T0259509

Inverse Problems and Inverse Scattering of Plane Waves

D. N. Ghosh Roy
SFA, Inc.
Largo, MD

L. S. Couchman
Naval Research Laboratory
Washington, DC

August 16, 2001

Preface

Let no one say that I have said nothing new;
the arrangement of the subject is new.

(From Pacsal's *Pensées* as quoted by G.M.L. Gladwell)

Printed and bound in the United Kingdom

Transferred to Digital Print 2011

Acknowledgement

The authors gratefully acknowledge the more than excellent help of Leslie Chaplin of SFA, Inc. at the Naval Research Laboratory in setting much of the manuscript in LATEX.

Contents

Chapter 1

Introduction

1.1 Direct and Inverse Problems

The terms *direct* and *inverse* problems are to be used in a relative sense depending on what is known and what is to be determined. A good example is an integral transform in which the calculations of the forward and the backward transform are entirely on an equal footing. Another example is the heat conduction problem in which the temperature distribution in a solid body is to be determined given appropriate boundary conditions and the initial data at a time $t = 0$. If the problem is to be solved for temperatures for times $t > 0$, it is usually considered to be a forward problem, forward meaning advancing in time. However, if for the same initial and boundary conditions, the sign of time t is reversed, then we have an inverse or backward heat conduction problem where we are to determine the past $(t < 0)$ temperatures that evolved into the initial data. Similarly, when a *known* electromagnetic field illuminates a slit of a *given* geometry, the determination of the *Fraunhoffer diffraction pattern* is considered to be the direct problem. Contrarily, determining the electromagnetic field on the slit *given* the Fraunhoffer diffraction pattern is the corresponding inverse problem. Numerous other examples can be cited. In any case, for the sake of fixing terminology, we will say that if $Af = g$ is a direct problem and $A^{-1}g = f$ its inverse, then g will be considered to be the *data* on the basis of which the *object* f is to be inferred.

The resolution of either problem requires some knowledge of the other. Usually, in a direct problem, the operator A and the object f are considered to be known so well that the quantity Af can always be calculated (at least in principle) in order to obtain g. Contrarily, in an inverse situation, where $f = A^{-1}g$, two problems arise. First, as a norm rather than as an exception, g is known only approximately, with errors which are experimental and/or numerical. Secondly, even if A is well-established, its inverse may be quite difficult to obtain. Good examples are the matrices. Given any matrix, it is always possible to use it as A. However, the determination of its inverse may be altogether a different

matter. The consequence of this is that the effect of a slight perturbation δg in
the data g can be dramatic in an inverse problem unlike in a direct calculation.
There an uncertainty δf in f will produce a comparable uncertainty δg in the
solution.

Moreover, inverse problems are mathematical in nature. There is no experimental solution to an inverse problem (Glasco, 1984). In solving any inverse
problem, it is first necessary to build a mathematical model based on whatever
information is available about the problem, solve it, compare the solution to the
given data and, finally, make as good a judgment as is possible regarding the
recovery of the desired quantity.

The inverse problems are essentially diagnostic in nature. All natural phenomena being causal, one not only perceives their manifestations, but can also
induce them under controlled conditions. For the most part, the causes behind
these manifestations remain inaccessible to direct observation or experimenta-
tion. To extract the hidden sources of the natural and biological phenomena
from their manifestations is the *leitmotif* of inverse problems. In these problems,
therefore, one proceeds backward from the data **g** to the source **f** of that data.
The problem is thus one of diagnosis. The widespread applicability of inverse
problems is, therefore, not surprising. Applications include not only the esoteric
and abstract, but also entirely mundane and down-to-earth problems.

1.1.1   Two Broad Divisions of Inverse Problems

Depending upon whether or not the data exists physically (experimentally or
numerically), the inverse problems fall into two broad classes: *direct* and *indirect*.
In the direct type of problems, inversion is obtained on the basis of data that
exists physically as outputs of measuring instruments. In the second case, on
the other hand, the data does not exist physically, but only as design criteria
to be achieved. Let us present some examples of both types.

A highly important class of inverse problems of the direct type is the *inverse
scattering* of plane waves from material objects. Here the data physically exists
in the form of the scattered fields. The inverse problem is to determine the
scatterer (e.g, the refractive index of a medium, shape and/or the composition
of an obstacle) from the fields it scatters in response to the plane wave/waves
incident on it. The problem is governed by the *Helmholtz equation* (Colton and
Kress, 1992) and will be discussed in detail in the latter half of the book.

A second example of an inverse problem of the direct variety is *Abel's equation* (Gorenflo and Vessella, 1991; Stakgold, 1979) given by

$$g(x) = \int_a^b \frac{F(y)}{(x-y)^\alpha}dx, \quad 0 < \alpha < 1. \tag{1.1}$$

Abel's equation (1.1) was originally derived in order to solve the *tautochrone*
problem. In this problem a bead is threaded through a string tied between two
endpoints a and b which form the limits of integration. The data **g** consist of
the time intervals taken by the bead to slide down various distances along the

thread. From this data, Abel tried to reconstruct the shape of the string F. Equation (1.1) which is a *Fredholm integral equation of the first kind* (Cochran, 1972; Wing, 1992) is the basic equation in astrophysics (Craig and Brown, 1986), plasma spectroscopy (Griem, 1997), gravimetry and instrumentation (Zidarov, 1980; Glasco, 1984) and imaging (Bakushinsky and Goncharsky, 1994), as well as many other areas of the physical sciences.

As the third example, we consider an *integral geometric* problem (Romanov, 1974). The basic inverse problem here is to recover a function f(x), $x \in R^n$, given its integrated values on the hyperplanes. The most well-known of these problems is *computed tomography* (Deans, 1983; Natterer, 1986; Herman, 1980). A more recent and rigorous treatment of the subject is by Ramm and Katsevich (1996). The starting equation is *Radon's transform* $\tilde{f}(\theta, p)$ of f defined by the integral (written for two space dimensions)

$$\tilde{f}(\theta, p) = \int_{R^2, x \cdot \theta = p} f(x) \mathrm{d}x,$$

p being a real parameter. In two dimensions, a hyperplane is a line and the Radon transform $\tilde{f}(\theta, p)$ is just the line integral of f, the line being normal to an axis rotated by an angle θ from the horizontal and intersecting it at a distance p from the origin (Figure 1.1). $\tilde{f}, \forall \theta \in [0, 2\pi]$ and $p \in R^1$ constitute the data called the *projections*. The function f is recovered from the projections through the *inverse Radon transform* (Gel'fand, Graev and Vilenkin, 1966; Helgason, 1980).

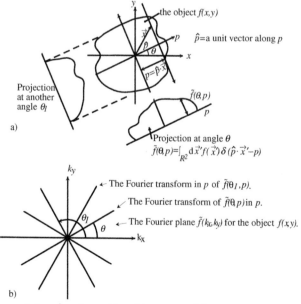

Figure 1.1 – A schematic of tomographic reconstruction of a two-dimensional function $f(x, y)$. (a) The tomographic projections of $f(x, y)$. (b) The Fourier Slice Theorem.

A number of interesting examples of direct inversions occur in heat conduction problems (Cannon, 1984; Ames and Straughan, 1997; Beck, et al.., 1985). Here we present an example from the propellant combustion problem (Kuo and Summerfeld, 1984; Ladouceur, 1990). In a direct problem of heat conduction, one determines the temperature distribution inside a solid body given either the temperature or heat-flux on the boundary. However, there are situations that involve hostile environment where sensors cannot be placed directly on the surface. Instead, the transient temperature is monitored at an interior point X. Let $\theta(X,t)$ be the temperature recorded by a thermocouple at X over a period of time. The inverse problem is: *given the time history $\theta(X,t)$ at a point X within a heat conducting body, determine the time history of the heat flux at its transient, receding boundary.* An important application of such an inverse heat conduction problem occurs in the burning of a solid propellant in a rocket where the ablating heat shield undergoes chemical reaction with the atmosphere. The problem is nonlinear and ill-posed. One way of resolving it is to approximate the heat flux and the transient burning profile mathematically, say, by a polynomial and solve the model problem. Compare the solution thus obtained with the data $\theta(X,t)$ given by the embedded sensor and keep updating till the desired accuracy is achieved.

Let us next consider some examples of inversions of the indirect type. These include the problems of control, design and synthesis. An archtypical example is the problem of shape optimization (Pironneau, 1984; Sokolowskii and Zolesio, 1992; Haug et al., 1986; Banicuk, 1990). In general, the problem is as follows. Suppose that the outcome of a certain physical problem depends upon the shape of some object. It is required that the shape be so designed as to make the outcome optimal (in a prescribed sense). As an example, consider that an optimal wing shape is to be designed for an airplane (Pironneau, 1984). Let S be a set of admissible (according to a certain criterion) shapes. A particular element from this set is to be so selected as to minimize drag.

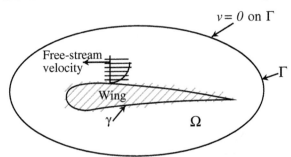

Figure 1.2 – A schematic of a wing-shape optimization problem.

Let the wing move with a free-stream velocity $\vec{v}_\infty$ in an unbounded domain as shown in Figure 1.2. Let Ω be a domain surrounding the wing and situated far away from the nonzero flow region so that the velocity on its boundary Γ is zero. The drag force is given by

$$\vec{F_{\mathrm{D}}} = \int_{\gamma} \sigma \cdot \vec{n} \;\; \mathrm{d}\gamma,$$

where

$$\sigma = \frac{\nu}{2}(\nabla \vec{u} + \nabla \vec{u}^{T}) - p\mathbf{I}$$

is the pressure tensor, $\mathbf{I}$ the unit tensor, $\vec{n}$ the outward normal to Γ and ν is the coefficient of kinematic viscosity. The integral is over the surface γ of the wing. *The inverse problem is to find the shape γ such that the work $\vec{F_{\mathrm{D}}} \cdot \vec{v}_{\infty}$ due to drag is minimal.* This involves determining the flow velocity $\vec{u}(x,t)$ via the *Navier–Stokes equation* (Schlichting, 1979)

$$\begin{aligned}
\dot{\vec{u}} + \vec{u} \cdot \nabla \vec{u} &= -\nabla p + \nu \Delta \vec{u}, & x &\in \Omega \\
\nabla \cdot \vec{u} &= 0, & x &\in \Omega, \quad t \in [0,T] \\
\vec{u}(x,0) &= \vec{u}_0(x), & x &\in \Omega \\
\vec{u}(x,t) &= 0, & x &\in \Gamma \quad \forall x, t.
\end{aligned}$$

There are other variations of the problem that include designs of structures such as buildings, bridges, automobiles (Banichuk, 1990; Haug *et al.*, 1986; Bennett and Botkin, 1986), design of materials with memory (Zochowski, 1992), design of electromagnets (Pironnaeu, 1984), and so on.

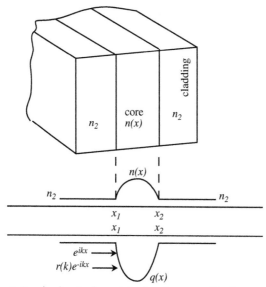

Figure 1.3 – A physical model of a planar optical waveguide and the associated inverse problem.

As a second example of the indirect type inversion, consider the problem of designing a planar optical waveguide with prescribed propagation constants

(Jordan and Lakshmanswami, 1989; Ge *et al.*, 1994). The refractive-index profile in the inhomogeneous core region of a planar waveguide (Figure 1.3) is to
be so designed as to propagate only a set of modes having prescribed propagation constants. Consider the propagation of a TE mode described by the wave
equation

$$\Delta \vec{E} + \nabla(\vec{E} \cdot \epsilon_r \nabla \epsilon_r) + c^{-2}\epsilon_r \ddot{\vec{E}} = 0, \tag{1.2}$$

where $\epsilon_r = \epsilon_x \epsilon_0^{-1}$ is the relative permittivity of the inhomogeneous core region that supports the guided mode. Representing the TE mode by the time-
harmonic wave

$$\vec{E} = \hat{y}\psi(k,x)e^{i(\beta z - \omega t)},$$

where β is the longitudinal wavenumber and ω the frequency, the wave problem
of Eq. (1.2) can be reduced to a Schrödinger-type problem

$$\left.\begin{array}{l} \psi'' + [k^2 - q(x)]\psi = 0 \\ k^2 = k_0^2 n_2^2 - \beta^2 \\ q(x) = k_0^2[n_2^2 - n^2(x)]. \end{array}\right\} \tag{1}$$

The parameters are shown in Figure 1.3. Given the reflection coefficient of the
waveguide, the solution of system (1) above obtained via the *Gel'fand–Levitan–
Marchenko equation* (Ablowitz and Segur, 1981; Lamb, 1980) which is

$$K(x,t) + R(x+t) + \int_{-t}^{x} K(x,y)R(x+y)\mathrm{d}y = 0, \quad t \le x, \quad x > 0. \tag{1.3}$$

R is the reflection coefficient which may consist of both discrete and continuous
components

$$R(t) = (2\pi)^{-1}\int_{R^1} r(k)e^{-ikt} \quad \mathrm{d}k - i\sum_{p=1}^{p} r_p e^{-ik_p t}.$$

The integral part is the Fourier transform of the reflection coefficient $r(k)$ and
represents the continuous spectrum corresponding to the unguided modes. The
sum over the poles k_p which lie on the positive imaginary axis (as is typical of
the Schrödinger equation) corresponds to the discrete spectrum over the propagating guided modes. Solving for $K(x,t)$ in Eq.(1.3) and considering the characteristic $x = t$, the potential $q(x)$ is obtained according to the Gel'fand–Levitan
theory as

$$q(x) = 2K'(x,x).$$

An analogous problem occurs in the *synthesis* of scientific instruments, for
example, in the design of a glass plate in order to obtain a prescribed transmission coefficient (Tikhonov and Samarskii, 1963). A light beam of frequency
ω, wavenumber k and polarized parallel to the surface of the plate is incident
normally on it. The propagation is described by the system

$$E'' + k^2 n^2(x)E = 0, \qquad x \in [0, h]$$
$$E'(0) - ikn_0(E(0) - 2E_0) = 0$$
$$E'(h) + ikn_0 E(h) = 0.$$

E_0 is the amplitude of the wave, n_0 the refractive-index of the ambiance and h is the thickness of the glass. The transmission coefficient is defined by

$$T(\omega) = \frac{n(h)}{n_0} \left| \frac{E(h)}{E_0} \right|.$$

The inverse problem is to determine the inhomogeneous refractive-index $n(x)$ of the plate which will result in the desired transmission coefficient $T(\omega)$.[1]

A similar problem arises in the design of reflectors (Ohiker, 1989). The set up consists of an illuminating source, a reflector and a given surface. The source illuminates the reflector and the reflected radiation then creates a radiation pattern on the surface. Given the geometry, that is, the relative positions of the source, reflector and the illuminated surface, the inverse problem is to design the reflector such that a prescribed distribution of radiation on the surface is obtained.

As the final example of inverse problems of indirect type, consider an infinitely long circularly cylindrical solenoid being inductionally tempered (case hardened) by passing a current in its coils (Glasco, 1984). At time $t = 0$ when the current is applied, the initial magnetic induction H is zero. The magnetic field being uniform in the core region, the normal (that is, radial) derivative of H is also zero at the center of the cylinder. This also applies to the radial temperature gradient. The problem is to determine the amplitude of the current $I(t)$ which will induce heating in the cylinder in such a way as to produce a certain prescribed transient temperature distribution on the surface. The control is characterized by the prescribed temperature regime given by

$$\theta(t) = \Theta(R, t), \quad t \in [0, T].$$

The current desired $I(t)$ depends implicitly through $\theta(t)$. The inverse problem is to determine $I(t)$ from an *a priorily* prescribed surface temperature function $\theta(R, t)$. The current depends implicitly upon $\theta(r, t)$ via the initial-boundary data as well as through the surface temperature evolution $\theta(R, t)$.

There is a variation on this problem (Tikhonov and Samarskii, 1963) in which the cylinder is immersed in a homogeneous magnetic field H_0. At time $t = 0$, the magnetic field is turned off and, according to Faraday's Law, a current is induced in the cylinder. The cylinder then demagnetizes during which an interplay takes place between the induced current and the demagnetization. Specifically, the induced current acts as a brake on the demagnetization process. A possible inverse problem here is to design the coil, particularly, the parameter $n(\rho\sigma)^{-1}$, in order to minimize the braking action.

---

[1]Note that in the above two examples, the inverse determination of the potential and the refractive index is the same as reconstructing the differential equation itself via that of the coefficients of the equation. As such, this forms a very large class of inverse problems, namely, the reconstruction of a differential equation from a given data set (Deuflhard and Hairer, 1983).

1.2   The Basic Concepts

Inverse problems can rarely be solved analytically and exactly. In most situations, an approximate solution is the best that can be hoped for. Even this approximate solution may have to be obtained in some convoluted way. One of the primary reasons is that in the presence of errors, the connection between the data and the range of the operator is lost. Now f can be inverted only if the data g is its image under A. The problem then is to establish some sort of a connection between the erroneous data and the image of f. This will be made abundantly clear in the sequel. This naturally raises the question of what is meant by a *solution* of an inverse problem and how to interpret such a solution. We turn to these issues now.

1.2.1   The Approximate Nature of an Inverse Solution

Often, the operator A in an inverse problem is an integral operator. The equation $Af = g$ then takes the form

$$(Af)(x) = g(x) = \int_{\Omega} K(x,y)f(y)\mathrm{d}y, \qquad (1.4)$$

where $x, y \in R^n$ and Ω is a region in R^n. The kernel $K(x,y)$ and the function f are dictated by the physics of the problem. For example, in a spectroscopic experiment, $f(y)$ may be the intensity of the spectral emission from a radiation source (e.g., a high-temperature gaseous plasma) falling on the slit of a monochromator. The monochromator processes the incident radiation with the kernel $K(x,y)$ which is a convolution operator in this case. Now if the slit is a line delta-function, a smeared image of the slit will be obtained thereby broadening its width. Consider, for the sake of discussion, that $f(y)$ is a continuous function (as the falling radiation is very likely to be) and let $K(x,y)$ be continuous in y and C^1 in x. $K(x,y)$ is thus once differentiable in x and $A : C \to C^1$ takes a continuous function $f(y)$ and turns it into a continuously differentiable function $g(x)$. This is typical of the so-called *compact* operators about which we will have much to say later.

The inverse problem is to find the continuous function f from the information about a once continuously differentiable, line-integrated spectral data g and the operator A. In reality, however, the data which is more than likely to come from physical and/or numerical experiments will be invariably mixed with errors. One, therefore, does not have the exact data g, but its approximation $g^\epsilon = g + \epsilon$, where the error ϵ is arbitrary in nature. The function g^ϵ cannot, therefore, be assumed to be differentiable. Moreover, as mentioned above, in order for a solution f to exist, g must be its image under the operation by A. In other words, g must be in the *range* $R(A)$ of A. Now if we insist on recovering f in $C(\Omega)$, then this simply is not going to be possible with an arbitrary g^ϵ. We must, therefore, search for a meaningful approximate solution corresponding to the approximate data g^ϵ.

One way of obtaining an approximate solution is to scan the space X to which f belongs and try to find an element $\tilde{f}$ in this space such that $A\tilde{f}$ is as close as possible to g^ϵ. The measure of the closeness is the metric ρ_Y in the image space Y. We are, therefore, looking for that particular element $\tilde{f}$ for which

$$\rho_Y(A\tilde{f}, g^\epsilon) = \inf_{f \in X} \rho_X(Af, g^\epsilon). \tag{1.5}$$

Obviously, if f is the exact solution then $f = \tilde{f}$. However, the method is certainly not trouble-free. In the first place, there is no guarantee that a $\tilde{f}$ can be found in X. Secondly, even if such a $\tilde{f}$ does exist, it may not be unique. As a matter of fact, an infinite number of $\tilde{f}$ may indeed exist.[2] As an example, replace the function f in Eq. (1.4) by a perturbed function $f_1(y) = f(y) + f_0 \sin ky$. Then

$$\rho_X(f_1, f) = |f_0| \sup_\Omega |\sin ky| = |f_0|. \tag{1.6}$$

Since f is assumed to be in C, the supremum-norm in (1.6) is appropriate. Let g_1, g be the data corresponding to f_1, f, respectively, and consider the L_2-norm in Y. By definition of the L_2-norm

$$\|z\|_2 = \Big[\int |z|^2 dz \Big]^{\frac{1}{2}}.$$

The integral is over the domain of definition of z. We will explain the choice of this norm for Y momentarily. Then

$$\begin{aligned}
\rho_Y(g_1, g) &= \|g_1 - g\|_2 \\
&= |f_0| \Big\{ \int_D \Big[\int_{D'} K(x, y) \sin ky \ \ dy \Big]^2 dx \Big\}^{\frac{1}{2}}.
\end{aligned}$$

Let f_0 be an arbitrary number and allow k to approach infinity. Now by assumption, the kernel $K(x, y)$ is summable. That is,

$$\int_D \int_{D'} |K(x, y)| dx dy < \infty$$

over every compact set of the product space $D \times D'$. For summable operators, the well-known *Riemann–Lebesgue Theorem* (Stakgold, 1979) tells us that the integrals of the type

$$\int K(x, y)\psi(ky) dy \to 0, k \to \infty.$$

---

[2]In the literature this is sometimes expressed by the statement that the solution space has too large a diameter. Further restrictions must be imposed on the space of functions to be recovered through *a priori* information. The situation is analogous to solving an initial or a boundary value problem of a differential equation without the initial and the boundary conditions in which case there may be an infinite number of solutions. However, once these conditions are specified, a concrete solution may result.

This means not only a nonunique $\tilde{f}$, but also instability in the solution.

The situation does not improve if we adopt the L_2-norm for the solution. For the sake of illustration, let the dimension of the space X be one and let y vary from a to b. Then in this norm

$$
\begin{aligned}
\|f - f_1\|_2 &= \left\{ \int_a^b |f(y) - f_1(y)|^2 \mathrm{d}y \right\}^{\frac{1}{2}} \\
&= |f_0| \left\{ \int_a^b \sin^2 ky \, \mathrm{d}y \right\}^{\frac{1}{2}} \\
&= |f_0| \left[\frac{b-a}{2} - \frac{1}{2k} \sin k(b-a) \cos k(b-a) \right].
\end{aligned}
$$

Again for arbitrarily small changes in g, arbitrarily large changes can occur in the solution as k goes to infinity.

An important fact to remember is that an inverse solution depends on the triple (sometimes called the *Gel'fand triple*) (A, X, Y) as a whole. The dependence on A reflects the fact that the resolution of an inverse problem requires either the full or the partial knowledge of the direct problem. That the members of the triple (A, X, Y) are not independent of each other can be seen as follows (Bertero, 1989; Kirsch, 1996). Let X and Y be Banach spaces. In addition, let the bounded linear operator A map the Banach space X continuously on the entire space Y (the *onto* mapping) in a one-to-one fashion. Then by the *Open Mapping Theorem* of linear analysis which guarantees that the inverse operator $A^{-1} = L$ is continuous and, therefore, bounded. The existence of the inverse operator thus depended on the choice of the spaces X and Y.

Before concluding this section a comment is due regarding the norm of the data. The measuring instruments that record the experimental data do not generally measure derivatives and the information about the difference $|g_1'(x) - g_2'(x)|$ of the derivatives of the data is not available in general. The norms that are appropriate for the measuring instruments are either L_1 or L_2. That is, either they measure the magnitude or the mean-square value. Our control over the space Y is thus limited and in practice, the space Y is usually considered to be the Hilbert space L_2 of the square integrable functions.

We saw that in finding an approximate solution to an inverse problem for data which is only approximately known, we can run into two problems: nonuniqueness (infinite number of solutions) and instability (large spurious oscillations in the solutions for only a slight change in the data). Next we take a closer look at some simple physics behind this problem.

1.2.2   The Smoothing Action Of An Integral Operator

Equation (1.4) demonstrates that the high-frequency components of f which are responsible for the object's sharp features are annihilated by the operator A (by the Riemann–Lebesgue Theorem) to below arbitrarily small levels of noise and other errors. As a consequence, only the low frequencies remain which results in a smoothed version of the image of the object. In other words, the integral operator is a smoothing operator and the high-frequency information about the object is irretrievably lost (see Craig and Brown, 1986, for an explicit demonstration of this point). Since the inverse operation attempts to restore the system to its original state, the small high-frequency perturbations in the data are amplified. As a consequence, oscillatory components of arbitrarily large amplitudes are introduced into the solution. These large-amplitude oscillations are entirely spurious and physically meaningless if there are perturbations in the data that are due to noise or other systemic errors.

This point is further demonstrated by considering the example of a convolution operator. Again for the sake of simplicity, let us consider only one space dimension and a convolutional integral operator. Thus

$$g(x) = A * f = \int_{R^1} K(x - y)f(y)\mathrm{d}y. \tag{1.7}$$

The integral is over the entire real axis and $*$ denotes convolution. Taking the Fourier transform on both sides of Eq. (1.7) and recalling that the Fourier transform of a convolution is an algebraic product of the transforms of the individual functions (Morse and Feshbach, 1953), we have

$$\hat{f}(k) = \left(\frac{1}{2\pi}\right)^{-\frac{1}{2}} \frac{\hat{g}(k)}{\hat{K}(k)}. \tag{1.8}$$

Suppose that the data has some perturbation $\delta g(x)$. Then the corresponding perturbation in the Fourier transform in Eq. (1.8) becomes

$$\delta\hat{f}(k) = \left(\frac{1}{2\pi}\right)^{-\frac{1}{2}} \frac{\hat{\delta g}(k)}{\hat{K}(k)}. \tag{1.9}$$

Now the Fourier transform $\hat{K}(k)$ must vanish as $k \to \infty$ in order for the energy to remain finite. However, this is not necessarily so for the noise spectrum $\hat{\delta g}(k)$ which may approach a finite limit as $k \to \infty$. The well-known *white noise* (Papoulis, 1965) is a clear example. Equation (1.9) then shows how a small variation in δg can induce large oscillations in the solution.

Another important example of the smoothing action of an integral operator is furnished by the scattering of plane waves by an obstacle. The scattering is described by an operator equation $F\psi^{\mathrm{sc}} = \psi^\infty$, where F is the far-field scattering operator, ψ^{sc} is the scattered field and ψ^∞ is the scattered field at a large distance form the scatterer. $F\psi^{\mathrm{sc}}$ is a covolution: $F\psi^{\mathrm{sc}} = G^0 * \psi$, where the free-space Green's function G^0 is given by

$$G^0(x, y) = \frac{e^{ik|x-y|}}{4\pi|x - y|},$$

in which $x \in R^n$ is the point at which ψ^{sc} is calculated and $y \in \Omega$ is a point of the scatterer. If $|x - y| > 0, \forall x, y$, Green's function G^0 is infinitely differentiable in both x and y. In particular, for $|x|$ far away from $|y|$, the scattered field is not only smooth but is actually an analytic function in all its arguments. These points are discussed in detail in the later chapters. This means that the high-frequency information is severely smoothed out in the scattering amplitude. Consequently, the determination of the scattered field from the scattering amplitude is problematic.

From all this it should be clear that unless some filtering is done in the high frequencies, instability is destined to set in. The filtering operation must depend on the kernel. The more sharply this function peaks in the k-space, the lower is the cut-off in frequency. Consequently, a kernel that is sharply peaked in the k-space and thus is relatively flat in the x-space, should result in enhanced smoothness. Contrarily, a kernel that is flat in the k-space and therefore, sharply peaked in the x-space, is less severe in restricting high frequencies. The cut-off also depends upon the noise level and, since a high noise level requires lower cut-off in the k values, the kernels of either type tend to degrade the image quality.

Sometimes, stability can be restored through the available *a priori* information about the solution. We consider this in the following section.

1.2.3   The Role of *a priori* Knowledge

The variational solution (1.5) was found to lead not only to nonuniqueness, but to instability as well. It is the latter that poses the most difficult problem in data inversion. A way out of this difficulty is to restrict the space of the solutions in order to avoid such pathologies. As an example, consider the equation

$$g(x) = \int_0^1 \sin n\pi y \sin n\pi x f(y) \, \mathrm{d}y. \qquad (1.10)$$

Let us suppose that $f \in L_2[0,1]$. It is clear that the integral operator A cannot be onto $L_2[0,1]$. It is certainly not one-to-one since all values of $n \in Z^+$ are in the null-space of the operator because $\sin n\pi z$ vanishes for these values of n. Moreover, the function $g(x)$ can not be arbitrary, but must be of the form const. $\sin n\pi x$. As discussed above, an inversion for f using Eq. (1.10) may produce large variations in the solution even though the corresponding variations in the data may be vanishingly small. Now let us assume that we so restrict the object function that its derivative, f', is uniformly bounded. This at once excludes functions such as $f_j(y) = \left(2\pi y^{-1}\right)^{\frac{1}{2}} \sin n\pi y$ from being object functions since their derivatives $f_j'(y)$ are not uniformly bounded. The problem of instability can thus be avoided. Note that the *a priori* information that f' must be uniformly bounded has resulted in the restriction of the object space from $L_2[0,1]$ to a compact subspace of $C[0,1]$.

Notice that restricting the space of solutions is equivalent to the knowledge of some *a priori* information about the object function and *vice versa*. This information usually comes in the form of bounds either on the function itself or on its smoothness, i.e., its differentiability. The bounds do not have to be sharp. It is important to point out that the *a priori* information must be justified either by the physics of the problem or by the mathematical consistency. As Sabatier (1990) points out, these are actually the missing pieces of information. We will see later that a backward heat conduction problem can be stabilized by knowing the bound on the temperature *a priorily* and this may have to be no more precise than simply the melting point of the solid. Another example (Baltes, 1978) is provided by the inverse determination of the stellar diameter from the measurement of the degree of coherence of the emission as a function of the angular spacing. The *a priori* information here has many pieces including a uniformly bright circular disk of zero coherence, and these pieces of information are the consequences of the already established physical knowledge. The role the *a priori* knowledge plays in the inverse solutions will be illustrated by the examples of Chapter 2.

1.2.4   Ill- and Well-posed Problems

Three difficulties were observed thus far in solving an inverse problem: i. the solution may not exist. ii. The solution may be nonunique, and iii. Even if a solution does exist, it may not change smoothly with the data. Nonexistence of a solution means nonexistence in a certain specified class of functions. If this class is altered, the nonexistence may disappear. The nonuniqueness means that the object is not mapped by the integral operator in a one-to-one fashion. The third condition states that a slight variation in the data may introduce arbitrarily large, spurious and physically meaningless components into the solution. In the parlance of the linear operator theory, these translate into the following. The data may not be in the range of the operator, implying nonexistence. The operator may not be one-to-one (also called *injective*) and finally, the inverse of the operator may not be continuous which, for a linear operator, is the same as not being bounded. Mathematically, the continuity is explained as follows. Let $Af = g$, $f \in X, g \in Y$. Let $\{f_n\}_1^\infty$ be a sequence of functions in X such that limit $n \to \infty f_n = f$. Then if limit$_{n \to \infty} Af_n = Af$ in Y for every member of the sequence, then A is continuous at $f \in X$. If, in addition, A is continuous for all $f \in X$, then it is continuous in X.

Let us next compare all this with the classical notion of a solution which was formulated by Hadamard at the turn of the century (1923). In reference to the inverse operator $L = A^{-1}$, these are:

1. For every $f \in X$, there must be at least one $g \in Y$ such that $Lg = f$. This is the problem of existence. 2. For every $f \in X$, there must be at most one $g \in Y$ such that $Lg = f$. This is the problem of the uniqueness. 3. Let there be a sequence $\{g_n\} \in Y$ such that $Lg_n \to Lg = f \in X$. Then it is required that there be a sequence $g_n \to g \in Y$. This is the requirement of stability.

In terms of A these conditions are 1. For every $g \in Y$, there must be at least one $f \in X$ such that $Af = g$. 2. For every $g \in Y$, there must be at most one $f \in X$ such that $Af = g$. 3. Let there be a sequence $\{f_n\} \in X$ such that $Af_n \to Af \in Y$ as $n \to \infty$. Then it must be required that $f_n \to f \in X$.

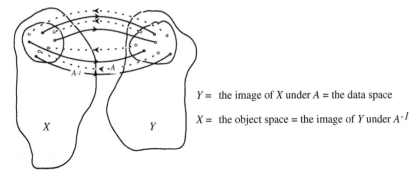

$Y =$ the image of X under $A =$ the data space

$X =$ the object space = the image of Y under A^{-1}

Figure 1.4 – A symbolic illustration of well-posedness. The
solid line is the action of the forward operator A and the dotted lines correspond to the inverse operation by A^{-1}. Small
variations in the data (•) result in correspondingly small variations (o) in the solution.

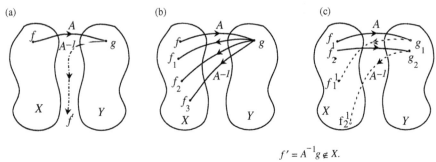

$$f' = A^{-1}g \notin X.$$

Figure 1.5 – A schematic illustration of ill-posedness. (a) nonexistence $f' = A^{-1} g \notin X,\ g \in Y$. (b) Nonuniqueness:
$A^{-1}g \rightarrow f_1, f_2$, and f_3. (c) instability: a small *distance* between two data points g_1 and g_2 produces a large difference
between f_1^1 and f_2^1 where $f_i^1 = A^{-1}g_i$.

If any of Hadamard's three conditions is violated, the problem is considered to be *ill-posed*.[3] Otherwise, it is *well-posed*. Figures 1.4 and 1.5 illustrate
the concept of well and ill-posedness schematically. When the set of the admissible solutions is large, the problem is said to be *strongly ill-posed*. If the
set is narrow, then it is *weakly ill-posed*. An ill-posed problem can be made
well-posed primarily in two ways, either by restricting the set of the admissible
solutions on the basis of available *a priori* information, or by projecting the
approximate data on the range of the unperturbed operator, and then inverting
this processed data. It is assumed that the restricted set S_R contains the true
solution. It is then proved that the solution obtained in the set S_R is unique
and varies continuously with the processed data. Also it is usual to show that

---

[3] Also called *improperly* posed or *incorrectly* posed.

the approximate solution goes over to the true solution in the limit that the error in the data tends to zero. This is primarily a mathematical requirement because the data in practice is bound to be contaminated with errors. However, an algorithm having this property guarantees that better accuracy can be obtained by improving the conditions of the experiment that would reduce errors (Bertero, 1989). **By an inverse problem having a solution is meant this approximate solution in the set** S_R.

The problem of existence is the least troublesome, whereas the most formidable is the problem of the continuous dependence of the solution on the data. The instability is reflected in small perturbations in the data not leading to correspondingly small perturbations in the solution. If these perturbations truly belong to the object, then the large-amplitude, high-frequency oscillations are the actual features of the object. However, since these are buried in the noise level in the data, the amplification will certainly contain noise and there is no way to isolate the actual from the spurious, noisy components. As these must necessarily be filtered out, the recovered object will be smooth.

The investigation of ill-posedness is vital for inverse solutions at least for the following reason. As pointed out earlier, an inverse problem cannot be solved experimentally. A reconstruction algorithm must be given. Stated otherwise, providing a reconstruction algorithm is an integral part of solving an inverse problem. Now there is hardly an inverse problem of any significance that can be solved without the benefit of numerical calculations. Since the numerics must perforce involve approximate data due to machine accuracy, error in the input data, limitations due to finite mesh size, and so on, their robustness depends critically on the ill-posedness in the formulation of the problem.

Let us make one further remark. Kernels that are convolutional (as in Laplace and Fourier transforms), summable (the Riemann–Lebesgue theorem holds) or *weakly singular* (typical of integral operators), are usually *compact*, and have smoothing properties that lead to the violation of Hadamard's third condition. Since these types cover most of the inverse problems in practice, it is clear that the inverse problems are almost always ill-posed. The converse is, however, not true. An example is the numerical differentiation of a function. This is an ill-posed problem and is not necessarily of an inverse type. Other examples include the calculation of a Fourier series with inexact coefficients, solving a system of linear equations the matrix of which is near-singular, and so on. The class of ill-posed problems is, therefore, larger than the class of inverse problems. However, the latter is probably the most important and significant of the ill-posed problems.

Thus far we have discussed inverse problems and their solutions in the setting of an infinite-dimensional function space. That is, the object and the data were considered to be functions. In reality, the data cannot be collected continuously, in an uninterrupted manner, and must perforce be discrete. At the same time, the solution must also be discrete since it is obtained numerically almost without exception. Therefore, the original continuous problem, $Af = g$, in an infinite-dimensional function space is reduced to solving a system of linear equations $\mathbf{A_n f_n} = \mathbf{g_n}$, where $\mathbf{f_n}$ and $\mathbf{g_n}$ are vectors in R^n and the operator $\mathbf{A}_n$ is a matrix operator. The inverse solution is formally given by $\mathbf{f_n} = \mathbf{A_n^{-1}g_n}$. Two problems arise immediately. First, the data and the solution are now vectors in finite-dimensional Euclidean spaces and may have no connection whatsoever with the function spaces.[4] It then behooves us to investigate the relation between the solution of the finite-dimensional approximation to an infinite-dimensional problem. Second, since the original problem $Af = g$ is ill-posed, its finite-dimensional approximation $\mathbf{A_n f_n} = \mathbf{g_n}$ must necessarily show the ill-posedness in its behavior if the discretization parameter n is allowed to approach infinity. The ill-posedness of the original continuous problem now manifests through the *ill-conditioning* of the matrix operator $\mathbf{A}_n$. These questions will be discussed in Chapter 5 where discrete inverse problems are treated.

---

[4]For example, R^n has nothing to do with function spaces, say, C or C^1.

Chapter 2

Some Examples of Ill-posed Problems

2.1 Introduction

In Chapter 1 we discussed the nature of inverse problems, their ill-posedness, instability and how *a priori* knowledge about the solutions can bring about stability. The mathematical formulation of these questions will be detailed in the sequel. However, the power of a mathematical analysis is better appreciated if first illustrated by examples, and this is the aim of the present chapter. Selected from various disciplines, the examples are designed with two specific objectives: to pinpoint how the ill-posedness creeps in and how the *a priori* information about the solution (which does not appear in the mathematical formulation of the problem itself) helps in the stabilization. Since these are the most vital considerations in inverse solutions, we have purposefully discussed the examples in more details than is otherwise necessary.

2.2 Examples

2.2.1 Example 1. The Cauchy Problem for the Backward Heat Equation

Consider the one-dimensional problem of heat conduction described by

$$\left. \begin{aligned} &\dot{u} + u" = 0, \\ &u(0,t) = u(L,t) = 0 \qquad t \in (0,T), x \in (0,L) \\ &u(x,0) = \frac{C}{n}\sin\left(\frac{n\pi x}{L}\right). \end{aligned} \right\} \qquad (2.1)$$

in a rod of length L the two ends of which are kept fixed at zero temperature. $u(x,t)$ is the temperature at a point x in the rod at time t. The initial temperature distribution at time $t = 0$ is given as a sine wave. Note that this is a heat equation backward in time. In the forward case, t is greater than zero and then the heat operator is not $\dot{u} + u''$, but $\dot{u} - u''$. The Cauchy problem for the heat equation is to determine the past temperature distributions in the rod which evolved into the sinusoidal distribution of Eq. (2.1) at time $t = 0$. As it stands, solving Eq. (2.1) is not different from solving a forward heat conduction problem, but with a backward heat operator. This is sometimes referred to in the literature as *solving a backward heat equation forward in time*.

The unique solution of the Cauchy problem is easily obtained using the method of the separation of variables and the solution is given by (Churchill, 1963):

$$u(x,t) = \frac{C}{n} \sin\left(\frac{n\pi x}{L}\right) e^{\frac{n^2 \pi^2 t}{L^2}}. \tag{2.2}$$

Note that had it been a forward heat equation, the exponential factor in Eq. (2.2) would have been $e^{-\frac{n^2 \pi^2 t}{L^2}}$. Following Hadamard, we consider a sequence of problems Eq. (2.1) one for each $n = 1, 2, \cdots$. We are, of course, interested in knowing if the solution satisfies the three conditions of Hadamard's well-posedness, particularly, if it depends continuously on the data. It is necessary to specify a function space for the solution and the measure or the metric in which its continuous dependence on the data is to be ascertained. We assume that $u \in C(X)$, $X = (0, L) \times (0, T)$, and that $Y = C(R^1)$ both equipped with supremum norm.

Now, $\text{limit}_{n \to \infty} u_n(x, 0) = u_\infty(x, 0) \to 0$. But from Eq. (2.2) it follows that

$$\underset{n \to \infty}{\text{limit}} \sup_{x \in (0,L)} \rho_X(u_n(x, t), u(x, 0)) \to \infty.$$

Thus although the data vanishes asymptotically, the solution grows without bound.

Let us next look at the problem in a more general setting and consider a different measure for continuous dependence (Ames and Straughan, 1997). Let us consider the same Cauchy problem in three space dimensions and write

$$\left.\begin{array}{ll} \dot{u} + \Delta u = 0, & \text{in } \Omega \times (0, T), \Omega \subset R^3 \\ u(x, t) = 0, & \text{on } \Gamma \times [0, T) \\ u(x, 0) = f(x), & x \in \Omega. \end{array}\right\} \tag{2.3}$$

The domain $\Omega \subset R^3$ is smooth with boundary Γ, and Δ is the Laplace operator. This time we consider a functional $F(t)$ defined by $F(t) = \|u\|_2^2$. Differentiating F with respect to t, using the heat equation in Eq. (2.3) and from Green's first identity (see Section 6.3), we find that

$$F\ddot{F} - (\dot{F})^2 \geq 0. \tag{2.4}$$

Actual calculation shows that

$$F\ddot{F} - (\dot{F})^2 = 4\left[\|u\|_2^2 \|\dot{u}\|_2^2 - \left(\int_\Omega u\dot{u}dx\right)^2\right].$$

The inequality (2.4) can also be expressed as

$$[\ln F(t)]_{tt} \geq 0,$$

the subscript "tt" implying second derivative in time. $\ln F(t)$ is, therefore, a convex function and F is said to have *logarithmic convexity* (van Tiel, 1984). It is well-known from the theory of functions (Titchmarsh, 1939) that a convex function $f(t)$ is bounded by

$$f(t) \leq a(t, t_1, t_2) f(t_1) + b(t, t_1, t_2) f(t_2), \tag{2.5}$$

where

$$a(t, t_1, t_2) = \frac{t_2 - t}{t_2 - t_1}, \qquad b(t, t_1, t_2) = \frac{t - t_1}{t_2 - t_1}. \tag{2.6}$$

t_1, t_2 being the end points and $t \in [t_1, t_2]$. If $f(t)$ in Eq. (2.5) in replaced by $\ln F(t)$, then

$$F(t) \leq [F(t_1)]^a [F(t_2)]^b.$$

From this and the definition that $F(t) = \|u\|_2^2$, we obtain

$$\|u(t)\|_2^2 \leq \|f\|_2^{2(1-\frac{t}{T})} \|u\|_2^{2\frac{t}{T}}, \qquad t \in [0, T). \tag{2.7}$$

Invoking the continuity of $F(t)$, it follows that if $F(t)$ vanishes at any point of the interval $[0, T)$, then it must vanish identically in the entire interval. Since F is the square of the L_2-norm of u, then u must also vanish identically in $[0, T)$. Looking at the boundary-initial value problem, Eq. (2.3), we see that this can happen only if the function $f(x)$ is zero. Thus the homogeneous problem has only the trivial solution. Then from Friedholm's alternative, the problem has a unique solution.

Having dispensed with the question of uniqueness, we would next like to determine if the solution depends continuously with the data. We have seen above that in the space of continuous functions with a supremum norm, the solution was unstable. Now let u be the difference between two solutions u_1 and u_2 of the Cauchy problem with the function $f(x)$ as the difference in the Cauchy data corresponding to these two solutions. From inequality (2.7) the problem may appear to be stable. That is, as if $u \to 0$ if $f \to 0$. But, unfortunately, it is not the case since there is no guarantee that the product $\|f\|^{2(1-\frac{t}{T})} \|u\|_2^{2\frac{t}{T}}$ will be small even if f is small. However, if $\|u(t)\|_2^2 \leq M^2$, for some constant M, then

$$\|u(t)\|_2^2 \leq \|f\|^{2(1-\frac{t}{T})} M^{\frac{2t}{T}}, \qquad t \in [0, T). \tag{2.8}$$

Inequality (2.8) does indeed show that u goes to zero if f vanishes thereby insuring stability. We can also rewrite Eq. (2.8) as

$$\sup_{t \in [0,T)} \|u_1(\cdot, t) - u_2(\cdot, t)\|_2 \leq C \epsilon^\alpha, \tag{2.9}$$

where C is a constant and

$$\|f_1 - f_2\|_2 < \epsilon,$$

and α is a constant index. According to Eq. (2.9) the solution depends continuously on the data f, but in the sense of Hölder. Recall the definition of Hölder continuity at this point. Let A map $U \to V$, U and V being linear vector spaces. Let $u_1, u_2 \in U$, and $f_1, f_2 \in V$. Then the mapping is *Hölder continuous* if

$$\sup_{u_1, u_2 \in \tilde{U}} \|u_1 - u_2\|_{\tilde{U}} < M\epsilon^\alpha,$$

whenever

$$\|f_1 - f_2\|_{\tilde{F}} < \epsilon.$$

M and α are positive constants and $\tilde{U}, \tilde{F}$ are subsets of U and F, respectively. Moreover, the solution $u(\cdot, t)$ is called *Hölder stable* if

$$\sup_{0 \le t \le T} \|u_1(\cdot, t) - u_2(\cdot, t)\|_{\tilde{U}} < C\epsilon^\alpha,$$

whenever

$$\|u_1(\cdot, 0) - u_2(\cdot, 0)\|_{\tilde{U}} < \epsilon,$$

where $\alpha \in [0, 1]$, and $\|\cdot\|_t, \|\cdot\|_0$ are the norms defined on the solution at time t and initially, respectively, and C is a positive constant independent of ϵ. Note that this is equivalent to saying that the solution depends continuously on the initial data for the interval $[0, T)$.

The above conclusions can be put in the theorem below.

Theorem 2.1 *Consider the Cauchy problem (2.3) for heat conduction. Let the Cauchy data be in $L_2(\Omega)$, and choose Hölder continuity as the measure of continuous dependence. Then the Cauchy problem is stable and the solution depends Hölder continuously on the initial data over the interval $[0, T)$.*

The example just presented reveals several important points. First, the stability depends upon the norm chosen to measure the continuity of the dependence of the solution on the data. Secondly, the imposition of a suitable *a priori* bound(s) on the solution ($\|u(t)\|_2^2 \le M^2$ in this case) can lead to stabilization and finally, the bounds need not be precise (their estimates are frequently possible from the physics of the problem). For example, M above can be simply the melting point temperature of the rod.

It is interesting to mention here that in the past, attempts were made to solve problems backward in time such as calculating the Earth's magnetic field in previous times. Most famous is Lord Kelvin's attempt to calculate the age of the Earth by extrapolations backward in time. These are, of course, improperly posed problems. In this context, a letter from Maxwell to Kelvin (see Garber *et al.* 1995) is most interesting. In this letter, Maxwell explicitly refers to the instability as t becomes negative.

2.2.2 Example 2. The Cauchy Problem for the Laplace Equation

The Cauchy problem for the Laplace equation can be considered to be an elliptic version of the parabolic backward heat equation of Example 1 and poses considerably more difficulty for analysis. Historically, this is that celebrated problem introduced by Hadamard (1923) at the turn of this century and has remained since the archetypical starting point for the twentieth century research in improperly posed problems. Hadamard considered an infinite sequence of Cauchy problems ($n = 1, 2, \cdots$) for the Laplace equation in a rectangular strip $(0 < x < a) \times (0 < y < b)$:

$$\Delta u(x, y) = 0$$
$$u(x, 0) = f(x) = \tfrac{\sin nx}{n}, \quad u_y(x, 0) = 0.$$

As $n \to \infty$, $u(x, 0) \to 0$. Therefore, the solution of the limit problem is zero. One can consider u in the above problem to be the deviation from the limit solution by a small perturbation in the data (i.e., large n). Again as in the previous example, consider $Y = C(R^1)$ to be the data space, $X = C(\overline{\Omega})$ the space of the solutions with supremum norm:

$$\rho_X(u_1, u_2) = \sup_{x \in (0, a)} |u_1(x, y) - u_2(x, y),$$

which, in this case, is

$$u_1(x, y) = \frac{\cosh ny \sin nx}{n^2}.$$

Thus

$$\rho_X(u_1, u_2) = \sup_{x \in (0, a)} \left| \frac{\cosh ny \sin nx}{n^2} \right|$$

becomes unbounded as $n \to \infty$. Again as in the previous example, the data tends to vanish, but the solution grows without bound indicating ill-posedness. The Cauchy problem for the Laplace equation considered on the pair of spaces $X = C(\overline{\Omega}), Y = C(R^1)$ is, therefore, ill-posed.

Following Flavin and Rionero (1995), let us consider the Cauchy problem in the plane:

$$\left. \begin{array}{ll} \Delta u = 0, & 0 < x < a, 0 < y < b \\ u(x, 0) = u(x, b) = 0 \\ u(0, y) = f(y), & u_x(0, y) = g(y). \end{array} \right\} \tag{2.10}$$

The Laplace equation and the first boundary conditions on y can be combined to yield a conservation statement

$$\int_0^b (u_x^2 - u_y^2) = E(\text{constant}). \tag{2.11}$$

Moreover, define a functional F:

$$F(x) = \int_0^b u^2 dy + C(x + x_0)^2, \tag{2.12}$$

where $C \geq 0$ and x_0 are constants to be determined. Differentiating F in Eq. (2.12) twice, integrating by parts and using the first boundary conditions in Eq. (2.10) gives

$$F''(x) = 2 \int_0^b (u_x^2 + u_y^2) dy + 2C. \tag{2.13}$$

Combining Eq. (2.13) with the conservation equation (2.11) yields

$$F''(x) = 4 \int_0^b u_x^2 dy + 2(C - E). \tag{2.14}$$

From Eqs. (2.12), (2.13) and (2.14), it follows that

$$
\begin{aligned}
FF'' - (F'')^2 &= \{ \int_0^b u_x^2 dy + C(x + x_0)^2 \}\{ 4 \int_0^b u_x^2 dy + 4C \} \\
&\quad -4\{ \int_0^b u u_x dy + C(x + x_0) \}^2 - 2(C + E)F,
\end{aligned}
$$

which can also be written as

$$FF'' - (F'')^2 \leq -2(C + E)F, \tag{2.15}$$

where Schwartz's inequality was used in reducing the first set of terms. Arguments similar to those in Example 1 show that if the data f and g are identically zero, then the homogeneous Cauchy problem of Eq. (2.10) has only a trivial solution and by Fredholm's alternative the Cauchy problem (2.10) has a unique solution.

We show, as before, that the stability is restored if we assume again that $|u| \leq M$, M being a constant. Let us first consider that the constant E in the conservation relation (2.12) is greater than zero. From the definition of the functional F in (2.12) and the inequality (2.15), it follows that

$$FF'' - (F'')^2 \geq -\frac{(2 + \epsilon)F^2}{(x + x_0)^2}, \tag{2.16}$$

where $\epsilon = 2E\beta^{-1}$. Noting that

$$[\ln(x + x_0)]'' = -\frac{(2 + \epsilon)}{(x + x_0)^2},$$

and

$$[\ln F]'' = F^{-2}[FF'' - (F'')^2],$$

Eq. (2.16) can be expressed as

$$[\ln F(x + x_0)^{-(2+\epsilon)}]" \geq 0.$$

The function $\tilde{F}(x) = F(x + x_0)^{-(2+\epsilon)}$ is, therefore, logarithmically convex and satisfies the same logarithmic convexity expression of Eqs. (2.5) and (2.6) only with $F(x)$ replaced by $\tilde{F}(x)$. Thus, proceeding as before we have, for $E > 0$, the following inequality: Hölder

$$F(x)(x + x_0)^{-(2+\epsilon)} \leq \left[F(0)x_0^{-(2+\epsilon)}\right]^{1-\frac{x}{a}} \left[\{M^2 + C(a + x_0)^2\}(a + x_0)^{-(2+\epsilon)}\right]^{\frac{x}{a}}.$$

In terms of the norm of the solution u, the above inequality satisfies a Hölder condition, namely

$$\|u(x)\|_X^2 (x + x_0)^{-(2+\epsilon)} \leq \|u(0)\|_X^2 x_0^{-(2+\epsilon)} \Phi(M, a, x, x_0)^{\frac{x}{a}},$$

where the function Φ includes (M, a, x, x_0) in its arguments. Therefore, $\|u(x)\|_X^2$ depends continuously on the data $\|u(0)\|_X^2$.

Next consider the case when $E < 0$. In this case, choosing C to be zero and from the inequality (2.15), we have

$$FF"' - (F"')^2 + 2EF \geq 0.$$

Again assuming that $|u| \leq M$, and using the expression (2.6) for logarithmic convexity results in

$$F(x) \leq M^{\frac{2x}{a}} F(0)^{1-\frac{x}{a}}. \tag{2.17}$$

In terms of the solution, Eq. (2.17) reduces to

$$\|u(x)\|_Y^2 \leq M^{\frac{2x}{a}} \|u(x)\|_Y^{2(1-\frac{x}{a})}.$$

Again the Hölder continuity appears.

In summary, therefore, we can state the following theorem.

Theorem 2.2 *The Cauchy problem Eq. (2.10) for Laplace's equation in a strip in two space dimensions has a unique solution. Provided that the L_2-norm of the solution $\|u\|_2 < M$, M being a constant, the Cauchy problem is stable. That is, the solution depends Hölder continuously on the initial data.*

In dimensions higher than two, the problem turns out to be quite involved and we refer the interested reader to Payne (1960) and Ames and Straughan (1997).

2.2.3 Example 3. The Laplace Transform

As our next example, we consider a special equation of the first kind; the *Laplace transform* (Widder, 1971; Morse and Feshbach, 1953). Next to the Fourier transform, the Laplace transform is perhaps the most important of integral transforms, especially in solving transient initial value problems. However, neither the ease with which the Fourier transform can be inverted nor its infinite information capacity is shared by the Laplace transform. Although closed form expressions exist for its inverse, they are usually difficult to apply in practice for numerical reasons. In view of the practical importance of inverting a Laplace transform, its discussion here was deemed worth while.

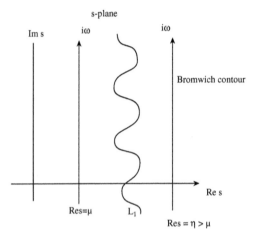

Figure 2.1 – The complex domain for the Laplace transform, showing the contour of Bromwich.

Let us first define the transform. Let $f(t)$, $0 \leq t \leq \infty$ be a continuous function of slow growth. This means that there are constants C and α such that $|f(t)| \leq Ce^{\alpha t}$. Also define a function

$$
\begin{aligned}
g(t) &= e^{-\mu t} f(t), \ t \geq 0, \ \mu > \alpha \\
&= 0 \qquad t \leq 0.
\end{aligned}
$$

The function $g(t) \in L_2(R^1)$. The Laplace transform $\tilde{f}(s)$ of $f(t)$ is essentially the Fourier transform $\hat{g}(\xi)$ of $g(t)$ and is defined by

$$
(\mathcal{L}f)(s) = \tilde{f}(s) = \int_0^\infty e^{-st} f(t) \, \mathrm{d}t, \qquad (2.18)
$$

where $\mathcal{L}$ is the Laplace transform operator and the transform variable $s = \mu + i\zeta, \mu, \zeta$ real. The inverse transform which is essentially the inverse Fourier transform of $\hat{g}(\xi)$ multiplied by $e^{\mu t}$ is given by

$$
f(t) = (2\pi i)^{-1} \int_L e^{st} \tilde{f}(s) \, \mathrm{d}s \qquad (2.19)
$$

The contour L in Eq. (2.19), known as the *Bromwich contour*, (Widder, 1971) is a vertical line at Re $s = \mu$, $\mu > \alpha$ in the complex s-plane. This is shown in Figure 2.1. The Laplace transform is an analytic function of s in the half-plane

Re $s > \alpha$. The line of integration can, therefore, be replaced by any path (such as L_1 in the figure) in this half-plane.

An alternative, although not very well-known expression for the inverse Laplace transform is that of Post and Widder (1946) given by

$$f(t) = \lim_{n \to \infty} \left\{ \left[\frac{(-1)^n}{n!} \tilde{f}^{(n)} \left(\frac{n}{t} \right) \right] \left(\frac{n}{t} \right)^{n+1} \right\}. \tag{2.20}$$

The Post–Widder inversion formula avoids integration in the complex plane, but requires derivatives of the transform of all orders. Since the transform is usually known only at a finite number of discrete points and numerical differentiation is, in general, an ill-posed problem (see Example 4 that follows), substantial difficulties can be expected in its practical implementation.

Let us show that inverting a Laplace transform is an ill-posed problem even though the direct operator $\mathcal{L}$ is well-defined. Let us introduce the *Mellin transform* (Titchmarsh, 1939) $(\mathcal{M}f)(\xi)$ of $f(t)$

$$(Mf)(\xi) = \int_0^\infty f(t) t^{i\xi - \frac{1}{2}} \, dt \tag{2.21}$$

and its inverse[1].

$$f(t) = (2\pi)^{-1} \int_{R^1} (Mf)(\xi) t^{-(i\xi + \frac{1}{2})} d\xi. \tag{2.22}$$

It is known that the transform $\mathcal{M}$ is an isometry from $L_2(0, \infty) \to L_2(-\infty, \infty)$. In other words, the transformation is norm preserving:

$$\|f\|^2_{L_2(R^+)} = (2\pi)^{-1} \|\mathcal{M}f\|^2_{L_2(R^1)}. \tag{2.23}$$

Also from the definition of the Γ-function (Abramowitz and Stegun, 1964), namely

$$\Gamma(z) = \int_0^\infty t^{z-1} e^{-t} dt, \quad \text{Re } z > 0,$$

it follows that the Mellin transform of the exponential e^{-t} is a Gamma function in the variable $(i\xi + \frac{1}{2})$, that is,

$$[Me^{-t}](\xi) = \Gamma\left(i\xi + \frac{1}{2} \right). \tag{2.24}$$

---

[1] Other definitions of the Mellin transform occur in the literature. Morse and Feshbach (1953), for example, define the transform as

$$(Mf)(\xi) = \int_0^\infty f(t) t^{i\xi - 1} \, dt$$

and the inverse by

$$f(t) = (2\pi i)^{-1} \int_{-i\infty - \sigma}^{i\infty + \sigma} x^{-\xi} (Mf)(\xi) d\xi, \quad \sigma > \sigma_0, \ x > 0.$$

It is also known (Abramowitz and Stegun, 1964) that

$$\Gamma\left(i\xi + \frac{1}{2}\right)\Gamma\left(\frac{1}{2} - i\xi\right) = \left|\Gamma\left(i\xi + \frac{1}{2}\right)\right|^2$$

$$= \frac{\pi}{\cosh(\pi\xi)} < \pi. \qquad (2.25)$$

Now consider the Laplace transform defined in Eq. (2.18)

$$\tilde{f}(s) = \int_0^\infty e^{-st} f(t)\, dt.$$

Taking the Mellin transform (Eq. (2.21)) on both sides of Eq. (2.18), we obtain

$$(M\tilde{f})(\xi) = (Me^{-st})(\xi)(Mf)(-\xi) = \Gamma(i\xi + \frac{1}{2})(Mf)(-\xi). \qquad (2.26)$$

where Eq. (2.24) was used. From Eqs. (2.26) and (2.25) and the isometry, Eq. (2.23), we have

$$\|\tilde{f}\|_2^2 \leq \pi \|f\|_2^2.$$

From this equation and the definition of the operator norm, it follows at once that the linear operator of the Laplace transform is bounded, the bound being precisely π. The Laplace transform is, therefore, well-defined.

Let us next calculate the inverse Laplace transform. From Eqs. (2.19) and (2.26) and the inverse Mellin transform, Eq. (2.22),

$$f(t) = (2\pi)^{-1} \int_{R^1} \frac{(M\tilde{f})(-\xi)}{\Gamma(\frac{1}{2} - i\xi)} t^{-(i\xi + \frac{1}{2})}\, d\xi, \qquad (2.27)$$

where relation (2.24) was used again. In view of the bound (2.25), it follows that the inverse operator $\mathcal{L}^{-1}$ cannot be bounded and, consequently, cannot be continuous. The inversion of the Laplace transform is, therefore, ill-posed.

The ill-posedness can be viewed from another angle. Being symmetric and Hilbert–Schmidt, the operator $\mathcal{L}$ has a complete orthogonal set of the eigenfunctions $\{\phi_n\}_1^\infty$ with real eigenvalues $\{\lambda_n\}_1^\infty$ which can be arranged in a decreasing order

$$|\lambda_{n+1}| \leq |\lambda_n|.$$

The function $f(t)$ can then be written as

$$f(t) = \sum_{n=1}^\infty \frac{\tilde{f}_n}{\lambda_n}\phi_n(t). \qquad (2.28)$$

Equation (2.28) reveals the ill-posedness. Note that although formally a complete solution, not all of the series (2.28) is useful in practical computations since $\lambda_n \to 0$ as $n \to \infty$. The series must be truncated after a certain number of terms N, the so-called *generalized Shannon number*. Thus Eq. (2.28) can be broken down into two separate series

$$f(t) = \left\{ \sum_{n=1}^{N} + \sum_{n=N+1}^{\infty} \right\} \frac{\tilde{f}_n}{\lambda_n} \phi_n(t). \qquad (2.29)$$

Therefore, an infinite amount of information must perforce remain undetermined. This is in contrast to the Fourier transform where the spectrum tends to a constant as n goes to infinity thereby resulting in an infinite information capacity (McWhirter and Pike, 1978).

McWhirter and Pike actually calculated the spectrum of $\mathcal{L}$ and the results are

$$\phi_\omega^+(\tau) = (\pi)^{-\frac{1}{2}} \left[\cos(\theta) \tau^{-\frac{1}{2}} \cos\left(\omega \ln \tau\right) + \sin(\theta) \tau^{-\frac{1}{2}} \sin\left(\omega \ln \tau\right) \right]$$

and

$$\phi_\omega^-(\tau) = (\pi)^{-\frac{1}{2}} \left[\sin(\theta) \tau^{-\frac{1}{2}} \cos\left(\omega \ln \tau\right) + \cos(\theta) \tau^{-\frac{1}{2}} \sin\left(\omega \ln \tau\right) \right].$$

The subscript ω symbolizes a continuous spectrum and the superscript $\pm$ correspond to the fact the eigenfunctions depend independently on t^s and t^{s-1}. The eigenvalues are given by

$$\lambda_\omega^\pm = \pm \sqrt{a^2 + b^2}, \qquad (2.30)$$

where

$$a = \operatorname{Re}\left(\Gamma\left(\frac{1}{2} + i\omega\right) \right)$$

$$b = \operatorname{Im}\left(\Gamma\left(\frac{1}{2} + i\omega\right) \right)$$

$$\theta = \frac{1}{2} \tan^{-1}\left(\frac{b}{a}\right).$$

In terms of the spectrum, the inverse transform takes the form

$$f(\tau) = \left\langle \frac{\phi_\omega^+}{\lambda_0^+}, \langle \phi_\omega^+, g \rangle \right\rangle + \left\langle \frac{\phi_\omega^-}{\lambda_0^-}, \langle \phi_\omega^-, g \rangle \right\rangle, \qquad (2.31)$$

where the inner product integrals are over ω and s. The ill-posedness of the inverse Laplace transform is manifested in the fact that beyond a certain upper limit Ω of ω, the integrals cease to exist. This limit Ω depends on the magnitude of the error in the data.

Note that in Eq. (2.29), the well- and the ill-conditioned parts are distinctly identified, which was not the case in either Eq. (2.27) or Eq. (2.20). In addition, the inversion formula, Eq. (2.31), explicitly points out that stability can be achieved if *a priori* information is available regarding Ω. It should be pointed out that such a separation can also be effected in the Post–Widder formulation,

Eq. (2.20). This follows from the fact that such separation can be indicated in the numerical derivatives.

It is interesting to note the following. From Eq. (2.30) and using Eq. (2.25)

$$|\lambda_\omega|^2 = \left|\Gamma\left(\frac{1}{2} + i\omega\right)\right|^2 = \frac{\pi}{\cosh(\pi\omega)}.$$

The spectrum of the Laplace operator, therefore, decays rapidly for large ω. On the other hand, for the Fourier transform over the positive half-line, the Mellin transform of a function such as $\cos(\beta s\tau)$, β finite, is given by

$$M(\cos(\beta s\tau)) = \beta^{-(\frac{1}{2}+i\omega)} + \Gamma\left(\frac{1}{2} + i\omega\right)\cos\left(\frac{1}{4}\pi + \frac{i\omega\pi}{2}\right),$$

resulting in the spectrum

$$|\lambda_\omega|^2 = (2\beta)^{-1}\left|\Gamma\left(\frac{1}{2} + i\omega\right)\right|^2 \left[\cosh^2\left(\frac{\omega\pi}{2}\right) + \sinh^2\left(\frac{\omega\pi}{2}\right)\right] = \frac{\pi}{2\beta},$$

which clearly does not decay to zero. This reflects the infinite information capacity of the Fourier transform.

A special type of Laplace transform is the so-called *wave-field transform* (Lavrentiev, 1967; Filatov, 1984) $u(q)$ of a function $f(t)$ defined by

$$f(t) = (2\sqrt{\pi t^3})^{-1}\int_0^\infty qe^{-\frac{q^2}{4t}}u(x, q)dq.$$

This is a Laplace transform between variables t and q. It has found interesting applications in electromagnetic prospecting of the Earth using EM-waves of very low-frequency (< 1 MHz) employed specifically for better ground penetration. At such low frequencies, the EM-field is diffusive, the second-order time derivative in Maxwell's equation being neglected in favor of the first-order. The wave-field transform is used in order to convert the diffusive EM-equation to a fictitious wave equation so that the known techniques for the latter can be applied for inversion of the slowness of the fictitiously propagating waves. The slowness thus determined is then connected to the diffusion coefficient in a known way from which the Earth's electromagnetic parameters can be derived. For further details, the reader is referred to Lee *et al.* (1989).

2.2.4   Example 4. Numerical Differentiation

An archetypical example (although not necessarily an inverse problem) of ill-posedness and the effect of *a priori* information on restoring stability is the numerical differentiation of a function. Consider the equation

$$(Af)(x) = g(x) = \int_0^x f(x')dx', \ 0 \le x \le 1, \tag{2.32}$$

the solution of which is $f(x) = g'(x)$. The differentiation problem will be discussed in various function spaces.

Let us begin with $A : C[0,1] \to C[0,1]$ with the supremum norm $\| \cdot \|_\infty$. Let $f(y)$ be perturbed to a new function $\tilde{f}(y) = f(y) + f_0 \sin ky$, f_0 being an arbitrary number. Then

$$\tilde{g}(x) = \big(g(x) + f_0 k^{-1}\big) - f_0 k^{-1} \cos kx.$$

We see again that $\tilde{g} \to g$ as $k \to \infty$, whereas $|\tilde{f} - f| \to |f_0|$, by the Riemann-Lebesgue theorem. The problem of differentiation in the pair of spaces $(C[0,1], C[0,1])$ is, therefore, ill-posed.

Let g be in C^1. We still have $|\tilde{f} - f| \to |f_0|$ as $k \to \infty$. But now

$$\|\tilde{g} - g\|_{C^1} \leq \sup_{0 \leq x \leq 1} \big[|\tilde{g} - g| + |(\tilde{g} - g)'| \big] \to |f_0|$$

as $k \to \infty$. Hence the problem is well-posed between spaces $(C[0,)], C^1[0,1])$.

Next consider that $f \in L_2(0,1)$ and measure g in L_2-norm. Then

$$\|\delta f\|_2 = \frac{f_0}{\sqrt{2}} [1 - (2k)^{-1} \sin 2k]^{\frac{1}{2}}.$$

At the same time

$$\|\delta g\|_2 = \frac{f_0}{\sqrt{2k}} [3 + k^{-1}(\sin 2k - 4\sin k)]^{\frac{1}{2}}.$$

Thus for a finite, but arbitrarily large f_0, $\|\delta g\|_2 \to 0$ while $\|\delta f\|_2 \to |f_0$ as $k \to \infty$. The problem is again ill-posed.

Let us look at the problem from a different angle and calculate the L_2-norm of f. Thus we have

$$
\begin{aligned}
\|f\|_2^2 &= \int_0^1 f(y)^2 dy \\
&= -\int_0^1 f'(y) \left\{ \int_0^y f(y') dy' \right\} + \left[f(y) \int_0^y f(y') dy' \right]_0^1 \\
&= -\int_0^1 f'(y) \left\{ \int_0^y f(y') dy' \right\} + f(1)g(1) \qquad (2.33)
\end{aligned}
$$

Assume *a priorily* that $|f'| \leq M_1$ is bounded, $f(1) = 0$ and $\|g\|_2 \leq \epsilon$. Then applying Schwartz's inequality to Eq. (2.33) obtains

$$\|f\|_2 \leq \sqrt{M_1}\epsilon. \qquad (2.34)$$

In other words, the problem of differentiation is well-posed between spaces X_1 and $L_2(0,1)$, where $X_1 = \{u : u \in H^1(0,1), u(1) = 0\}$.

Suppose that we calculate the norm $\|f'\|_2$. This gives

$$
\begin{aligned}
\|f'\|_2^2 &= \int_0^1 f'(y)^2 dy \\
&= -\int_0^1 f(y)f''(y)dy + [f(y)f'(y)]\Big|_0^1 \\
&\leq \|f\|_2 \|f''\|_2.
\end{aligned}
$$

Substituting $\|f'\|_2$ thus calculated into Eq. (2.33) gives

$$\|f\|_2^2 \leq \|g\|_2 \sqrt{\|f\|_2} \sqrt{\|f''\|_2}.$$

If it is assumed that $\|f''\|_2 \leq M_2$, i.e. the second derivative of f is also bounded, then we finally have

$$\|f\|_2 \leq \epsilon^{\frac{2}{3}} M_2^{\frac{1}{3}}. \tag{2.35}$$

The numerical differentiation problem (2.32) is again well-posed between X_2 and $L_2(0,1)$ with

$$X_2 = \{u : u \in H^2(0,1), u(1) = 0\}.$$

In the above last two cases, f was in a space Y (X_1 and X_2) which happened to be embedded in a larger space X which was $L_2(0,1)$ such that $\|\cdot\|_Y \leq \|\cdot\|_X, \|f\|_Y \leq M$ being bounded. It is a general result (Franklin, 1974; Cullam, 1979; Kirsch, 1996) that in such a case $\|f\|_2 \leq \epsilon^j M^k, j, k$ depending upon Y. In general, the exponent j over ϵ cannot be increased and the estimate is called *asymptotically sharp*.

Finally, consider differentiation in a discrete setting. Let the interval $[0,1]$ be subdivided into equal spacing of size $2h$. Then

$$2h \sum_{j=1}^i f[(2j-1)h] = g(2hi), \quad i = 1, 2, \cdots N$$

or in matrix notation

$$\mathbf{g} = \mathbf{A}\mathbf{f},$$

where $\mathbf{A}$ is lower-triangular: $(\mathbf{A})_{ij} = H(i-j)$, H is unity if $i \geq j$, and zero otherwise. Approximating g' by a forward-difference formula gives

$$f(2i+1)h = \frac{g(2i+2)h - g(2ih)}{2h} + O(h^2)$$

to second-order in h.

Let us calculate the *condition number* of $\mathbf{A}$ (see Appendix A.5.1). Assuming that the norm of $\mathbf{A}$ is given by

$$\|\mathbf{A}\|_\infty = \max_i \sum_{j=1}^n |(\mathbf{A})_{ij}|,$$

the condition number is

$$\text{cond}(\mathbf{A}) = \|\mathbf{A}\|_\infty \|\mathbf{A}^{-1}\|_\infty = N^2 \sim O(h^{-2}).$$

The system thus becomes ill-conditioned as $h \to 0$. This means that the mesh cannot be refined *ad infinitum* and some optimal step-length must be obtained.

The situation remains the same if the integral in Eq. (2.32) is reduced by quadrature, that is, if we write the integral as

$$g(x_i) = \sum_{j=1}^{N} w_{ij} f(x_j), \qquad j = 1, 2, ..., N$$

with weight functions w_{ij}. The quadrature sum may also run into the same problem as the mesh is refined indiscriminately. The resulting matrix may become ill-conditioned and the sum may oscillate violently with the refinement of the discretization (Varah, 1983).

In any given case, an optimum discretization may be obtained if appropriate *a priori* information about f is available. The only information used so far about f was its boundedness or equivalently, that of the first derivative of g. Let us now suppose that $|f'| \leq M$, $\forall y \in (0, 1)$, that is, the first derivative of f is bounded. By approximating f with a first-order accurate difference, it can be easily seen that the error in this case is of the order of hM. Thus the additional information that f' is bounded has introduced stability against finer discretization.

We now carry on the analysis a little further by considering data with error, namely, $g^\epsilon : |g^\epsilon - g| \leq \epsilon > 0$. If a first-order accurate difference is used, then by the Taylor expansion around y we obtain

$$\begin{aligned}
|f^\epsilon(y) - f(y)| &\leq \frac{|g^\epsilon(y+h) - g^\epsilon(y)|}{h} + |f(y)| \\
&\leq hM/2 + \frac{2\epsilon}{h}.
\end{aligned} \qquad (2.36)$$

Minimizing the right-hand side of Eq. (2.36), h is determined to be $2M^{-\frac{1}{2}}\epsilon^{\frac{1}{2}}$. Substituting back into Eq. (2.36) yields

$$|f^\epsilon(y) - f(y)| \leq 2\sqrt{M\epsilon}. \qquad (2.37)$$

The convergence then proceeds at the rate of $\sqrt{M\epsilon}$. The condition number in this case is M/ϵ and the error can no longer be unbounded if the *a posteriori* choice of $h = 2M^{-\frac{1}{2}}\epsilon^{\frac{1}{2}}$ is used.

Let us next use a second-order central-difference approximation

$$f^\epsilon(y) = \frac{g^\epsilon(y+h) - g^\epsilon(y-h)}{2h}. \qquad (2.38)$$

Now taking the norm of the difference $f^\epsilon(y) - f(y)$, using Eq. (2.38) and after a little manipulation, we obtain

$$
\begin{aligned}
|f^\epsilon(y) - f(y)| &\leq \frac{|g^\epsilon(y+h) - g^\epsilon(y-h)|}{2h} + |f(y)| \\
&\leq \frac{|g(y+h) - g(y-h)|}{2h} + |f(y)| + \frac{\epsilon}{h} \\
&\leq \frac{h^2 M}{6} + \frac{\epsilon}{h}.
\end{aligned}
\qquad (2.39)
$$

Again minimizing the right-hand side of Eq. (2.39), h is found to be $(3\epsilon M^{-1})^{\frac{1}{3}}$ which when substituted back into Eq. (2.39) yields

$$
|f^\epsilon(y) - f(y)| \leq 2^{-\frac{1}{2}} (3M^{\frac{1}{2}}\epsilon)^{\frac{2}{3}}.
\qquad (2.40)
$$

The rate of convergence is now proportional to $(M^{\frac{1}{3}}\epsilon^{\frac{2}{3}})$. The new condition number is $(M/3\epsilon)^{\frac{2}{3}}$. Again since the optimal fineness of discretization is determined *a posteriorily*, the unbounded error propagation is eliminated, the resolution being determined solely by the level of error ϵ in the data and the *a priori* information about the solution. This is in conformation with the fact that for $\mathbf{A}f = \mathbf{g}$, the error in the solution is directly correlated to that in the data implying continuous dependence of the solution on the latter. The results also show that the rate of convergence and stability depend upon the smoothness of the function being recovered. This again illustrates what was said in Chapter 1, namely, the ill- or well-posedness is a function of the Gel'fand triple as a whole. It is also interesting to consider the discrete results Eqs. (2.37) and (2.40) with the corresponding continuous quantities given by Eqs. (2.34) and (2.35) which are the same.

2.2.5   Example 5. Inverse Source Problem

The inverse source problem runs as follows. Let $f(x)$, $x \in R^3$, describe a source of radiation (e.g., a current distribution in an antenna, vibrations on the surface of a body immersed in a fluid, rotating propeller blades of an aircraft, and so forth) inside a bounded domain $\Omega \in R^3$. The emitted radiation (electromagnetic, acoustic or elastic) is detected on a measuring surface in the space surrounding Ω. The problem is to recover the source distribution f from the knowledge of the radiation field on the measurement surface. The inverse source problem (ISP) is a paradigm of ill-posedness and, naturally, has been investigated extensively (Baltes, 1978; Anger, 1994; Anikonov *et al.*, 1997). An important class of ISP is *inverse diffraction* and its close variant *near-field holography* (references cited below).

The emission from a time harmonic source satisfies the *reduced wave* or the *Helmholtz equation* (Chapter 6)

$$
(\Delta + k^2)\psi = -f, \ f \in \Omega \subset R^3,
$$

where $k = 2\pi\lambda^{-1}$ is the wavenumber of the emitted radiation of wavelength λ and f is the source distribution (or rather its Fourier transform in time). Moreover, ψ must satisfy the radiation condition at infinity:

$$\lim_{r\to\infty}(\psi_r - ik\psi) \sim o(r^{-1}), \quad r = |x|.$$

The radiation condition ensures that the emitted waves propagate away as outgoing spherical waves in order to eliminate the existence of any source of radiation at infinity (see Chapter 6 for details). The emission is described by the following integral equation of the first kind

$$\psi(x) = \int_\Omega G^0(x, x')f(x')\mathrm{d}x', \qquad (2.41)$$

where

$$G^0(x, x') = \frac{\mathrm{e}^{ik|x-x'|}}{4\pi|x - x'|}$$

is the free-space Green's function which reduces to an outwardly propagating spherical wave at infinity according to

$$\lim_{|x|\to\infty} G^0(x, x') \sim \frac{\mathrm{e}^{ik|x|}}{4\pi|x|}\left[\mathrm{e}^{-ik\hat{x}\cdot x'} + O(|x|^{-1})\right], \qquad (2.42)$$

where $\hat{x} = x|x|^{-1}$ denotes the direction of propagation and

$$\psi^\infty(\hat{x}) = \lim_{|x|\to\infty} \psi(x)$$

the *radiation amplitude*. Then from Eq. (2.41) and the asymptotic form of Eq. (2.42) of Green's function, we have

$$\psi^\infty(\hat{x}) = (Af)(\hat{x}) = \frac{1}{4\pi}\int_\Omega \mathrm{e}^{-ik(\hat{x}\cdot x')}f(x')\mathrm{d}x', \qquad (2.43)$$

the integral operator $A : L_2(\Omega) \to L_2(\hat{\Omega}_3)$ being compact. $(\hat{\Omega}_3)$ is the unit sphere in R^3. We can, therefore, expect difficulties and the following discussion will make it abundantly clear.

The source is assumed to have a compact support. The integral in Eq. (2.43) may, therefore, be considered to be over R^3. $\psi^\infty(\hat{x})$ then becomes the Fourier transform of f over a sphere of radius k in the Fourier space, the so-called *Ewald sphere* (Luneberg, 1966; Langenberg, 1987; Carrion, 1987). Moreover, $\psi^\infty(\hat{x})$ is an analytic function and from discussions in Chapter 1 this implies ill-posedness.[2] $L_2(\Omega)$ can be uniquely decomposed into the *direct sum* of $N(A)$ (the nullspace of A) and its orthogonal complement $N(A)^\perp$, i.e., $L_2(\Omega) = N(A) \oplus N(A)^\perp$, $\oplus$ denoting the direct sum (see Chapter 3, Figure 3.2). This decomposition means that any function in a Hilbert space can be uniquely

---

[2]It will be shown in Chapter 6 that any solution of Helmholtz's equation in the exterior is an analytic function in all its variables.

decomposed into two components, one in $N(A)$ and the other in $N(A)^\perp$. In ISP, $N(A)$ consists of those sources for which $\psi^\infty(\hat{x})$ is identically zero over the Ewald sphere, and in $N(A)^\perp$ are the sources which produce nonzero radiation. The nonradiative sources in $N(A)$ cannot be reconstructed from the measured data revealing at once the inherent nonuniqueness of ISP[3] Mathematically, the reconstruction is done *via* the *adjoint* operator A^* giving

$$(A^*\psi^\infty)(x) = \frac{1}{4\pi} \int_{\hat{\Omega}_3} e^{ik\hat{x}'\cdot x}\psi^\infty(\hat{x}')d\hat{x}'. \tag{2.44}$$

The range of A^* coincides with $N(A)^\perp$ (see Chapter 3).

In order to see the ill-posedness somewhat more explicitly, let us use the following well-known expansion (Jackson, 1975)

$$e^{ik|x|(\hat{x}\cdot\hat{x}')} = 4\pi \sum_{\ell=0}^{\infty} \sum_{m=-\ell}^{\ell} i^\ell j_\ell(k|x|)Y_\ell^m(\hat{x})Y_{\ell m}^*(\hat{x}'), \tag{2.45}$$

where Y_ℓ^m is a *spherical harmonic* of order ℓm and the superscript * denotes complex conjugation. The spherical harmonics are discussed in Appendix A.8.1. $j_\ell(k|x|)$ is the spherical Bessel function of order ℓ and argument $k|x|$. Substituting the expansion in Eq. (2.45) into Eq. (2.44) yields

$$(A^*\psi^\infty)(x) = \sum_{\ell=0}^{\infty} \sum_{m=-\ell}^{m=\ell} \psi_{\ell m}^\infty \tilde{\phi}_{\ell m}, \tag{2.46}$$

where

$$\psi_{\ell m}^\infty = \int_{\hat{\Omega}_3} \psi^\infty(\hat{x}')Y_{\ell m}^*(\hat{x}')$$

and $\tilde{\phi}_{\ell m}(x) = i^\ell j_\ell(k|x|)Y_{\ell m}^*(\hat{x})$. Equation (2.46) shows that the functions $\tilde{\phi}_{\ell m}$ span the space $N(A)^\perp$ since they are in the range of A^* (see section 3.4). We can, therefore, expand $f(x) \in N(A)^\perp$ in terms of the set $\{\tilde{\phi}_{\ell m}(x)\}$ as

$$f(x) = \sum_{\ell=0}^{\infty} \sum_{m=-\ell}^{m=\ell} f_{\ell m}\tilde{\phi}_{\ell m}(x)$$

yielding

$$(Af)(\hat{x}) = \sum_{\ell=0}^{\infty} \sum_{m=-\ell}^{m=\ell} f_{\ell m}A\tilde{\phi}_{\ell m}(x).$$

Furthermore, inserting Eq. (2.45) into Eq. (2.43) gives

$$(Af)(\hat{x}) = \sum_{\ell=0}^{\infty} \sum_{m=-\ell}^{m=\ell} \lambda_\ell f_{\ell m}Y_{\ell m}^*(\hat{x}'),$$

---

[3]An interesting discussion as to how to specify nonradiating sources is given in Gamliel *et al.* (1989).

where

$$\lambda_\ell = \int_0^R r^2 j_\ell(kr) dr.$$

The source was assumed to be confined within a ball of radius R. Comparing the above two expressions of $(Af)(\hat{x})$, we obtain

$$A\tilde{\phi}_{\ell m} = \lambda_\ell Y_{\ell m}.$$

Defining

$$\phi_{\ell m} = \frac{\tilde{\phi}_{\ell m}}{\sqrt{\lambda_\ell}},$$

we finally obtain an eigenvalue equation for A, namely

$$A\phi_{\ell m} = \mu_\ell Y_{\ell m}, \tag{2.47}$$

where $\mu_\ell = \sqrt{\lambda_\ell}$. Similarly, noting that the spherical harmonics constitute the basis set for $L_2(\hat{\Omega}_3)$, Eqs. (2.43) and (2.45) lead to the eigenvalue problem for the adjoint operator A^* which is

$$A^* Y_{\ell m} = \sqrt{\lambda_\ell}\phi_{\ell m}. \tag{2.48}$$

The pair of equations (2.47) and (2.48) form the basis of what is known as the *singular value decomposition* (SVD) of A, σ_ℓ being the singular values. A detailed discussion of SVD will appear in Chapters 3 and 5. It is clear that the faster the singular values vanish, the more rapid is the loss of information contained in the high spatial frequencies of f, and the more severe is the smoothing. This becomes more evident if we write

$$\psi^\infty(\hat{x}) = \sum_{\ell=0}^\infty \sum_{m=-\ell}^{m=\ell} \psi_{\ell m}^\infty Y_{\ell m}(\hat{x})$$

and compare with the R.H.S. of Eq. (2.48). This gives

$$\psi_{\ell m}^\infty = (-\mathrm{i})^\ell \sigma_\ell^2 f_{\ell m}$$

which is connected to the normal operator A^*A. It is clear that small values of σ_ℓ implies less information about f in the data. In the language of partial wave expansion (Messiah, 1958; Roman, 1965), the emission *cross-section* becomes small meaning that less energy reaches the detectors. This results in ill-posedness. The faster the singular values vanish the stronger is the ill-posedness.

We next discuss an important class of inverse source problems, namely, inverse diffraction and near-field holography (Williams, 1999; Bleistein and Cohen, 1977; Devaney and Wolf, 1973; Hoenders, 1978).

Inverse Diffraction and Near-Field Holography

In inverse diffraction and near-field holography, the fields due to some distribution of sources are measured on a surface (planar, cylindrical or spherical) away from the sources. The determination of the emitted fields on the measurement surface, given the sources, is the problem of forward diffraction, whereas the recovery of the source distribution by propagating the measured fields backward to the sources constitutes the inverse problem. The latter is plagued with the problem of the so-called *evanescent* or *inhomogeneous* waves which decay rapidly with the propagation distance. It is analogous to the phenomenon of *skin depth* in electrodynamics (Jackson, 1975). A detailed account of these waves, their mathematics and applications in diverse physical problems can be found in the monograph of Caviglia *et al.* (1992). The waves that reach the detectors and can, therefore, be used for gathering information about the source are called the *homogeneous* waves. Below we will demonstrate how ill-posedness arises, its connection with the inhomogeneous waves and how stability can be restored via appropriate *a priori* information about the source. We consider only the planar geometry here. More involved cylindrical and spherical geometries are discussed in a recent monograph by Williams (1999).

Let us begin with the concept of the *angular spectrum* (Goodman, 1968; Luneberg, 1966; Langenberg, 1987; Lalor, 1968) which is central to our discussion. Let $\psi(x)$ be the field (due to some source f) on a planar surface Γ, say, parallel to the xy-plane, at a certain height $z > 0$. Then for $x \in \Gamma$, $\psi(x) = \psi(x, y, z) = \psi(x_\perp; z)$, where $x_\perp = (x, y)$. The coordinate z being constant, we can Fourier transform $\psi(x)$ in x and y to obtain

$$\psi(x_\perp; z) = \int_{R^2} \hat{\psi}(k_\perp; z) e^{ik_\perp \cdot x_\perp} \, dk_\perp, \tag{2.49}$$

and the inverse Fourier transform is

$$\hat{\psi}(k_\perp; z) = \frac{1}{4\pi^2} \int_{R^2} \psi(x_\perp; z) e^{-ik_\perp \cdot x_\perp} \, dx_\perp, \tag{2.50}$$

where $k_\perp = (k_x, k_y)$ is the Fourier transform vector in the plane and $\hat{\psi}(k_\perp; z)$ is the Fourier transform of $\psi(x)$ in $x_\perp$. The spectrum $\hat{\psi}(k_\perp; z)$ is given the name *angular spectrum* and its determination is tantamount to the solution of the forward propagation problem. Using Eqs. (2.49) and (2.50), the following propagation equation between planes at z and z' can be obtained, namely

$$\psi(x_\perp; z) = \frac{1}{4\pi^2} \int_{R^2} \int_{R^2} \hat{\psi}(x'_\perp; z') e^{-i\left[k_\perp \cdot (x'_\perp - x_\perp) + k_z(z' - z)\right]} \, dx'_\perp dk_\perp. \tag{2.51}$$

$k_\perp + k_z$ is the total wavevector k. Note that Eq. (2.51) is independent of whether $z > z'$ or $z < z'$ as long as $z, z' \geq 0$. It tells us that each Fourier component in the spectrum $\hat{\psi}(k_\perp; z')$ on a plane at z' propagates to a plane at z with a wavevector k_z and that the field distribution $\psi(x; z)$ on z is the superposition of these waves. Thus a whole gamut of plane waves propagates out of the plane at z' toward a plane at z with all possible $k_\perp$-vectors, so that the tips of the total wavevectors cover the entire hemisphere of $\hat\Omega_3$. Moreover, it follows from Eqs. (2.49)–(2.51) that the angular spectrum at z is related to that at z' by

$$\hat\psi(k_\perp; z) = \hat\psi(k_\perp; z')e^{ik_z(z-z')}.$$

This implies that each frequency component of the angular spectrum $\hat\psi(k_\perp; z')$ reaches z carrying information about this particular frequency alone. This is important for inverse propagation as will appear later.

Now, since $k = k_\perp + k_z$, it follows that

$$k_z = \sqrt{k^2 - |k_\perp|^2}.$$

Thus if $|k_\perp| > k$, then k_z is imaginary and the propagation decays as $e^{-k_z z}$ along z producing evanescent waves.[4] The waves for which $|k_\perp| < k$, in which case k_z is real, are the homogeneous waves. Pictorially, if one draws a circle of radius k, called the *radiation circle*, then $|k_\perp|$ lies outside (inside) this circle for the inhomogeneous (homogeneous) waves. Accordingly, the wave speed is greater (homogeneous) or less (inhomogeneous) than that in the medium. In the literature, these pass for *supersonic* and *subsonic* waves, respectively (Williams, 1999).

At this point let us introduce the well-known *Weyl representation* (Weyl, 1919; Banos, 1966) of a spherical wave in terms of the plane waves which is

$$\frac{e^{ikr}}{r} = \frac{ik}{2\pi}\int_{R^2}\frac{1}{k_z}e^{i(k_\perp \cdot r_\perp + k_z|z-z'|)}dk_\perp, \qquad (2.52)$$

where $r = |x - x'|$ and $r_\perp = x_\perp - x'_\perp$. Assuming that $z > z'$ in Eq. (2.51) and using the representation Eq. (2.52) we obtain the celebrated *Rayleigh integral formula* (Rayleigh, 1897) of propagation from z' to z

$$\psi(x_\perp; z) = 2\int_{R^2}\psi(x'_\perp; z')\big(G^0(x_\perp, x'_\perp; z \leftarrow z')\big)_{z'}\,dx'_\perp, \ z > z'. \qquad (2.53)$$

The subscript z' on Green's function implies differentiation with respect to that variable. We have also explicitly indicated the direction of propagation by the arrow inside Green's function. The integral in Eq. (2.52) describes propagation in the forward direction. The integral in Eq. (2.53) is continuous and converges if $z \geq z'$.

It is in reversing the direction of propagation from $z' \to z, z > z'$ in the Rayleigh formula (2.53) where the problem lies. In order to locate the problem,

---

[4]In order to appreciate the rapidity with which these waves attenuate, consider, as an example, $k_x = 2k$ and $k_y = k$. This means that the periodicity in the x-coordinate is $\Delta x = 2^{-1}\lambda$ and in the y-direction $\Delta y = \lambda$. The wavenumber k_z^2 is then $-4k^2$. In propagating over a distance equal to one wavelength, the waves attenuate by a factor of $e^{-4\pi z\lambda^{-1}}$ which is about 3.5×10^{-6} of its initial value. If the propagation distance is $10^5\lambda$ (roughly a length of 5 cm at the optical frequencies) this factor is an astronomical number of $10^{-5\times10^5}$.

let us go back to Eq. (2.51) and interchange z and z', z still being greater than z'. We then obtain the equation

$$\psi(x'_\perp; z') = \int_{R^2} \hat{\psi}(x_\perp; z) \left[\frac{1}{4\pi^2} \int_{R^2} e^{-i\left[k_\perp \cdot (x_\perp - x'_\perp) + k_z(z - z')\right]} \, dk_\perp \right] dx'_\perp.$$

The above equation can be written as a sum of two integrals, namely

$$\psi(x'_\perp, z') = \frac{1}{r\pi^2} \left[\int_{R^2} \psi(x_\perp, z) K_>(x_\perp - x'_\perp, z - z') dx_\perp \right.$$
$$\left. + \int_{R^2} \psi(x_\perp, z) K_<(x_\perp - x'_\perp, z - z') dx_\perp \right], \qquad (2.54)$$

where

$$K_>(x_\perp - x'_\perp, z - z') = \int_{R^2, |k_\perp| > k} e^{ik_\perp \cdot (x_\perp - x'_\perp) + |k_z|(z - z')} \, dk_\perp, \qquad (2.55)$$

and

$$K_<(x_\perp - x'_\perp, z - z') = \int_{R^2, |k_\perp| \le k} e^{i[k_\perp \cdot (x_\perp - x'_\perp)] - k_z(z' - z)} \, dk_\perp. \qquad (2.56)$$

The integral in Eq. (2.55) is over the subsonic and in Eq. (2.56) over the supersonic waves. Since k_z is real for the latter, the integral in Eq. (2.56) is well-behaved. However, the subsonic part of the integral Eq. (2.55) (where k_z is purely imaginary) diverges since the integration variable $k_\perp$ ranges over the entire R^2-plane. Here then we encounter the difficulty in back propagating the forward-propagated waves by using Rayleigh's type of propagation formula.

Now Eq. (2.54) involves a fourfold iterated integrals (two over $dk_\perp$ and two over $dx_\perp$) and we have expressed a fourfold integral of the form

$$\int_{R^2 \times R^2} f(x)g(y) dx dy, \quad x, y \in R^2$$

as

$$\int_{R^2} f(x) \left[\int_{R^2} g(y) dy \right] dx.$$

We have thus interchanged the integration variables in an iterated integral expression. Recall that *Fubini's theorem* (Hille, 1964) permits us to do so provided that each of the integrals exists. Since one of the integrals, Eq. (2.55), is divergent, Fubini's theorem does not hold and the interchange of the integration variables is illegitimate. This was well pointed out by Shewell and Wolf (1968), but without reference to Fubini's theorem which provides the correct mathematical interpretation.[5]

---

[5]Our primary concern in this example is to illustrate the origin of the ill-posedness and how additional information about the source helps in restoring stability. We, therefore, do not derive the inverse Rayleigh integral here. Discussions about the inverse diffraction formula appear in the references already cited.

Let us look at the physics involved (see also Shewell and Wolf (1968) for a clear exposition). Now in the forward propagation, the subsonic or the short wavelength spectrum undergoes decay. The shorter the wavelength, the faster is the decay. Because the inverse propagation tries to restore this wave at the source, it amplifies it, and the higher the decay, the stronger is the amplification. If the amplification balances the decay of the signal exactly, then, of course, correct restoration would take place. The reality, however, is otherwise, and destroys this delicate balance by introducing small perturbations which are always present in the form of noise. These small high-frequency perturbations are then amplified by the inverse propagator. We again encounter the archetypical maláise of ill-posedness: small perturbations in the data inducing arbitrarily large changes in the solution. Again the stability can be restored *via* proper *a priori* information regarding the high-frequency content of the source. In other words, the spectrum must be bandlimited based on physical considerations. This means that the region of integration in Eq. (2.55) is finite and no divergence occurs. As to the details regarding various filtering processes, we refer the reader to Williams (1999).

Let us next look at the problem from the perspective of the direct and inverse operators. We again assume that the sources are contained in a ball of radius R with boundary Γ_R, the field distribution on which is $\psi(R)$. The direct problem is to solve the reduced wave equation

$$(\Delta + k_0^2)\psi(x, k_0) = 0$$

with boundary data $\psi(R, k_0)$ on Γ_R plus the radiation condition at infinity and k_0 is the wavenumber in the exterior region. It is well-known (Colton and Kress, 1992) that outside Γ_R, the solution can be written as

$$\psi(x, k_0) = \sum_{\ell=0}^{\infty} \sum_{m=-|\ell|}^{|\ell|} \alpha_{\ell m} \frac{h_\ell^{(\ell)}(k_0|x|)}{h_\ell^{(1)}(k_0 R)} Y_{\ell m}(\hat{x}). \tag{2.57}$$

For large ℓ, the ratio

$$\frac{h_l^{(1)}(k_0|x|)}{k_0 R} \sim O\left(\frac{R}{|x|}\right) \tag{2.58}$$

(Abramowitz and Stegun, 1964). Therefore, if $|x| > R$, the expansion (2.57) converges, the rate of convergence being faster the farther away $|x|$ is from Γ_R. The coefficient $\alpha_{\ell m}$ in the expansion (2.57) is just the inner product $\langle \psi|_{\Gamma_R}, Y_{\ell m}\rangle_{\hat{\Omega}_3}$, where $\psi|_{\Gamma_R} = \psi(R, k_0)$, the value of ψ on the boundary Γ_R. More explicitly,

$$\alpha_{\ell m} = \frac{k_0}{4\pi} i^\ell \int_\Omega j_\ell(k_0|x'|f(x')Y_{\ell m}(\hat{x})dx'. \tag{2.59}$$

For every fixed field-point x, the expansion is unique and clearly depends continuously on the data through α_{lm} in view of Eq. (2.59). The direct problem is, therefore, well-posed.

Let us next look at the far-field asymptotics of Eq. (2.57). For $|x| \to \infty$, the Hankel function $h_l^{(1)}(k_0|x|)$ behaves as (Abramowitz and Stegun, 1964)

$$\lim_{|x| \to \infty} h_l^{(1)}(k_0|x|) = (-\mathrm{i})^{l+1} \frac{e^{\mathrm{i}k_0|x|}}{k_0|x|} \left[1 + O(|x|^{-1})\right]. \qquad (2.60)$$

Using $\alpha_{\ell m}$ and the asymptotic expression (2.60) for $h_\ell^{(1)}$ into the expansion (2.57), and after a little algebraic manipulation, we obtain

$$\lim_{|x| \to \infty} \psi(x, k_0) = \psi^\infty(\hat{x}, k_0) = \int_{\hat{\Omega}_3} K(\hat{x}, \hat{x}')\psi(R, k_0)\mathrm{d}\hat{x}', \qquad (2.61)$$

where

$$K(\hat{x}, \hat{x}') = k_0^{-1} \sum_{l=0}^{\infty} \sum_{m=-|l|}^{|l|} (-\mathrm{i})^{l+1} \left[h_l^{(1)}(k_0 R)\right]^{-1} Y_{lm}(\hat{x}) Y_{lm}^*(\hat{x}').$$

The kernel $K(\hat{x}, \hat{x}')$ is square integrable. It is, therefore, Hilbert-Schmidt and the integral operator in Eq. (2.61) is compact. Moreover, the eigenvalues λ_l, given by

$$\lambda_l = k_0^{-1}(-\mathrm{i})^{l+1} \left[h_l^{(1)}(k_0|x|)\right]^{-1}$$

are nonzero and vanish only at infinity. Since λ_l tends to zero, the solution cannot depend continuously with the data. The recovery of the source field distribution from the far-field diffraction pattern is thus ill-posed in spite of the uniqueness of the solution.

Next, suppose that $|x|$ is not very far away from Γ_R. From Eq. (2.60) it is clear that if $R|x|^{-1}$ is not too small, the series in Eq. (2.57) does not fall off rapidly. This only reflects the presence of the evanescent waves. The success of the near-field holography in many areas of practical applications (Williams, 1999) is due to the fact that it uses data quite close to the surface and, therefore, can afford a higher cut-off in frequencies.

2.2.6   Example 6. An Example from Medical Diagnostics

In this example, we present an interesting inverse problem in medical diagnostics. It is a biomagnetic ISP and its objective is the diagnosis of the dynamic activities and anomalies of the brain. As of now, the technique has been applied primarily to the cerebral cortex region, the uppermost layer of the brain which is a 2–4 mm thick sheet of gray tissue and contains approximately 10^{24}–10^{26} neurons.[6] These cells are active in a vast signal processing network containing

---

[6] The cortex region is as large as about 2500 cm^2. In order to accommodate this massive area within the cranial cavity formed by the skull, this sheet of gray matter is folded over and again in a complicated way. It is this that gives rise to the convoluted appearance of the cerebral cortex familiar from the anatomical pictures of the brain in medical texts. The various areas of the cortex are responsible for processing signals from and directing the functions of

some 10^{14} switching connections called *synapses*. If an area of the brain is activated by a stimulus (sound, a picture or the tactile somatosensory perceptions), the neurons in the stimulated area become minuscule current dipoles. These are primarily intramembrane and intracellular currents which are generated by the diffusion of ions down the concentration gradients set up by the stimulus. The neural current thus generated produces a weak magnetic field in accordance with the laws of electrodynamics. A typical strength of a current dipole representing the synchronous, coherent activities of possibly tens of thousands of neurons in a particular site is of the order of 10 nAm (10^{-9} Ampere-meter) and the neuromagnetic field produced is of the order of 0.1 pT (10^{-13} Tesla) which is roughly 10^{-8} or 10^{-9} times that of the Earth's magnetic field (Hämäläinen *et al.*, 1993, and Malmivuo and Plonsey, 1995). The neuronal magnetic fields are measured outside the brain by biomagnetometers called SQUIDs or *superconducting quantum interference devices*. A recording of the spatio-temporal neuromagnetic field of the brain is called a *magnetoencephelogram.*[7] and the diagnostic technique *magnetoencephelography* or MEG in short (see the review by Williamson and Kaufman, 1989).

The inverse problem is: given a set of magnetoencephelograms, obtain a *neuromagnetic image* of the brain, a three-dimensional image of the current sources that produced the recorded magnetoencephelograms. The image is indicative of brain functions. In MEG it is possible to follow the neural activity on a millisecond scale thereby providing unique insights into the dynamic behavior of the brain compared to a much slower response of at best one second in other imaging modalities such as magnetic resonance imaging (MRI) and computed tomography (Mosher *et al.* 1997).

The problem is ill-posed on two counts. First, the basic formulation is via an integral equation of the first kind and secondly, the solution is nonunique, i.e., the integral operator has a nontrivial nullspace. This means that analogous to the nonradiative sources in the inverse diffraction problem of the last example, there are magnetically silent current sources in MEG that do not produce magnetic fields outside. It is a long-standing result (Helmholtz, 1853) that a current distribution inside a conductor cannot be retrieved uniquely from the information about the electromagnetic (EM) fields outside (Malmivuo and Plonsey, 1995). The nature of the *a priori* information in MEG will be discussed later after the fundamentals are presented.

In MEG, the time variations of the electric (**E**) and magnetic (**B**) fields are

---

various parts of the body. For example, the primary motor cortex directs the movement of a specific part of the body. There is an area in the frontal lobe of the cortex where the neurons convey the signals of muscular activity. Similarly, there are primary visual and auditory areas in the cortex as well as large areas for controlling those body parts which need finer and more accurate control of movement such as the lips, and so on. Besides these, there are fissures in the cortex (much like canyons in a mountain) in between the lobes of the brain. These are called the *Sylvan* and the *Rolandic* fissures. The moment vectors of the neuronal dipoles in these fissures are roughly tangential to the approximately hemispherical lobes. This fact has particular significance for imaging the dynamic activities of the brain (as we will see momentarily), especially if the head can be considered spherical in shape.

[7]The first such fields were detected by Cohen (1968).

considered to be negligible in a region of the size of the head and the starting point is thus the quasi-static system of Maxwell's equations (Jackson, 1975)

$$
\begin{aligned}
\nabla \cdot \mathbf{E} &= \frac{\rho}{\epsilon_0}, \\
\mathbf{E} &= -\nabla \phi \\
\nabla \cdot \mathbf{B} &= 0, \\
\nabla \times \mathbf{B} &= \mu_0(\mathbf{J}^{\mathrm{P}} - \sigma \nabla \phi),
\end{aligned}
$$

where ϕ is a scalar *potential* and the current density is written as

$$
\mathbf{J} = \mathbf{J}^{\mathrm{P}} - \sigma \nabla \phi, \tag{2.62}
$$

where ρ is the charge density. The permeability of the tissue of the head is assumed to be that of the free-space, i.e., $\mu = \mu_0$ (Hämäläinen *et al.*, 1993). The component $\mathbf{J}^{\mathrm{P}}$ of the total current vector $\mathbf{J}$, known as the *primary* or *impressed* current, is the actual source of bioelectricity. It arises out of the conversion of biochemical energy into electricity at highly localized sites inside or in the vicinity of the cells. The mechanisms behind the primary current are, therefore, entirely microscopic in nature and involve details at cellular levels. It is this highly localized current source $\mathbf{J}^{\mathrm{P}}$ that contains information about the neural activities associated with brain functions and is the quantity to be imaged by MEG. The rest of the current $\mathbf{J}^v$, called the *volume* current, flows passively through and is distributed over the head tissue.

The current $\mathbf{J}$ and the magnetic field $\mathbf{B}$ are related by the well-known law of *Biot–Savart* (Jackson, 1975)

$$
\mathbf{B}(\vec{x}) = \frac{\mu_0}{4\pi} \int_D \left[\mathbf{J}^P(\vec{x}') - \sigma(\vec{x}')\nabla\phi(\vec{x}') \times \mathbf{R} \right] \frac{1}{R^3} d\vec{x}', \tag{2.63}
$$

where $\mathbf{R} = \vec{x} - \vec{x}'$ and $R = |\mathbf{R}|$ is the distance between the point $\vec{x}$ where the field is monitored and the source position $\vec{x}'$. D is the region containing the sources. The head is usually modeled as consisting of a few homogeneous regions, the material parameters undergoing changes only at their interfaces. Let N be the number of the interfaces including the nonconducting region outside the head. Applying the divergence theorem and Green's second identity (see Sections 6.2 and 6.3) to Eq. (2.63), the following two relations are obtained for the field $\mathbf{B}$ and the scalar potential ϕ (Sarvas, 1987; Hämäläinen *et al.*, 1993; Williamson and Kaufman, 1989; Tripp, 1983):

$$
\mathbf{B}(\vec{x}) = \mathbf{B}_0(\vec{x}) - \frac{\mu_0}{4\pi} \sum_{j=1}^{m}(\sigma_j^- - \sigma_j^+) \left[\int_{\Gamma_j} \phi(\vec{x}')\,(\hat{n}_j(\vec{x}') \times \mathbf{R}) \frac{1}{R^3} d\Gamma_j \right] \tag{2.64}
$$

and

$$
\phi_0(\vec{x}) = \frac{(\sigma_j^- - \sigma_j^+)}{2}\phi(x) + \frac{1}{4\pi}\sum_{j=1}^{m}(\sigma_j^- - \sigma_j^+)\left[\int_{\Gamma_j} \phi(\vec{x}')(\hat{n}_j(\vec{x}') \cdot \mathbf{R})\frac{1}{R^3} d\Gamma_j \right],
$$

$$
\tag{2.65}
$$

where $+$ and $-$ refer to outer and inner sides of an interface, respectively, and $\hat{n}_j$ is the unit outward normal to the j-th interface. The term $\mathbf{B}_0$ is due only to the primary current dipole and is given by the Biot–Savart Law as

$$\mathbf{B}_0(\vec{x}) = \frac{\mu_0}{4\pi} \int_D (\mathbf{J}^P(\vec{x}') \times \mathbf{R}) \frac{1}{R^3} d\vec{x}'.$$

Similarly, the primary potential $\phi_0(\vec{x})$ is given by

$$\phi_0(\vec{x}) = \frac{1}{4\pi} \int_D (\mathbf{J}^P(\vec{x}') \cdot \mathbf{R}) \frac{1}{R^3} d\vec{x}'.$$

It is assumed that all surfaces and the potential are sufficiently smooth for the divergence theorem and Green's identity to apply.

Simplifications arise in two cases. First, if there exists only one current dipole $\mathbf{Q}$ at a location $\vec{x_Q}$. In this case

$$\mathbf{J}^P(\vec{x}) = \mathbf{Q}\delta(\vec{x} - \vec{x_Q}),$$

where δ is *Dirac's delta*. In this case, the primary field and potential reduce to

$$\mathbf{B}_0(\vec{x}) = \frac{\mu_0}{4\pi}(\mathbf{Q} \times \mathbf{R}) \frac{1}{R^3}$$

and

$$\phi_0(\vec{x}) = \frac{1}{4\pi}(\mathbf{Q} \cdot \mathbf{R}) \frac{1}{R^3}.$$

A second simplification results if the head is modeled as a sphere. Under this assumption, the magnetic field outside the head is given by (Sarvas, 1987; Geselowitz, 1970)

$$\mathbf{B}(\vec{r}) = \frac{\mu_0}{4\pi F}(F\mathbf{Q} \times \vec{r}_Q - \mathbf{Q} \times \vec{r}_Q \cdot \vec{r}\nabla F),$$

where $\vec{r}$ is the position of the detector and F is a scalar function of $\vec{r}, \vec{r}_Q$ and the radius of the head. From the above equation, it is seen at once that the radial part of the current does not affect the measured field.

It is also readily seen from Eqs. (2.63) and (2.64) that if the MEG sensors are radially oriented, then the volume current contributions disappear and the detector sees only the desired primary source. If $\vec{r}_Q$ is the position of the dipole inside the head, then the radial component of the field is given by

$$\mathbf{B}_r(\vec{R}) = \frac{\mu_0}{4\pi}\hat{r}\{\vec{R} \times \vec{r}_Q \cdot \mathbf{Q}\}\backslash R.$$

Moreover, if the dipole is at the center of the spherical head, then $\mathbf{B}_r$ disappears altogether. Therefore, the dipoles that are oriented radially cannot be imaged. At this point recall the Rolandic and Sylvian fissures that were mentioned at the beginning of this example: they show up here. These fissures are more or less tangential to the lobes and the neuronal dipoles in them are the only dipoles that can be imaged and the dynamics of the corresponding cortex activities revealed.

An MEG forward problem is solved by first solving Eq. (2.64) for the interface potentials ϕ and then obtaining the field **B** from Eq. (2.63) with the potentials thus obtained. Only in the case of a spherically symmetric head, can the solution be obtained in one step. The steps are reversed when solving the inverse problem and a solution is obtained usually by optimizational or statistical means. In practice, the inversion is carried out by solving an integral equation of the first kind of the form

$$B_j(\vec{x}) = \int_D \mathbf{L}(\vec{x}, \vec{x}') \cdot \mathbf{J}^{\mathbf{P}} d\vec{x}'.$$

The kernel **L** is called the *lead field* which is determined by the physical modeling of the problem. For details of the construction of the lead field and the numerical inversion of the above equation in MEG, we refer the reader to the already cited references of Sarvas (1987) and Mosher *et al.* (1997).

The inverse MEG problem is clearly nonunique. In the case of a spherical head, this nonuniqueness was discussed above. The question of the nonuniqueness in MEG and its resolution are discussed rigorously from a mathematical standpoint by Fokas *et al.* (1996). However, in practice, some *a priori* information is incorporated in order to obtain a stable MEG inversion. Primarily, this consists of assuming rather specific physical models of the neuronal source distributions such as an equivalent current dipole; dipoles of fixed orientation, but variable positions; dipoles fixed at locations, but with variable orientations, and so on (see Williamson *et al.* and Kaufman, 1989). Alternative approaches include combining MEG with other imaging modalities such as MRI, emission tomography (Gullberg *et al.*, 1999); combining MEG with EEG; introducing physiological constraints such as the requirement that the dipoles point outward, the activation sequence be continuous, and so on (Mulmivuo and Plonsey, 1995) and, of course, the time-honored empirical approach of physician recognition of the correlation between certain signals and their sources.

Finally, it is worth pointing out that the formulation presented above is also applicable to EEG or *electroencephelography* which involves measuring the potential difference due to the primary current source instead of the magnetic field. As a matter of fact, the integral equation for the calculation of the potential already provides the basic formulation for solving the EEG problem. It is for this reason that the problem is sometimes referred to as E/MEG in order to indicate that the formulation presented can be suitably adopted for both the EEG as well as the MEG problem.

2.2.7   Example 7. A Nonlinear Problem

The examples given above all involved linear equations such as the Laplace, Helmholtz, linear differential equations, and so forth. At this point, the reader may wonder if the occurrence of instability and its mediation through additional information about the problem are characteristics of linear problems alone. In our final example, therefore, we discuss a nonlinear problem with a view to dispelling such a notion. We choose the well-known *Burger's equation* (Gustafsson, et al. 1995; Drazin and Johnson, 1989). It is a nonlinear evolution equation with Navier-Stokes type nonlinearity and is one of the most important equations in shock dynamics in fluids. The Cauchy problem for Burger's equation is given by

$$\dot{u} + uu' = \nu u'', -\infty < x < \infty, t \geq 0. \tag{2.66}$$

$$u(x,0) = f(x).$$

As usual, $\dot{u}$ stands for differentiation with respect to time and the u' with respect to x. $u = u(x,t)$ and ν is the coefficient of kinematic viscosity. The function f is considered to be smooth, i.e., $f \in C^\infty$. However, it also suffices to assume that f is a function the derivatives of which (in x) are uniformly bounded.

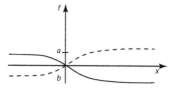

Figure 2.3 – An illustration of strictly monotonic functions. The solid line (—) is for $a > 0 > b$, whereas the dotted line (- - -) corresponds to $a < 0 < b$.

Let us first consider the fluid to be inviscid ($\nu = 0$) and show how instability arises in this case. To begin with, let f be a strictly monotonic function such that

$$\lim_{x \to -\infty} f(x) \to a \text{ and } \lim_{x \to -\infty} f(x) \to b. \tag{2.67}$$

Either $a > 0 > b$ or $a < 0 < b$. A hyperbolic tangent function is an example. The function f described by Eq. (2.67) is schematized in Figure 2.2. It is customary (see Drazin and Johnson, 1989) to look upon u in the nonlinear term uu' in Eq. (2.66) as a coefficient and as such as a given function. Equation (2.66) can be solved by the *method of characteristics* (Whitham, 1974). The equation for a characteristic is given by

$$\dot{x}(t) = u, \quad x(0) = x_0.$$

If $f(x_0)$ is the value of f at the point x_0, then according to the method of characteristics,

$$u(x,t) = f(x_0) = \text{constant} \tag{2.68}$$

along the characteristic originating at x_0 and the characteristics are linear, being the straight lines

$$x(t) = f(x_0)t + x_0.$$

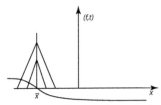

Figure 2.3 – A schematic illustration of charactersitcs and
their interwsections. In the figure $a > 0 > b$.

Let us assume that $a > 0 > b$. By Eq. (2.67) there exists a point $\tilde{x}$ such that

$$f(x_0) > 0 \text{ for all } x_0 < \tilde{x}, \text{ and } f(x_0) < 0 \text{ for all } x_0 > \tilde{x}.$$

Therefore, the left and right propagating characteristics will intersect and this intersection is shown in Figure 2.3. Now by Eq. (2.68), u is constant along a characteristic. Therefore, if two characteristics intersect, then the solution is not going to be unique. Even more. Since the solutions along the two characteristics are different, a huge gradient will be set up leading to instability.

To give an example, consider the following problem:

$$\dot{u} = \sin x u', \quad -\pi < x < \pi, t \geq 0$$

$$u(x, 0) = \sin x.$$

The characteristic near $x = 0$ behaves as x. Now u is strictly zero at $x = 0$, whereas it can be arbitrarily large in the vicinity of the origin for large values of time. Thus discontinuities are expected as $t \to \infty$. Numerical calculations (Gustafsson *et al.*, 1995) also verify this quite well. This is the phenomenon of shock formation.

In order to see explicitly how a shock can be formed, let us consider the following initial condition:

$$u(x, 0) = \begin{cases} 1 & \text{if } x \leq 0 \\ 1 - x & \text{if } 0 \leq x \leq 1 \\ 0 & \text{if } x \geq 1. \end{cases}$$

The characteristics are then

$$x(t) = \{u(x_0)t + x_0, t\}$$

and the values of the solution along the characteristics are:

$$u(x, t) = \begin{cases} 1 & \text{if } x \leq t \\ \frac{1-x}{1-t} & \text{if } t \leq x \leq 1 \\ 0 & \text{if } x \geq 1, \end{cases}$$

$0 \leq t \leq 1$. Note that the solution breaks down for $t \geq 1$ since the characteristics cross and a shock forms.

Let us now add the viscous term $\nu u''$ and return to Burger's equation (2.66). In the viscous case, it can be shown (Gustafsson, et al. 1995) that the derivatives of u are uniformly bounded by

$$\left\| \frac{\partial^{(p+q)} u(\cdot, t)}{\partial_x^p \partial_t^q} \right\|_\infty \leq C(p, q) \nu^{-(p+q)}. \tag{2.69}$$

The constant C does not depend upon time. Since there are bounds on the derivatives now, the solution cannot blow up and the instability is mitigated. Thus again, the additional information about the problem (viscous fluid and bounded derivatives) regularized the solution.

Some intuition can be gained about the viscosity term. Note that in the presence of the term $\nu u''$, Burger's equation acts somewhat like the heat equation.[8] The viscous solution, therefore, is infinitely differentiable notwithstanding the nonlinearity. The bound given in Eq. (2.69) is thus not surprising. Moreover, since viscosity introduces dissipation into the system, it prevents the crossing of the characteristics.

Finally, let us determine the order of the instability (Evans, 1991). By this we mean how fast the inviscid solution can grow. Toward that, let us write the solution of Eq. (2.66) explicitly. Define

$$w(x, t) = \int_{-\infty}^x u(y, t) dy \tag{2.70}$$

and

$$h(x, t) = \int_{-\infty}^x g(y) dy,$$

where we have written $g(y)$ for $u(x, 0)$. In terms of w and h, Eq. (2.66) becomes[9]

$$\dot{w} + \frac{1}{2} w'^2 = \nu w'' \text{ in } R^1 \times (0, \infty)$$

$$w = h \text{ on } R^1 \times \{t = 0\}.$$

The solution w is given by

$$w(x, t) = -2\nu \log \left(\frac{1}{\sqrt{r\pi\nu t}} \int_{R^1} e^{-\frac{L(x, y)}{2\nu}} \right),$$

where

$$L(x, y) = h(y) + \frac{|x - y|^2}{2t}.$$

Recovering u from w by inverting Eq. (2.70), we obtain

$$u(x, t) = \frac{\int_{-\infty}^\infty \frac{|x-y|}{t} e^{-\frac{L(x,y)}{2\nu}} dy}{\int_{-\infty}^\infty e^{-\frac{L(x,y)}{2\nu}} dy}. \tag{2.71}$$

---

[8] As a matter of fact, as early as 1906, the connection between Burger's and heat equations was already pointed out by Forsythe (see Ablowitz and Clarkson, 1991).
[9] Relations of the form $\dot{z} + F(z') = 0$ are known as the conservation laws and play an important role in the study of evolution equations.

Let us introduce the following lemma which can be found in Evans (1991).

Lemma *Let $\phi, \psi : R^1 \to R^1$ be two continuous functions such that ϕ grows at most linearly and ψ grows at least quadratically. Also assume that there exists a point $x^* \in R^1$ such that $\psi(x^*) = \min_{x \in R^1} \psi(x)$. Then*

$$\underset{\delta \to 0}{\text{limit}} \, \frac{\int_{-\infty}^{\infty} \phi(y) e^{-\frac{\psi(x,y)}{\delta}} \, dy}{\int_{-\infty}^{\infty} e^{-\frac{\psi(x,y)}{\delta}} \, dy} = \phi(x^*).$$

Applying this lemma to Eq. (2.71) at once shows that the solution to Burger's equation grows at most linearly as the viscosity term goes to zero.

Chapter 3

Theory of Ill-posed Problems in Function Spaces

3.1 Introduction

In general, inverse problems are ill-posed and a straightforward solution $f = A^{-1}g$ of an ill-posed problem $Af = g$ can produce nonsense. The problem must somehow be made *well-posed* before its approximate, yet meaningful and stable solutions can be obtained. Briefly speaking, the mathematics of ill-posed problems aims at developing methods for solving them approximately, but in such a way that the solutions obtained change continuously with the data. This program goes under the general name *regularization*. As pointed out in Chapter 1, ill-posedness is a function of the triple $\{A, X, Y\}$ and the well-posing of the problem naturally involves manipulating this triple on the basis of whatever *a priori* information may be available about the solution. Now the operator A is given and the image space Y usually cannot be changed since it should remain sufficiently broad in order to accommodate the inexact data. Therefore, it is primarily the object space X that can be manipulated in accordance with the *a priori* knowledge of the problem, i.e., whatever missing pieces of information about the system are available on physical and mathematical grounds.

There are several approaches to the solution of an ill-posed problem each with its own subtleties, variations and limitations. The most important are: (i) generalized solutions; (ii) singular value expansions; (iii) the theory of regularization; (iv) the iterative approaches and (v) mollification. The method of generalized solutions is essentially a generalization of the theory of ill-conditioned matrix operators (Groetsch, 1977) to functions spaces. The singular value expansion uses the spectral properties of the so-called *normal* operators associated with the ill-posed problem and is a highly important tool in dealing with com-

pact operators which occur more than frequently in inverse problems and cause
substantial mathematical difficulties. Of the above approaches, the theory of
regularization (Tikhonov and Arsenin, 1977) is perhaps the most important and
versatile and has two variations. In the methods of the first group (which may
be collectively called *Tikhonov's type of well-posedness*), the emphasis is on the
behavior of a class of sequences of approximate solutions. This approach of
establishing well-posedness via sequences of approximate solutions is important
from a practical point of view. Since for an ill-posed equation $Af = g$, the simple
inversion $f = A^{-1}g$ is not generally viable, suitable definitions of its approximate solutions must be given. These in turn are usually obtained as minimizers
(both constrained and unconstrained) of certain appropriately constructed functionals. Numerically, the minimization proceeds through iteratively generated
minimizing sequences. It is, therefore, important to ascertain that the approximate solutions do not stray too far from the true solution. Essentially, there are
three methods for constructing stable approximate solutions within Tikhonov's
variety. These are (i) the *quasisolutions*; (ii) the *method of residuals*, and (iii)
regularization, also called the *classical theory* (Engl *et al.*, 1996). The first two
are variational in character while the last mentioned directly minimizes a functional. In the second variation, the existence of a solution is assumed at the
outset and attention is given to its continuous dependence on the data. This
approach is of the type of *Hadamard*. Our focus here is primarily on the first
approach which is more in keeping with the materials presented in this book. Of
the iterative approaches, the most important are *Landweber's iterations* and the
method of *conjugate gradients*. The technique of mollification stands somewhat
apart from the rest. It uses a mollifier and its convergence properties in the
limit that the width of the mollifier goes to zero.

The organization of the chapter is as follows. In Section 3.2, a fundamental
result in the mathematics of ill-posed problems is presented, namely, the theorem of Tikhonov. The quasisolution of Ivanov (1963) (see also Glasco, 1984)
is discussed in Section 3.3. Section 3.4 deals with the generalized solutions and
Section 3.5 is devoted to singular value expansion and its significance for compact operators. Section 3.6 develops Tikhonov's theory of regularization in its
general form while a special case of this theory, involving a functional of the
first order is discussed in Section 3.7.

The mathematical theory of ill-posed problems is a vast subject. There exist copious definitions and theorems which at times differ only slightly from
one another. When combined with myriad nuances and variations, the subject
may appear to be overwhelming to a reader encountering it for the first time.
The literature on it is naturally extensive and still growing. Clearly, it is impossible to discuss the topic in any great depth in a work such as this, where
the breadth and scope are limited. We, therefore, present only the essentials
of the subject. The present discussion is based on the work of Tikhonov and
Arsenin (1977), Glasco (1984), Groetsch (1984), Lavrentiev *et al.*(1986), Engl
and Groetsch (1987), Bertero (1989), Kress (1989), Colton and Kress (1992),
Kirsch (1996), Bakushinsky and Goncharsky (1994), Lebedev *et al.*(1996), and
Engl*et al.*(1996).

3.2 Tikhonov's Theorem

Let $A : X \to Y$ be a linear, continuous, one-to-one mapping of a metric space X onto another metric space Y. The inverse A^{-1} thus exists, but is not supposed to be continuous everywhere in Y. Let $\hat{X} \subset X$ be a subset of X and $\hat{Y} = \{g \in Y | Af = g, \forall f \in \hat{X}\}$ be the image of $\hat{X}$ under A. Moreover, it is assumed that the exact solution f_e of the equation $Af = g$ is in $\hat{X}$. The objective of Tikhonov's theorem is to demonstrate that if $\hat{X}$ is compact in X, then $A^{-1} : \hat{Y} \to \hat{X}$ is continuous on $\hat{Y}$. The proof is schematized in Figure 3.1.

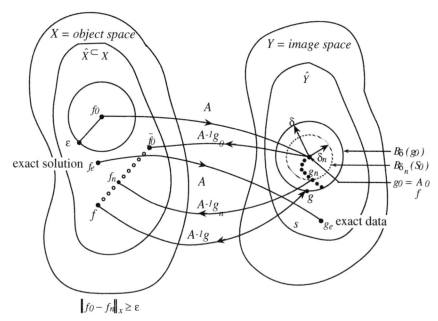

Figure 3.1 – A schematized illustration of the proof of Tikhonov's theorem. $\hat{X} \subset X$ and $\hat{Y} = A\hat{X}$ by assumption. The exact solution $f_e \in \hat{X}$.

Consider an arbitrary point $f_0 \in \hat{X}$ and $g_0 = Af_0$ is its image in $\hat{Y}$. Let $B_\delta(g_0)$ be a ball of radius δ centered around g_0 and $\{g_n\}_1^\infty$ be a sequence in $B_\delta(g_0)$ such that $g_n \to g_0, n \to \infty$. Consider an arbitrary n such that $\|g_n - g_0\|_Y \le \delta_n, \delta_n \le \delta$, and the corresponding sequence $\{A^{-1}g_n\}_1^\infty = \{f_n\}_1^\infty \in \hat{X}$. Let $\hat{X}$ be compact and assume that $\|f_0 - f_n\|_X \ge \epsilon, \forall n, \epsilon > 0$, i.e., the sequence does not converge to f_0. Now since $\hat{X}$ is compact by assumption, there exists a subsequence $f_{n_k} \to \tilde{f}_0$. Corresponding to this subsequence is the image subsequence $\{g_{n_k}\} \in \hat{Y} \to \tilde{g}_0 \in \hat{Y}, A\tilde{f}_0 = \tilde{g}_0$. However, the sequence $\{g_n\} \to g_0$ and, therefore, $\tilde{g}_0 = g_0$. This implies that $A\tilde{f}_0 = g_0 = Af_0$. Since by assumption A is one-to-one, $\tilde{f}_0$ must be identical to f_0. This tells us that if $\|g_0 - g\|_Y \le \delta$, then $\|f_0 - f\|_X = \|A^{-1}(g_0 - g)\|_X \le \epsilon$. That is, the inverse mapping A^{-1} is continuous at g_0. Since g_0 is arbitrary, A^{-1} is continuous in

$\hat{Y}$. This leads to the celebrated theorem of Tikhonov (Tikhonov and Arsenin, 1977) which is stated below.

Theorem 3.1 (Tikhonov) *Let a linear operator A map a metric space X into another metric space Y in a one-to-one fashion. Let $\hat{X} \subset X$ be a subset of X and $\hat{Y} = \{g \in Y | Af = g, \forall f \in \hat{X}\}$ be the image of $\hat{X}$ in Y. If $\hat{X}$ is compact, then its inverse $A^{-1} : \hat{Y} \to \hat{X}$ is continuous in $\hat{Y}$.*

Two natural corollaries follow from Theorem 3.1.

Corollary 3.1. Let $g_e \in \hat{Y}$ be the exact data corresponding to the exact solution f_e. Let $f_e \in \hat{X}$. Then $A^{-1}g_e = f_e$.

Corollary 3.2. Let $\{g_n\}_1^\infty$ be a sequence in $\hat{Y}$ and $\left\{A^{-1}g_n\right\}_1^\infty = \{f_n\}_1^\infty$ the corresponding sequence in $\hat{X}$. Then if $\lim_{n\to\infty} g_n \to \tilde{g}$ in the norm of Y, then $\lim_{n\to\infty} f_n \to \tilde{f}$ in the norm of X. Moreover, $A\tilde{f} = \tilde{g}$.

Next, suppose that the data is known only approximately and we have $g^\epsilon = g_e + \epsilon$ instead of g_e, the quantity ϵ being the error. However, assume that $g_e \in \hat{Y}$, i.e., the error does not take the data out of the image space $\hat{Y}$. Now by Tikhonov's theorem, A^{-1} is continuous on $\hat{Y}$. Therefore, $f^\epsilon = A^{-1}g^\epsilon \in \hat{X}$ is unique. Moreover, by the same theorem, f^ϵ changes continuously with g^ϵ and hence $f^\epsilon \to f_e$ as $\epsilon \to 0$. The following theorem due to Lavrentiev (1967) then follows.

Theorem 3.2 (Lavrentiev) *The problem of solving the ill-posed operator equation $Af = g, f \in \hat{X}$ and $g \in \hat{Y}$ is well-posed on $\hat{X}$ if (i) it is known a priorily that a solution exists and is in $\hat{X}$, (ii) the solution is unique, and (iii) infinitesimally small variations in the data do not take it out of the data space $\hat{Y}$.*

If the conditions of Tikhonov's theorem are met, the problem is said to be *well-posed in the sense of Tikhonov*. Well-posedness according to Lavrentiev's theorem 3.2 is called *conditional*. When the subset $\hat{X}$ is distinguished in X, the problem is said to be inserted into the *class of well-posedness* or into the *correctness class*. Note that the well-posedness in the sense of Tikhonov differs from the classical well-posedness of Hadamard in two aspects: (1) $Af = g$ is not required to hold for an arbitrary data in Y and (2) the inverse operator A^{-1} is not required to be bounded over the entire Y, but only over that subspace of it which is the image of the correctness class.

A large number of inverse solutions is essentially a trial-and-error process in which a model for the forward problem is formed on the basis of whatever knowledge is available about the problem. Its solution is then compared with the data and the model is further refined. The entire process is repeated until the best agreement is obtained with the data in some well-defined sense. Tikhonov's theorem tells us how to carry out this program efficiently and intelligently by narrowing the region of trial solutions and guarantees an approximate solution. The integral equation

$$\int_0^1 K(x, y) f_j(y) \mathrm{d}y = g_j(x)$$

of Example 4 in Chapter 2 illustrates this point. There it was found that if the search was confined to the functions in $C[a, b]$ in general, the trial-and-error process could not be successful. In order to obtain an approximate, stable solution, the trial solutions had to be restricted to a compact subspace of $C[0, 1]$. The search, therefore, was narrowed down to those functions f in $C[a, b]$ for which $\max |f| \leq M$ and also $\max |f'| \leq M_1$.

Note that Tikhonov's theorem says nothing about how to remedy the fact that the range of the operator A does not exhaust Y. Moreover, by restricting the solution space to $\hat{X}$, it restricts the data space Y to $\hat{Y}$. It was mentioned previously that the data space should remain as broad as possible in order to accommodate erroneous data. In addition, the theorem assumes A to be injective. Tikhonov's theorem simply formulates the conditions under which the inverse operator A^{-1} is guaranteed to be continuous and, consequently, guarantees the fulfillment of the critical third condition of Hadamard's well-posedness. Note also that if the theorem applies, then the inversion can be carried out by any classical means, that is, any classical method can be used to solve for f in the equation $Af = g$.

The natural question is, of course, how to find a correctness class. In practice, the physics of the situation provides the answer. Consider, as an example, the inverse problem of determining the boundary of a scattering obstacle from the scattering data. Now any reasonably smooth boundary in R^3 can be parameterized by a Fourier series:

$$\xi(\theta, \phi) = \sum_{l=0}^{\infty} \sum_{m=-l}^{l} \xi_{lm} Y_{lm}(\theta, \phi),$$

where (θ, ϕ) are the angular coordinates, ξ_{lm} the Fourier coefficients in the parameterization and Y_{lm} the spherical harmonics of order lm. Recovering the Fourier coefficients is thus equivalent to recovering the boundary. In practice, only a finite number of the coefficients can be considered. The correctness class is, therefore, a certain bounded and closed domain of a Euclidean space and is thus compact. Another example is the geophysical prospecting by gravimetric measurements. The inverse problem here is to determine the sources of the gravitational field from its observations on the surface of the Earth. The quantity to be recovered here is a 4N-dimensional vector $\mathbf{V} = \{m_i, x_i, y_i, z_i\}, i = 1, 2, ..., N$, where m_i is the i-th mass located at coordinates $\{x_i, y_i, z_i\}$. From the observations, known geological data and interpretations of available results, it is known *a priorily* that $0 \leq |z_i| \leq H$ and $0 \leq |x_k|, |y_k| \leq L$. Thus the vector $\mathbf{V}$ is restricted to be an element of the finite-dimensional space R^{4N} which is certainly compact. Example 4 of Chapter 2 provides another example. Many others can be cited from a variety of inverse problems.

3.3 Regularization on a Compactum: The Quasisolution

The conditions laid down in the previous two theorems are by far the simplest. In the next stage of complexity, the restriction that the data be in $\hat{Y}$ is relaxed to be an arbitrary element of Y. In this case Tikhonov's theorem does not apply and the inverse of A^{-1} is not defined. Ivanov (1963) suggested a method for *finding* a solution in this situation and his solution is called the *quasisolution* defined below. The operator A is still considered injective. Notice the emphasis on the word *finding*, the reason for which will be explained momentarily.

Definition 3.1. (Quasisolution) An element $\tilde{f} \in \hat{X}, \hat{X} \subset X$ a compactum, is called a *quasisolution* of the ill-posed problem $Af = g, f \in \hat{X}, f \in X, g \in Y, f_e \in \hat{X}$, if it minimizes the variational problem

$$\|A\tilde{f} - g\|_Y = \inf_{f \in \hat{X}} \|Af - g\|_Y. \tag{3.1}$$

A method for finding an approximate solution is thus prescribed. If $\hat{X}$ is a compactum, the norm functional of Eq. (3.1) attains a minimum in $\hat{X}$ for any fixed g. A quasisolution, therefore, exists for an arbitrary data in Y. Clearly, finding a quasisolution is a problem in constrained optimization: minimize the functional $\rho_Y(Af, g)$ for any g in Y subject to the constraint $\tilde{f} \in \hat{X}$. This is a difficult problem in itself. Observe that we are required to find the *global* minimum and there is no satisfactory method for doing so. This is why the word *finding* in the previous paragraph was emphasized. Although the global minimum is the *desideratum*, the local minima are more than likely to occur in which case a set of quasisolutions will be obtained.

Two problems arise at this point: (i) how to solve Eq. (3.1) and (ii) the relation between a quasi-solution and the exact solution. The first part of the question can be answered thus. Since $\hat{X}$ is assumed to be compact, then by Tikhonov's theorem, any element in the image space $\hat{Y}$ can be inverted. Therefore, if the arbitrary data g in Y can be put into a one-to-one correspondence with an element in $\hat{Y}$, and the inverse operator A^{-1} is applied to this element, then a quasisolution can be obtained. It is natural to use a projection operator $P_{Y \to \hat{Y}}$ for the purpose. Thus $P_{Y \to \hat{Y}}g = \hat{g} \in \hat{Y}, g \in Y$. This results in a unique quasisolution $\tilde{f} = A^{-1}\hat{g} = (A^{-1}P_{Y \to \hat{Y}})g$. Obtaining the inverse operator $R = A^{-1}P_{Y \to \hat{Y}}$ is the program of the quasisolution method. If $P_{Y \to \hat{Y}}$ is continuous on Y, then the quasisolution will be a stable approximate solution to the ill-posed problem $Af = g$.

The second question, the relation between a quasisolution and the desired solution, can only be answered indirectly. For any method of solving an ill-posed problem, it is customary to show that the approximate solution it yields converges to the true solution as the error in the data goes to zero (see p. 14). The relation between a quasisolution $\tilde{f}$ and the true solution f_e is revealed by showing that $\tilde{f}$ tends to f_e as the error in the data goes to zero. This is also a confirmation that convergence to the true solution depends continuously on the data. Indeed, if $Af^\epsilon = g^\epsilon$, then

$$
\begin{aligned}
\|A(\tilde{f} - f_e)\|_Y &= \|A(\tilde{f} - f^\epsilon) - A(f_e - f^\epsilon)\|_Y \\
&\leq \|A(\tilde{f} - f^\epsilon)\|_Y + \epsilon = \inf_{f \in \hat{X}} \|A(f - f^\epsilon)\|_Y + \epsilon \leq 2\epsilon.
\end{aligned}
$$

This shows that a quasisolution converges to the true solution continuously with the data and hence is well-posed.

In order to go beyond the convergence in this weak sense, it is required to assume that there is a unique projection of an element $g^\epsilon \in Y$ on the set $\hat{Y}$, the image of the compactum $\hat{X}$. Let $P_{Y \to \hat{Y}}$ be this projection and $\{g_n^\epsilon\}$ a sequence of elements of Y converging to $\hat{g}$ in $\hat{Y}$ with $A^{-1}\hat{g} = \hat{f} \in \hat{X}$. Let $g_n = P_{Y \to \hat{Y}} g_n^\epsilon$ be the unique projection of g_n^ϵ on $\hat{Y}$ and by Tikhonov's theorem, there is a unique f_n satisfying the equation $Af_n = g_n$. Then

$$
\lim_{n \to \infty} \|Af_n - Af_n^\epsilon\|_{\hat{Y}} \leq \lim_{n \to \infty} \|A\hat{f} - g_n^\epsilon\|_{\hat{Y}} \to 0.
$$

Now

$$
\begin{aligned}
\lim_{n \to \infty} \|Af_n - A\hat{f}\|_{\hat{Y}} &\leq \lim_{n \to \infty} \|A\hat{f} - g_n^\epsilon\|_{\hat{Y}} \\
&= \lim_{n \to \infty} \|(A\hat{f} - Af_n^\epsilon) + (Af_n^\epsilon - Af_n)\|_{\hat{Y}} \to 0.
\end{aligned}
$$

Since the sequence $\{g_n\}$ is in $\hat{Y}$ and hence the problem is well-posed in the sense of Tikhonov and by Corollary 3.2, it follows that $\rho_X(f_n, \hat{f}) \to 0$ as $n \to \infty$ in the metric of X. Since the inverse operator is continuous on $\hat{Y}$, it follows that $\rho_X(\overline{f}, \hat{f}) \to 0$ as $\epsilon \to \infty$. Alternatively, it can be said that

$$
d_X(\overline{f}, \hat{f}) \leq \omega(2\epsilon),
$$

where ω is the modulus of continuity. It may be recalled that the modulus of continuity ω is defined as

$$
\omega(\epsilon) = \sup_{x_i, x_j \in X} [d_X(x_i, x_j); d_Y(Ax_i, Ax_j)] \leq \epsilon.
$$

All this can be summed up in the following theorem.

Theorem 3.3 *Let the solution of the equation $Af = g$ be well-posed in the sense of Tikhonov between the compactum $\hat{X} \subset X$ and its image $\hat{Y}$ under A. Let $P_{Y \to \hat{Y}}$ be a unique operator of projection from Y to $\hat{Y}$. Then the variational problem*

$$
\|A\tilde{f} - g\|_Y = \inf_{f \in \hat{X}} \|Af - g\|_Y, g \in Y,
$$

is well-posed between Y and $\hat{X}$.

The quasisolution essentially approximates the inverse $A^{-1}g^\epsilon, g^\epsilon \in Y$. Hence the well-posedness can be considered to be in the sense of Hadamard. Note that if the data is exact, that is, $g^\epsilon = g_e$, then $\tilde{f} = f_e$.

Finally, we comment briefly on the uniqueness of a quasisolution. Since it solves a constrained optimization problem, any consideration of uniqueness must assume that the compactum $\hat{X}$ is convex. We just quote the following theorem from Bakushinsky and Goncharsky (1994).

Theorem 3.4 *Let A be a linear, continuous, one-to-one mapping of a convex compactum $\hat{X} \subset X$ into a rigorously normed Banach space Y. Then a unique quasisolution exists for any $g^\epsilon \in Y$.*

Recall the following definition of rigorous norm.
Definition 3.2. A Banach space is *rigorously normed* or *rigorously convex* if the identity $\|x + y\| = \|x\| + \|y\|$ holds if and only if $y = \alpha x, \alpha$ being a real number.
A Hilbert space is always rigorously normed simply due to the existence of an inner product for which Schwartz's inequality holds. For rigorously normed functionals $F(z)$ in a Banach space, there is an important theorem.

Theorem 3.5 (Rocafellar) *Let X be a Banach space and $\hat{X} \in X$ be a convex, compact and closed set in X. Let $F(z)$ be a rigorously normed continuous functional defined on $\hat{X}$. Then $F(z)$ attains its exact lowest bound at a unique point $z^* \in \hat{X}$.*

Note that $\hat{X}$ must be compact and closed. There are sets that are compact but not closed, e.g., a disc with a part of the boundary excluded. In the context of the variational problem for the quasisolution, let us consider the minimization of the functional $\|Af - g^\epsilon\|_2$. If the space X is rigorously normed and $\mathbf{A}$ is a continuous, one-to-one mapping of $\hat{X}$ into $\hat{Y}$, then the functional $F(f) = \|Af - g^\epsilon\|_2$ is rigorously normed and hence according to the above Theorem 3.4 of Rocafellar, must attain its exact lowest bound at a unique point f^* in $\hat{X}$. Hence the following theorem.

3.4   Generalized Solutions

Before proceeding to the theory of regularization, let us introduce the concept of a *generalized solution*. The discussion in this section relies much on Bertero (1989). Thus far the operator A was assumed to be injective and defined on a compact subset of the object space. We now set aside these simplifications. Again $R(A)$ is not assumed to exhaust the data space Y. Not only can no unique solution be expected in this case, but a solution may not exist for an arbitrary data because any error in the data may take it out of the range of the operator. In a situation so general, only approximate solutions are possible of which one must be selected as the best in some well-defined sense. This is the motivation behind a generalized solution. Furthermore, it also stands to reason that as the simplifying assumptions of the previous two sections are restored, the generalized solution should converge to the previous solutions. This will indeed turn out to be the case.

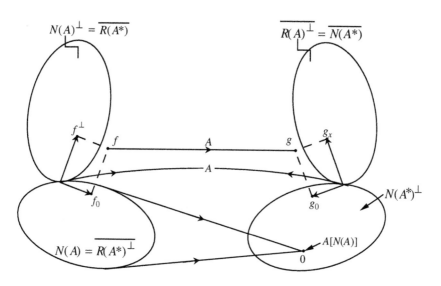

$$f = f_0 \oplus f_\perp, \quad f_0 \in N(A), \quad f_\perp \in N(A)$$

$$N(A)^\perp \oplus N(A) = X.$$

$$g = Af = g_0 \oplus g_\perp, \quad g_0 \in N(A^*)^\perp, \quad g_\perp \in N(A^*)$$

$$N(A^*) \oplus N(A^*)^\perp = Y$$

Figure 3.2 – A schematized illustration of various components of the object and image space and their relations to each other.

In what follows, A will be assumed to be linear and continuous between two Hilbert spaces X and Y which is frequently the case in practice. Therefore, X and Y can each be decomposed into two mutually orthogonal subspaces, namely, the nullspace $N(A)$ of A and its orthogonal complement $N(A)^\perp$. The following relations are known to be true from functional analysis

$$\overline{R(A)} = N(A^*)^\perp, \quad \overline{R(A^*)} = N(A)^\perp,$$

the overbar denoting closure. Then

$$X = \overline{R(A^*)} + N(A), \quad Y = \overline{R(A)} + N(A^*).$$

These are illustrated pictorially in Figure 3.2 for an easy visualization. It is also known (see, for example, McLean, 2000) that the following conditions are equivalent:

i. $R(A) = N(A^*)^\perp$.   ii. $R(A^*) = N(A)^\perp$.
iii. $R(A)$ closed in Y.   iv. $R(A^*)$ closed in X.

Under these conditions, the operator $N(A^*)^\perp \to R(A)$ has a bounded inverse.

Since $R(A)$ does not exhaust Y, the data may be out of the range $R(A)$ in Y and will have a component in the orthogonal complement $R(A)^\perp$ ($g_\perp$ in Figure 3.2). The aim is, therefore, to extend the inverse operator A^{-1} from $R(A)$ to $R(A) \oplus N(A^*) = R(A) \oplus R(A)^\perp$ which is dense in Y, i.e., its closure is Y. The idea is to project the erroneous, physical data g^ϵ in Y on its subspace $R(A)$ and then invert this projected data to an element in X as in the case of the quasisolution. The relation between the generalized and quasisolution will be explained at the end of the section. The solution thus obtained is given by $(A^{-1}P_{Y \to R(A)})g^\epsilon$, P being the operator of projection from Y to $R(A)$. If A is one-to-many then this would result in multiple solutions. We are interested in that particular solution (denote it by $\hat{f}$) which will result in the minimum distance between g^ϵ and the set $R(A)$. In other words, find $\hat{f}$ such that

$$\|A\hat{f} - g^\epsilon\|_Y = \inf_{f \in X} \|Af - g^\epsilon\|_Y, g^\epsilon \in Y. \tag{3.2}$$

The solution $\hat{f}$ of this variational problem is called the *least square solution* or LSS in short, of the ill-posed equation $Af = g^\epsilon$.

Definition 3.3 $\hat{f} \in X$ *is called a least square solution of the equation $Af = g^\epsilon$, A being linear and bounded between two Hilbert spaces X and Y, if it minimizes the distance between the range $R(A)$ of A and the perturbed data g^ϵ. That is, if*

$$\|A\hat{f} - g^\epsilon\|_Y = \inf_{f \in X} \|Af - g^\epsilon\|_Y.$$

Let ϕ be an arbitrary element in X. Then $A\phi \in R(A)$. What we are really looking for is the orthogonality condition

$$< A\hat{f} - g^\epsilon, A\phi >_Y = 0, \ \forall \phi \in X,$$

from which

$$< \phi, A^*A\hat{f} - A^*g^\epsilon >= 0.$$

Since ϕ is arbitrary

$$A^*A\hat{f} = A^*g^\epsilon. \tag{3.3}$$

Equation (3.3) is the *Euler–Lagrange equation*[1] (see, for example, Goldstein, 1980) for the solution of the variational problem (3.2). Noting that $PA = A$ and $(I - P)A = QA = \mathbf{0}, \mathbf{0}$ being the null operator, we have

$$\begin{aligned}\|A\hat{f} - g^\epsilon\|_Y &= \|A\hat{f} - Pg^\epsilon|_Y + \|Pg^\epsilon - g^\epsilon\|_Y \\ &= \|A\hat{f} - Pg^\epsilon\|_Y + \|Qg^\epsilon\|_Y.\end{aligned}$$

Hence it follows that $\hat{f}$ is a solution of the equation

$$A\hat{f} = Pg^\epsilon. \tag{3.4}$$

---

[1]Another, and more explicit, derivation of Eq. (3.3) will be given in Section 5.7.

Let us make a remark about the content of Eqs. (3.3) and (3.4). Symbolically, Eq. (3.4) states that $\hat{f} = (A^{-1}P)g^\epsilon$ and goes to show that the LS solution is obtained by solving $Af = g$ for f by replacing the physical data g^ϵ by its projection g on the range of A. The solution $\hat{f}$ given by either Eq. (3.3) or Eq. (3.4) minimizes the mean-square distance (in the norm of Y) between $R(A)$ and the physical data g^ϵ. This is why $\hat{f}$ is called the least-square solution (also known as *pseudosolution*). Both Eqs. (3.3) and (3.4) give us the recipe for obtaining a least-square solution. However, as Bertero (1989) points out, it is relatively easier to find the adjoint than the projection operator.

Equation (3.4) makes clear the fact that in order for a least-square solution to exist the projected data must be in the range of the operator. A LS solution, therefore, exists for an arbitrary data g^ϵ if and only if $Pg^\epsilon \in R(A)$. Equation (3.3), on the other hand, shows that the LSS $\in R(A^*)$ and, consequently, in $N(A)^\perp$. From the equivalence statements on p. 57, it follows that in order to have a LSS for an arbitrary data, the range $R(A)$ must be closed. No LS solution is possible for an arbitrary data if the range is not closed. Since the range is always closed if the space is finite-dimensional, it follows that a least-square solution always exists if data is discrete or band-limited. As to the uniqueness of a LS solution, the nullspace $N(A)$ must be trivial. So a unique LSS exists if and only if: (i) $Pg^\epsilon \in R(A)$; $R(A) = \overline{R(A)}$, and (iii) $N(A) = \emptyset$. This can also be seen from the Euler–Lagrange equation (3.3). In this case, $N(A)$ must be empty so that $N(A^*A)$ is empty. This is because

$$< A^*Af, f >_X =< A^*(Af), f >_X =< Af, Af >_Y = \|Af\|_Y^2 = 0,$$

and, therefore, $Af = 0$ which in turn implies that f is identically zero since A is one-to-one. Thus $N(A^*A) = N(A)$. Similarly, $N(AA^*) = N(A^*)$.

However, our intention is to deal with a one-to-many operator for which $N(A)$ is nonempty. The solution of Eq. (3.3) or Eq. (3.4) is then nonunique up to an element of the nullspace. Therefore, all LS solutions comprise the set $N[F] = \{F \in X; F = \hat{f} + \phi, A\phi = 0\}$. Note that as long as $Pg^\epsilon \in R(A)$, the set $N[F]$ is nonempty and $N[F] = \hat{f} \oplus N(A)$. The set is also closed and convex as shown from the following considerations. Let there be two solutions f_1 and f_2 in $N[F]$ corresponding to some data g^ϵ. Then $A^*(Af_1) = A^*(Af_2) = A^*g^\epsilon$. Hence if λ is a real parameter, then

$$A^*\big(A[\lambda f_1 + (1 - \lambda)f_2]\big) = A^*g^\epsilon.$$

Thus, if f_1 and f_2 are any two elements of $N[F]$, then the combination $[\lambda f_1 + (1-\lambda)f_2]$ also belongs to $N[F]$. Hence $N[F]$ is closed and convex. Alternatively, $N[F]$ can be considered to be an element of the *quotient space* $Y/N(A)$ (Taylor, 1964).

Now in the absence of uniqueness, the best that can be hoped for is to select that element of $N[F]$ which has the minimum norm on the set. The convexity of $N[F]$ guarantees the existence of such a norm. Moreover, it will force uniqueness. This solution, denoted by f^+ is called a *generalized solution*.

Symbolically, $f^+ = A^+ g^\epsilon$, A^+ being the *generalized inverse* in a function space setting.[2] The minimum norm least square solution f^+ is shown in Figure 3.3.

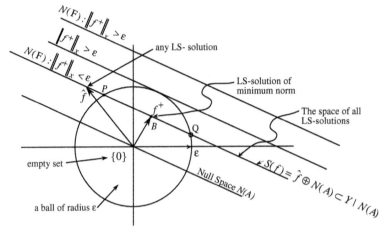

Figure 3.3 – A geometric illustration of the least-square (LS) solution $\hat{f}$, the generalized solution f^+, quasisolution $\tilde{f}$ and relations between them. If $\|f^+\|_X < \epsilon$, no unique quasisolution $\tilde{f}$ exists. All solutions on the line segment PQ are possible $\tilde{f}$s. If $\|f^+\|_X = \epsilon$, then $f^+ = \tilde{f}$ and $\tilde{f}$ is unique. If $\|f^+\|_X > \epsilon$ then $\tilde{f}$ is any solution on the surface of the sphere of radius ϵ.

The domain $D(A^+)$ of A^+ is the dense subspace $R(A) \oplus R(A)^\perp$ of Y and $N(A)^\perp$ is its range. Hence $A^+ : R(A) \oplus R(A)^\perp \to N(A)^\perp$ is an operator which takes in the corrupted data g^ϵ and produces a unique minimum norm LS solution $f^+ \in N(A)^\perp$ of the ill-posed equation $Af = g^\epsilon$. In Figure 3.3, the vector $\vec{AB}$ is perpendicular to the line through the origin representing the nullspace $N(A)$. Let $\tilde{A} = A|_{N(A)^\perp} : N(A)^\perp \to R(A)$ be the restriction of A to $N(A)^\perp$. Obviously, $N(\tilde{A}) = 0$ and $R(\tilde{A}) = R(A)$. $\tilde{A}^{-1}$, therefore, exists. Now $g \in Y = g_1 + g_2$, $g_1 \in R(A)$, $g_2 \in R(A)^\perp$, and $A^+ g \in N(A)^\perp$. Summing up all this into the following theorem, we have:

Theorem 3.6 *The generalized inverse A^+ of a linear bounded operator A between two Hilbert spaces X and Y uniquely extends $\tilde{A}^{-1}$ from $R(A) \to D(A^+) = R(A) \oplus R(A)^\perp$ and $\tilde{A}^{-1} : R(A) \to N(A)^\perp$ is the inverse of the restriction $\tilde{A}$ of A to $N(A)^\perp$. $\tilde{A} = A|_{N(A)^\perp} : N(A)^\perp \to R(A)$ of A.*

---

[2] A^+ is a generalization of the *Moore–Penrose* inverse in matrix theory (Groetsch, 1977). The Moore–Penrose inverse exists for an arbitrary matrix operator and is unique irrespective of the number of rows and columns and the rank. Otherwise, the problem of finding an inverse of an arbitrary $n \times m$ matrix depends upon n, m and the rank p of the matrix (Lanczos, 1961). Analogously, the generalized inverse, if it exists, exists for an arbitrary data and is unique. The distinction, however, is that unlike the Moore–Penrose inverse for a matrix, the generalized inverse in an infinite-dimensional function space may not necessarily exist for an arbitrary data.

As to A^+ being continuous on all of Y, it depends upon whether $R(A)$ is closed. If not, then A^+ is not continuous on Y and the problem of finding a generalized solution is ill-posed. On the other hand, if the range is closed, then $D(A^+) = Y$ and finding f^+ is a well-posed problem. In the former case, A^+ is found to be a *closed operator* (Roman, 1975; Stakgold, 1979). This is a class of unbounded operators which resembles bounded operators (analogous to compact operators resembling operators on a finite-dimensional space). If T is a closed operator having a domain $D(T)$, then for each sequence $\{x_n\} \in D(T)$,

$$\{x_n\} \to x, \ T\{x_n\} \to y \Longrightarrow x \in D(T), \ y = Tx.$$

We show that A^+ is indeed a closed operator. Consider a sequence $\{g_n\}$ in $D(A^+)$. The corresponding sequence in the object space is $\{f_n\} \in N(A^\perp)$. Since $N(A^\perp)$ is closed, $\{f_n\} \to f \in N(A^\perp)$. Now $A^*Af_n \to A^*Af \Rightarrow A^*Af_n \to A^*g_n \to A^*g \Longrightarrow A^*(Af - g) = 0 \Longrightarrow Af - g \in N(A^*) = R(a)^\perp \Rightarrow g \in R(A) \oplus R(A)^+$. Hence $g \in D(A^+)$. Therefore,

$$\{g_n\} \in D(A^+), \ A^+\{g_n\} \to f \Longrightarrow g \in D(A^+), \ f = A^+g.$$

The generalized inverse A^+ is, therefore, closed.

Let us next assume that A^+ is continuous and consider a sequence $\{g_n\}$ in $R(A)$ converging to g. Then $\{f_n\} = A^+\{g_n\}$ is in $N(A)^\perp$. Since $N(A)^\perp$ is closed, the limit f is in $N(A)^\perp$. But by continuity, $A\{f_n\} \to Af \in R(A)$. Hence g is in $g = Af$. The range $R(A)$ must, therefore, be closed Hence if A^+ is continuous on $D(A^+)$, then the range $R(A)$ is closed, that is, $R(A) = \overline{R(A)}$. Conversely, suppose that the range $R(A)$ is closed. Then A^+ is a linear operator defined on all of Y which is also closed. Then from the famous *closed graph theorem* (Taylor, 1964) A^+ is continuous on all of Y.

Let us combine all this into the following theorem.

Theorem 3.7 *Let $A : X \to Y$ be a continuous linear operator between two Hilbert spaces X and Y. Then the generalized inverse A^+ is continuous on Y if and only if the range of A is closed and vice versa. If the range $R(A)$ is not closed, then $A^+ : D(A^+) \to N(A^\perp)$ is a closed operator.*

When $R(A)$ is closed, A^+ is defined on the entire Y and produces a unique solution in $N(A)^\perp$. The problem of determining the solution of the equation $Af = g^\epsilon, f \in X, g^\epsilon \in Y$ is, therefore, well-posed in the classical sense although the original equation possesses none of these properties. The continuity of A^+ on Y is, however, a necessary but not a sufficient condition for the stability of the solution. The propagation of error from the data to the solution must still be controlled by the so-called *condition number* of the problem (see Appendix A.5.1). If $R(A)$ is not closed, the situation becomes complicated. The generalized inverse is then merely closed and not continuous on all of Y. This is entirely unsatisfactory if A is either a convolution or a compact operator for which the range is not closed unless it is finite-dimensional. This poses a significant problem since operators of these two types govern a large number of inverse problems. Recourse must then be made to the theory of regularization proper to which we will turn momentarily. But before leaving, it is worthwhile for us to summarize the results and conclusions obtained in this section.

3.4.1   Summary

An ill-posed problem $Af = g^\epsilon$ is to be solved, $f \in X, g \in Y, X, Y$ being Hilbert spaces. g^ϵ is an arbitrary data with errors. The operator A is not assumed to be either 1:1 or to exhaust Y. Then g^ϵ may not generally be in the range of A. The conditions for the classical well-posedness are thus not satisfied. All one can hope for is an approximate solution which is the best in some sense. This is given by the solution $\hat{f} \in X$ of the variational problem

$$\|A\hat{f} - g^\epsilon\|_Y = \inf_{f \in X} |Af - g^\epsilon\|_Y$$

and minimizes the distance between the set $Af, f \in X$ and the perturbed data g^ϵ. The variational solution $\hat{f}$ is called the *least-square or pseudosolution* of the ill-posed problem $Af = g^\epsilon$. $\hat{f}$ is obtained by projecting g^ϵ orthogonally onto the range $R(A)$ of A by a projection operator $P_{Y \to R(A)}$ and then inverting this projected data by A^{-1}. The LS solution is given either by the equation $A\hat{f} = Pg^\epsilon$ or via the *normal* equation $A^*A\hat{f} = A^*g^\epsilon$. The latter form is more suitable in practice.

A least-square solution always exists if Pg^ϵ is in $R(A)$ and is unique if and only if the null space $N(A)$ of A is empty. Otherwise, the solution is nonunique up to an element of $N(A)$. In that case multiple least-square solutions are possible out of which the one having the minimum norm is considered to be the best approximate solution. This minimum norm, least-square solution, denoted by f^+, is called the *generalized* solution of the ill-posed equation $Af = g^\epsilon$ and is a solution of the equation $A^+g^\epsilon = f^+, A^+ = (A^*A)^{-1}A^*$. The linear operator A^+ which is called the *generalized* inverse is a generalization in function space of the well-known *Moore–Penrose* inverse for matrix operators. Its domain $D(A^+)$ is the dense subspace $R(A) \oplus R(A)^\perp$ of Y and its range is $N(A)^\perp$, the orthogonal complement of the nullspace $N(A)$ of the operator A. A^+ is a unique extension of $\tilde{A}^{-1} : R(A) \to N(A)^+$ to $R(A) \oplus R(A)^+ \to N(A)^+$.

The generalized inverse is not necessarily continuous on its domain $D(A^+)$. It is continuous if and only if the range $R(A)$ is closed. In that case, a unique solution f^+ exists in $N(A)^\perp$ which is stable, depends continuously upon the data and the problem is well-posed in the classical sense although the original problem is not. In this case, a generalized solution exists for any arbitrary data. If $R(A)$ is not closed, then the most that can be said is that A^+ is a closed operator. In such situations, which occur for a convolution or a compact operator in an infinite-dimensional space, the problem of finding the generalized inverse is ill-posed and recourse must be taken to other means such as regularization.

3.4.2 Connection with Quasisolution

Definition 3.1 for Ivanov's quasisolution and Definition 3.3 for the LS solution are identical except for the fact that Definition 3.3 is a problem in unconstrained optimization as opposed to the quasisolution being determined by constrained optimization. In other words, a quasisolution is a *constrained least-square solution*, the solution being constrained to be in $\hat{X} \in X$. A quasisolution then is the solution to the problem

$$\|A\tilde{f} - g\|_Y = \inf\{\|Af - g\|_Y; \|f\|_X \le d_{\hat{X}}\},$$

$d_{\hat{X}}$ being the *diameter* of the correctness class $\hat{X}$. Usually, the solution is restricted to a ball of a constant diameter F. The generalized solution, on the other hand, imposes the constraint that the norm $\|f\|_X$ be minimum.

Figure 3.3 geometrically illustrates the various solutions, namely, least-square solution $\hat{f}$, the generalized solution f^+ and Ivanov's quasisolution $\tilde{f}$ and explains the connection between them. The region Ω in the figure is the region to which Ivanov's quasisolutions are constrained (i.e., the subspace $\hat{X}$ of Section 3.3) with $\epsilon = r_{\hat{X}}$ as the radius of Ω. Assume that f^+ exists and $\|f^+\|_X \le \epsilon$. Then there exists a unique $\hat{f}$ which coincides with f^+. If, on the other hand, $\|f^+\|_X > \epsilon$, then it will only be said at the moment that the quasisolutions must be constrained to lie on the sphere (two-dimensional in the figure) with a radius ϵ. A further discussion about the minimization on the sphere will lead to the discussion of Tikhonov's functional and is deferred till Section 3.6.

3.5 Singular Value Expansion

Before moving on to the theory of regularization of ill-posed problems, let us introduce the technique of the *singular value expansion* or SVE in short.[3] SVE explicitly reveals the physical reason behind the unboundedness of the inverse of a compact operator in an infinite-dimensional space and allows us to treat the ill-posedness of a compact operator in a reasonable manner. Furthermore, one can recover the generalized solution from SVE analysis.

Let $A : X \to Y$ be a compact linear operator between two Hilbert spaces X and Y, and AA^*, A^*A the *normal* operators associated with A which are compact since A is compact. Both $< A^*Af, f >_X$ and $< AA^*g, g >_Y$ are nonnegative. Next consider the eigenvalue problem $A^*A\phi = \lambda\phi$. Then $(AA^*)A\phi = \lambda A\phi$. Therefore, if ϕ is an eigenfunction of A^*A for an eigenvalue λ, then $A\phi$ is an eigenfunction of AA^* for the same eigenvalue. Similarly, if ψ is an eigenfunction of AA^* for an eigenvalue λ, then $A^*\psi$ is an eigenfunction of A^*A for the same eigenvalue. There exist well-established results (Stakgold, 1979) for these symmetric, nonnegative normal operators. These are:
i. They have the same spectrum.

---

[3]SVE is the generalization to infinite-dimensional function spaces of *singular value decomposition* or SVD for the matrix operators (Golub and Van Loan, 1989).

ii. The spectrum lies on the positive-half of the real axis.

iii. The number of eigenvalues is either finite or countably infinite.

iv. The multiplicities of the eigenvalues are finite.

v. In the latter case, only zero is an *accumulation point*, i.e., $\lim_{n\to\infty} \lambda_n \to 0$.

vi. If we count each eigenvalue the number of times as its multiplicity, then they can be arranged in a nondecreasing order: $\lambda_0 \geq \lambda_1 \geq \lambda_2 \cdots$.

vii. The eigenvalue problems are

$$\left.\begin{array}{l} Au_n = \mu_n v_n \\ A^* v_n = \mu_n u_n \\ A^* A u_n = \lambda_n u_n \\ AA^* v_n = \lambda_n v_n, \end{array}\right\} \quad (1)$$

where $\mu_n = \sqrt{\lambda_n} > 0$.

viii. $\{u_n\}_1^\infty$ and $\{v_n\}_1^\infty$ constitute the complete orthonormal sets of eigenfunctions of the operators A^*A and AA^*, respectively, corresponding to the spectrum $\{\lambda_n\}_1^\infty$. Note (Section 3.4) that $\overline{R(A^*A)} = N(A^*A)^\perp = N(A)^\perp$ and $\overline{R(AA^*)} = N(AA^*)^\perp = N(A^*)^\perp = \overline{R(A)}$. It then follows that $\{u_n\}$ span the space $N(A)^\perp$ and $\{v_n\}$ span the space $N(A^*)^\perp$. Since $N(A^*)^\perp = \overline{R(A)}$, the sequence $\{v_n\}$ forms an onthonormal basis for the closure of the space of the exact images.

The triple $\{\mu_n, u_n, v_n\}$ is called the *singular system* and $\{\mu_n\}_1^\infty$ the *singular values* for the one-to-many operator A.[4] The singular values must satisfy the relation

$$\sum_{i=1}^\infty \mu_i^2 = ||A||^2$$

and, therefore, must decay at a rate faster than $i^{-\frac{1}{2}}$. The singular function u_n is mapped onto the corresponding v_n by the operator A and the amplification of this mapping is given by the singular value μ_n, and *vice versa*. The behavior of the singular system cannot be arbitrary and depends on the nature of the kernel of the operator A. As a matter of fact, the singular system is unique for a given operator. The smoother (that is, more differentiable) the kernel is, the faster is the decay of the singular values. It is known (Smithies, 1937) that if the kernel has derivatives of order p, then the decay rate of the n-th singular value μ_n is $\sim O(n^{-p-\frac{1}{2}})$. Moreover, practice shows that the more a singular value decays, the more oscillatory the corresponding eigenfunction becomes. This bears a close relation to the Riemann–Lebesgue theorem.

From property viii above, the following expansions follow:

$$f = \sum_{j=1}^\infty f_j u_j = \sum_{j=1}^\infty <f, u_j>_X u_j \quad (3.5)$$

---

[4]The eigensystem $\{\lambda, \phi\}$, $\lambda \neq 0$, $\{\phi\}$ a complete orthonormal set of eigenfunctions, exists only for positive, self-adjoint operators. If the operator is not self-adjoint, an eigensystem does not have to exist. The singular value system plays the role of an eigensystem for a nonself-adjoint operator.

and

$$g = \sum_{j=1}^{\infty} g_i u_i = \sum_{j=1}^{\infty} < g, v_j >_Y v_j. \tag{3.6}$$

From Eq. (3.5) and the eigensystem (1) above

$$Af = \sum_{j=1}^{\infty} \mu_j < f, u_j >_X v_j. \tag{3.7}$$

Since $Af = g$, then comparing Eqs. (3.6) and (3.7) gives

$$< f, u_j >_X = \mu_j^{-1} < g, v_j >_Y \tag{3.8}$$

It follows at once that

$$f = \sum_{j=1}^{\infty} \frac{1}{\mu_j} < g, v_j >_Y u_j. \tag{3.9}$$

Equation (3.9) expresses the object function f in terms of the data function g and is essentially a representation of the inverse operator. Equation (3.7) is the *singular value expansion* of A and (3.9) of f. Similar SV expansions can be obtained for the adjoint operator A^* by interchanging the eigenfunctions u_j and v_j.

Equations (3.7) and (3.9) are interesting. Let us look at Eq. (3.7), which is the spectral expansion for the data function g in the form

$$\sum_{j=1}^{\infty} \mu_j < f, u_j >_X v_j = \sum_{j=1}^{\infty} < g, v_j >_Y v_j. \tag{3.10}$$

In view of the behavior of the singular system described above, Eq. (3.10) tells us that the higher-order spectral components in the object function f are damped out in the data since the component $< f, u_j >$ is multiplied by μ_j. The higher the order, the more is the damping. This clearly reveals the smoothing effect of the integral operator. Equation (3.9) describes the inverse of Eq. (3.7)– calculating the object function f from its image g. This shows that the higher-order spectral components in the data function g are amplified in the object since the component $< g, v_j >$ in the equation is divided by μ_j. The higher the order, the more amplified it is in f introducing greatly amplified high-frequency noise of an oscillatory nature.

It is thus clear that a square-integrable solution f cannot exist for an arbitrary data g because of the multiplication by the factor μ_j^{-1} in Eq. (3.9). The data must, therefore, be somewhat smoother than the object function if the series in this equation is to converge. From Eqs. (3.5) and (3.8), we must have

$$\|f\|_X^2 = \sum_{j=1}^{\infty} | < f, u_j >_X |^2 = \sum_{j=1}^{\infty} |\lambda_j^{-1}| < g, v_j >_Y |^2 < \infty. \tag{3.11}$$

For a compact operator, $\text{limit}_{j\to\infty} \lambda_j \to 0$. This implies that if the estimate (3.11) is to hold then the generalized Fourier coefficients $| < g, v_j >_Y |$ must vanish faster than the singular values μ_j. The bound (3.11) is known as *Picard's existence condition*. The condition says that beyond some value N of j in the summation in (3.9), the coefficients $< g, v_j >$ must start falling-off faster than μ_j^{-1} if the series in (3.9) is to converge.

Now if the series in Eq. (3.9) converges, then it converges in the mean-square sense if the kernel is square-integrable, whereas the convergence is uniform if the kernel is continuous (Smithies, 1958; see also Hansen, 1998). The data g must be in the range of A. Therefore, for a solution of the equation $Af = g$ to exist, the following two conditions must apply;

$$\left. \begin{array}{l} g \in N(A^*)^\perp \\ \sum_{j=1}^{\infty} \lambda_j^{-1} | < g, v_j >_Y |^2 < \infty. \end{array} \right\} (2)$$

Together they are known as *Picard's conditions*.

Note that Eq. (3.9) is of the minimum norm since any solution can be added to it from the null space. The generalized solution is thus recovered, namely

$$f^+ = \sum_{j=1}^{\infty} \mu_j^{-1} < g, v_j >_Y u_j = A^+ g. \tag{3.12}$$

Equation (3.12) is an explicit expansion of the generalized solution in terms of the singular system of the operator A. Operating by A on both sides of this equation gives

$$\begin{aligned} Af^+ &= \sum_{j=1}^{\infty} \mu_j^{-1} < g, v_j >_Y Au_j \\ &= \sum_{j=1}^{\infty} < g, v_j >_Y v_j \end{aligned} \tag{3.13}$$

upon using the eigenvalue problem (1). The relation (3.13) is a confirmation of what was mentioned earlier in Section 3.4, namely, a generalized solution exists if and only if $Pg \in R(A)$. From Eq. (3.12) it follows that

$$\|A^+\| = \sup_{\|v_k\|=1} \|A^+ v_k\|_X = \mu_k^{-1} \to \infty$$

as $k \to \infty$. The generalized inverse is thus unbounded. SVE, therefore, demonstrates explicitly how the inverse of a compact operator in an infinite-dimensional space becomes discontinuous. This further reflects the fact that the range of a compact operator cannot be closed if the space is infinite-dimensional. This is in accord with the equivalence relations of p. 57. In order to see it more clearly, suppose that Y is a Banach space and $R(A)$ is closed in Y. Then $R(A)$ itself is a Banach space. Now, by Banach's *Open Mapping Theorem*,

$\tilde{A}^{-1} : R(A) \to N(A)^{\perp}$ exists. Thus $A\tilde{A}^{-1} = I|_{R(A)}$. Since A is compact, it follows that $A\tilde{A}^{-1}$ is also compact implying that $I|_{R(A)}$ is compact. This cannot be true if the space is infinite-dimensional.

It is clear that if f^{+} is to remain bounded, then the SV expansion (3.12) must be truncated at some value n of j and one must be satisfied with this n-th order approximation of the solution

$$f_n^{+} = \sum_{j=1}^{n} \mu_j^{-1} < g, v_j >_Y u_j.$$

Therefore,

$$f_n^{+} - A^{+}g = \sum_{j=n+1}^{\infty} \mu_j^{-1} < g, v_j >_Y u_j$$

and

$$\|f_n^{+} - A^{+}g\|_X = \sum_{j=n+1}^{\infty} \mu_j^{-1} | < g, v_j >_Y | \to 0 \qquad (3.14)$$

as $n \to \infty$ by Picard's condition.

The determination of n at which the series is to be terminated is equivalent to some *a priori* knowledge about the solution. For noisy data g^{ϵ}, n depends on the error level ϵ, i.e., $n = n(\epsilon)$. It is interesting to calculate the norm of the difference between the n-th order SV decomposition of the generalized solution for the exact and noisy data. Denoting these by f_{en}^{+} and f_{en}^{+}, respectively, we obtain

$$
\begin{aligned}
\|f_{en}^{+} - f_{en}^{+}\|_X^2 &= \| \sum_{j=1}^{n} \mu_j^{-1} < g_e - g^{\epsilon}, v_j >_Y u_j \|^2 \\
&= \mu_n^{-2} | \sum_{j=1}^{n} < g_e - g^{\epsilon}, v_j >_Y u_j |^2 \le \epsilon^2 \mu_n^{-2}.
\end{aligned}
$$

Equation (3.14) shows that as $n \to \infty$, $f_{en}^{+} \to A^{+}g_e$. It is interesting to do the same for f_{en}^{+} in which case

$$
\begin{aligned}
\|f_{en}^{+} - A^{+}g\|_X &\le \|f_{en}^{+} - A^{+}g\|_X + \|f_{en}^{+} - f_{en}^{+}\|_X \\
&\le \|f_{en}^{+} - A^{+}g\|_X + \epsilon \mu_n^{-1}. \qquad (3.15)
\end{aligned}
$$

The first term on the R.H.S. of Eq. (3.15) is the error of truncation of the SV expansion after n terms and goes to zero as $n \to \infty$ according to Eq. (3.14). The second term is due to the error in the data and becomes unbounded for a fixed error level ϵ in the same limit. Therefore, given an error level in the data, there exists a trade-off here. We would like to choose a large value for n in order to minimize the truncation error. On the other hand, n must not be too large lest the data error should become too large. The trade-off depends on

ϵ and, therefore, strictly speaking, one should write $n(\epsilon)$ in place of n. All that can be said at this point is that given an ϵ, n must be chosen in such a way that the ratio $\epsilon \mu_n^{-1}$ goes to zero as $\epsilon \to 0$. It is also clear that the faster the rate at which the singular values decay, the more difficult it is to satisfy Picard's conditions. On the basis of this consideration, one can crudely characterize the severity of ill-posedness (Hoffman, 1986; Engl $et\ al.$, 1996). For example, if $\mu_n \sim O(n^{-\alpha})$, $\alpha \in R^+$, i.e. μ_n falls off no faster than a rational function of n, then the problem is said to be $mildly$ $(modestly)$ ill-$posed$. On the other hand, if $\mu_i\ O(e^{-\alpha i})$, $\alpha > 1$, then the problem is $severely$ ill-$posed$. If $\alpha \le 1$, some authors consider the problem to be $moderately$ ill-$posed$ The rate $\mu_n \sim O(n^{-1/2})$ is found to be appropriate for two-dimensional computerized tomography (CT). The problem is, therefore, only mildly ill-posed and this is well-known in the CT literature. On the other hand, if only a limited number of projection data is available, then the rate of decrease in μ is dramatic and the problem becomes severely ill-posed. The strong ill-posedness of the limited-angle CT is a well-established fact (Natterer, 1986; Louis, 1989).

3.6　Tikhonov's Theory of Regularization

In Section 3.4 we discussed generalized solutions of the ill-posed equation $Af = g$ in a Hilbert space setting. The generalized solution was obtained by minimizing the residual $\|Af - g^\epsilon\|_Y^2$ subject to the constraint that the norm of the solution was minimum. In this sense it is essentially a problem in constrained minimization. A generalization of this procedure would be to minimize a functional $T_\alpha(f)$ defined as

$$T_\alpha(f) = \|Af - g^\epsilon\|_Y^2 + \alpha \|f\|_X^2 \tag{3.16}$$

by appropriately choosing a real, positive parameter α, the minimizer $f \in X$. An even further generalization can be obtained by replacing $\|f\|_X^2$ in Eq. (3.16) by a functional $\Omega(f)$ of a general nature yielding

$$T_\alpha(f) = \|Af - g^\epsilon\|_Y^2 + \alpha \|\Omega(f)\|_X. \tag{3.17}$$

The functional $\Omega(f)$ is called $stabilizer$, $stabilizing$ or $smoothing$ $functional$. It is nonnegative, continuous and embodies any a $priorily$ available information about the solution. This information can be $qualitative$ as well as $quantitative$. The qualitative knowledge involves nearness and similarity information. These include considerations such as closeness of the exact solution to some known order of smoothness, existence of discontinuities, and so on. One strives for the maximum possible in this regard. As an example, if it is known a $priorily$ that the exact solution is close to being in, say, the Sobolev space of order 2, then the stabilizer will be expressed as $\|f\|_{H^2}$ in order to attain the maximum smoothness possible compatible with the available information. Glasco (1984) points out that if the true solution is known to be close to some functions in a certain class of smoothness, then the extra smoothness requirement introduces no substantial errors in applications. The quantitative knowledge deals with the information available with more certainty than just closeness and similarity. For example, in determining the density of heat sources $\rho(x,t)$, $x \in [0,a]$, $t \in [0,T]$ from the measurements of the temperature distribution over a line segment in $[0,a]$, it may be known that $\rho(x,0) = \rho(0,t) = $ constant.

$\Omega(f)$ can also be given in a weighted Sobolev space as

$$|<f,\phi>_X|^2 = \left|\sum_{j=0}^{\infty} \int_D w_j(x) f^j(x) \phi^j(x) \; dx\right|^2,$$

where $w_j(x)$ is a weight functions, $f(\phi)^j$ is the j-th order generalized derivative of $f(\phi)$, and the integral is over some regular domain D. As to how the above sum over j is to be truncated would depend upon the *a priorily* available information about the smoothness of f. Imposing the constraint $\Omega(f)$, of course, restricts the solution space. In a nutshell, *the objective of the theory of regularization is to construct stable, approximate solutions of the ill-posed problem $Af = g$ by incorporating any a priorily available information about the solution directly into the mathematical formulation itself instead of ad hoc guesses as afterthoughts.* This is achieved through the functional $\Omega(f)$.

Roughly speaking, to *regularize* an ill-posed problem in which the inverse operator A^{-1} is not continuous, is to modify the problem in such a way that the inverse mapping is continuous and the solution of this modified problem– the *regularized* solution–is an approximation of the actual solution. In practice, a single modified problem may not suffice and the unbounded inverse A^{-1} is approximated *via* a sequence of bounded (but not necessarily linear) operators. Thus regularizing an ill-posed problem means approximating the problem by a family of neighboring well-posed problems. Alternatively, it refers to finding an approximation to the discontinuous inverse operator A^{-1} in terms of a neighboring family of bounded operators. These operators are called *regularizing operators* and obtaining the solution by a family of regularizing operators is known as a *regularizing algorithm* or *regularization strategy*. This is achieved through the minimization of Tikhonov's functional, the process of which consists of constructing minimizing sequences acting as regularizing operators. A particular regularization algorithm yields a sequence of regularized solution $A_\alpha f_\alpha = g_\alpha$, where A_α is continuous and $g_\alpha \in R(A_\alpha)$. It will be shown later that minimizing Tikhonov's functional is equivalent to solving an integral equation of the second kind and equations of this kind are generally known to behave well. The basic idea behind regularization by minimizing Tikhonov's functional is that it would result in a suitably small residual and at the same time satisfy certain specified constraints so that the approximation is not too far from the desired solution.

The parameter α in Tikhonov's functional is important as it controls the numerical stability of the inverse operator A_α^{-1}. If this parameter is too small, then the effect of the stabilizer $\Omega(f)$ becomes negligible compared to the residual term $\|Af - g\|$ in which case the problem becomes unstable since the original problem is ill-posed. Contrarily, if α is unduly large so that $\Omega(f)$ dominates the residual, then the regularized operator A_α^{-1} will be too far away from the actual operator A. In this case the solution will be over-smoothed and quite stable, but the approximation may be unacceptable. Clearly, there is a trade-off between minimizing the residual and the stability of the solution. The regularization parameter α balances the two. More will be said on this later. However, the basic results are briefly:

$$\alpha = \alpha(\epsilon)$$
$$\text{limit}_{\epsilon \to 0}\, \alpha(\epsilon) \to 0$$
$$\text{limit}_{\epsilon \to 0}\, \frac{\epsilon^2}{\alpha(\epsilon)} \to 0$$
$$\text{and } \|f_\alpha - f\| \to 0 \text{ as } \epsilon \to 0.$$

These will be borne out in the discussions that are to follow.

3.6.1   The Regularizing Operator

Let us begin with the definition of a regularizing operator. There are basically two mutually related definitions: the usual $\epsilon - \delta$ definition of continuity and a parametric definition. Let A, f_e, g_e, f^ϵ and g^ϵ be as in the previous sections. It is agreed upon that the perturbed solution f^ϵ cannot be obtained by simply writing it as $A^{-1} g^\epsilon$. Instead let us assume that f^ϵ is given by $f^\epsilon = \mathcal{R} g^\epsilon, \mathcal{R} : Y \to X$ is some operator. If such an operator exists, it will be called a *regularizing operator*. We intend to define this operator and determine the conditions for its existence.

The $\epsilon - \delta$ Definition

Let $B_{\epsilon_1}(g_e)$ be a ball of radius $\epsilon_1 > 0$ in Y centered around the exact data g_e such that $\|g^\epsilon - g_e\|_Y \le \epsilon, \forall \epsilon \in (0, \epsilon_1]$.

Definition 3.4. (Tikhonov) *An operator* $\mathcal{R}_\epsilon : Y \to X$ *is called a* regularizing operator *if (i) it is defined on the ball* $B_{\epsilon_1}(g_e)$ *and (ii) for every* $\delta > 0$, *there exists an* $\epsilon(\delta), 0 < \epsilon(\delta) \le \epsilon_1$, *such that* $\|f^\epsilon - f_e\|_X \le \delta$, *where* $f^\epsilon = \mathcal{R}_\epsilon g^\epsilon$.

A slightly modified version of the above definition is

Definition 3.5. (Glasco) *An operator* $\mathcal{R}_\epsilon : \mathcal{R}_\epsilon g^\epsilon = f^\epsilon$ *is called a* regularizing operator *for the ill-posed problem* $Af = g$ *in the metric space* X *if (i)* $\mathcal{R}_\epsilon$ *is defined for every* $\epsilon, 0 < \epsilon \le \epsilon_1$, *and for arbitrary* $g^\epsilon \in Y$, *such that* $\|g^\epsilon - g_e\|_Y \le \epsilon > 0$, *and (ii)* $\text{limit}_{\epsilon \to 0} \|f^\epsilon - f_e\|_X \to 0$.
A similar definition is also given by Domanskii (1987).

Definition 3.6. (Domanskii) *An operator* $\mathcal{R}_\epsilon$ *mapping* $Y \to X, \epsilon \in (0, \epsilon_0)$, $\epsilon_0 > 0$, *is a regularizing operator if* $\forall f \in X, \exists f_0 \in X$, *such that*

$$\limsup_{\epsilon \to 0} \|f_0 - \mathcal{R}_\epsilon g\|_X : g \in Y, \|g - Af\|_Y \le \epsilon = 0.$$

The second part of Definition 3.5 (also Definition 3.6) is termed the *principle of regularization* by Glasco (1984) and reflects the physics of ill-posedness. It says that only those solutions that converge to the exact solution as the error goes to zero (see p. 15) are admissible. It, therefore, singles out only the solutions that vary continuously with the data out of the space V_ϵ of all solutions for which the residual $\|Af^\epsilon - g^\epsilon\|_Y$ may be small.

Maslov (1968) gives the following interpretation of a regularizing operator. Define a set

$$M(\mathcal{R}_\epsilon) = \{g \in Y | \exists f \in X, \, \mathcal{R}_\epsilon g \to f, \epsilon \to 0\} \Longrightarrow Af = g. \tag{3.18}$$

If such a set exists, then $\mathcal{R}_\epsilon$ is said to *converge to an element in* Y. According to Masolv, the convergence of $\mathcal{R}_\epsilon$ to an element in Y is equivalent to the existence of a stable, approximate solution. Moreover, such convergence guarantees a linear regularizer, that is, a linear $\mathcal{R}_\epsilon$. Hence the definition is:

Definition 3.7. (Domanskii) *A linear regularizer in the set* $M(\mathcal{R}_\epsilon)$ *defined in (3.18) is called a M-regularizer and the corresponding algorithm a M-regularizing algorithm for the ill-posed problem* $Af = g$ *and the solution* $\mathcal{R}_\epsilon g^\epsilon = f^\epsilon$ *an approximate solution.*
The term *M-soluble* is also used.

The Parametric Definition

In this definition, the family of the regularizing operators is parameterized by a so-called *regularizing parameter* $\alpha \in (0, \infty)$.

Definition 3.8. (Tikhonov) *An operator* $\mathcal{R}_\alpha : Y \to X$ *depending on a real parameter* $\alpha \in (0, \infty)$ *is called a regularizing operator for the ill-posed equation* $Af = g$ *in a neighborhood of the exact data* g_e *if (i)* $\mathcal{R}_\alpha$ *is defined* $\forall \alpha > 0, \forall g^\epsilon \in Y, \|g^\epsilon - g_e\|_Y \leq \epsilon, 0 < \epsilon \leq \epsilon_1$, *and (ii)* $\exists \alpha = \alpha(\epsilon)$ *such that* $\forall \delta > 0, \exists$ *a number* $\epsilon(\delta) \leq \epsilon_1$, *such that*

$$\|g^\epsilon - g_e\|_Y \leq \epsilon(\delta) \Longrightarrow \|f_\alpha^\epsilon - f_e\|_X \leq \delta, f_\alpha^\epsilon = \mathcal{R}_{\alpha(\epsilon)} g^\epsilon.$$

Definition 3.9. (Glasco) *An operator* $\mathcal{R}_\alpha$ *acting on* $g \in Y$ *is called a regularizing operator for the equation* $Af = g$ *provided (i) it is defined* $\forall \alpha(0 < \alpha \leq \alpha_1)$ *for some* α_1 *and arbitrary* $g^\epsilon, \|g^\epsilon - g_e\|_Y \leq \epsilon$, *and (ii)* $\alpha = \alpha(\epsilon)$, *a function of* ϵ *such that* $\lim_{\epsilon \to 0} \alpha(\epsilon) \to 0$, *and* $\lim_{\epsilon \to 0} \|f_\alpha^\epsilon - f_e\|_X \to 0, f_\alpha^\epsilon$ *as in Definition 3.7.*

The second condition is again the principle of regularization.

Definition 3.6 of Domanskii remains intact provided that the convergence set M is redefined as

$$M(\mathcal{R}_\alpha) = \{g \in Y | \exists f \in X : \mathcal{R}_\alpha g \to f, \alpha \to 0\}.$$

The regularization parameter α depends not only on ϵ, but also on g^ϵ and, consequently, on g_e and hence on f_e. The continuity is thus not uniform in α, but more on this later.

As we mentioned, the regularizing operator need not be linear. However, when it is, one speaks of a *linear regularizer*. A sufficient condition for $\mathcal{R}_\alpha$ to be linear and continuous in g can be given. Let f_0 be in X and its image $g_0 = Af_0$ in Y. Let $B_{\epsilon_0}(g_0)$ be a ball of radius ϵ_0 centered at g_0, $0 < \epsilon \le \epsilon_0$. Let $\mathcal{R}_\alpha : Y \to X$ depend on a parameter $\alpha > 0$. Suppose that $f_{0\alpha} = \mathcal{R}_\alpha g_0$ and let $g^\epsilon \in B_{\epsilon_0}(g_0)$, $\|g^\epsilon - g_0\|_Y \le \epsilon$, and $\mathcal{R}_\alpha g^\epsilon = f_\alpha^\epsilon$. Then

$$\|f_0 - f_\alpha^\epsilon\|_X \le \|f_{0\alpha} - f_\alpha^\epsilon\|_X + \|f_{0\alpha} - f_0\|_X$$

$$= \|\mathcal{R}_\alpha(g^\epsilon - g_0)\|_X + \|\mathcal{R}_\alpha g_0 - f_0\|_X. \tag{3.19}$$

Consider the modulus of continuity

$$\omega_\alpha(\epsilon) = \sup_{g_1,g_2 \in B_{\epsilon_0}(g_0)} \|\mathcal{R}_\alpha(g_1 - g_2)\|_X.$$

Let us further assume that

$$\mathrm{limit}_{\alpha \to 0} \mathcal{R}_\alpha(Af) = f. \tag{3.20}$$

In terms of $\omega_\alpha(\epsilon)$, Eq. (3.19) becomes

$$\|f_0 - f_\alpha^\epsilon\|_X \le \omega_\alpha(\epsilon) + \|\mathcal{R}_\alpha g_0 - f_0\|_X.$$

Now by the condition (3.20), α can be so selected that $\|\mathcal{R}_\alpha g_0 - f_0\|_X \le \omega_\alpha(\epsilon)$ resulting in $\|f_0 - f_\alpha^\epsilon\|_X \le 2\omega_\alpha(\epsilon)$. Moreover, Eq. (3.20) implies that $\omega_\alpha(\epsilon) \to 0$ as $\epsilon \to 0$. Therefore, for some $\delta > 0$, $\exists$ an $\tilde{\epsilon}(\delta) \in B_{\epsilon_0}(g_0)$ such that

$$\|f_0 - f_\alpha^{\tilde{\epsilon}}\|_X \le \delta, \ \forall \delta.$$

The continuous nature of $\mathcal{R}_\alpha$ is thus established and the linearity is a consequence of the assumption (3.20).

The regularizing operator $\mathcal{R}_\alpha$ is not uniformly bounded in α. In other words, a sequence $\{\alpha_j\}$ can be found for which the norm of $\mathcal{R}_{\alpha_j}$ becomes unbounded as $j \to \infty$. Indeed suppose the contrary and let $\mathcal{R}_\alpha\| \le C \ \forall \alpha > 0, C$ being a constant. Now for noise-free images to which Eq. (3.20) applies, this would imply that $\|\mathcal{R}_\alpha f\|_X \to \|f\|_X = \|A^{-1}g\|_Y \le C\|g\|_Y, \forall g \in Y \implies \|A^{-1}\| < \infty$. This is impossible if A is compact. Alternatively, $\mathrm{limit}_{\alpha \to 0} \mathcal{R}_\alpha A$ cannot converge to the identity operator I in the operator norm for a compact operator since I is not compact in an infinite-dimensional space.

The following theorem follows.

Theorem 3.8 (Tikhonov) *Let $L(X,Y)$ be the space of all continuous linear operators in X and Y. Then a family of operators $\{\mathcal{R}_\alpha\}, \alpha > 0, \mathcal{R}_\alpha \in L(X,Y)$, is a regularizing algorithm for A^{-1} if*

$$\forall f \in X, \mathrm{limit}_{\alpha \to 0} \mathcal{R}_\alpha Af = f. \tag{3.21}$$

Moreover, the convergence in Eq. (3.21) is not uniform in α.

If A is one-to-many then Eq. (3.21) should be replaced by

$$\forall f \in X, \text{limit}_{\alpha \to 0} \mathcal{R}_\alpha(Af) = \hat{f}, f = \hat{f} + f_0, \forall \in N(A).$$

Replacing f_0 by the exact solution f_e in Eq. (3.19), we obtain

$$\|f_\alpha^\epsilon - f_e\|_X \le \|\mathcal{R}_\alpha(g^\epsilon - g_e)\|_X + \|\mathcal{R}_\alpha g_e - f_e\|_X$$

$$\le \epsilon \|\mathcal{R}_\alpha\| + \|\mathcal{R}_\alpha g_e - f_e\|_X. \tag{3.22}$$

As $\alpha \to 0$, the second term in Eq. (3.22) goes to zero by Theorem 3.9. However, $\|\mathcal{R}_\alpha\|$ becomes unbounded because of the ill-posedness of the problem. Hence the regularization parameter $\alpha(\epsilon)$ must be so chosen that the error in Eq. (3.22) is minimized. As a matter of fact, $\epsilon \mathcal{R}_{\alpha(\epsilon)}$ must tend to zero as $\epsilon \to 0$. Suppose the contrary, i.e., $\text{limit}_{\epsilon \to 0} \epsilon \mathcal{R}_{\alpha(\epsilon)} \ge C > 0$ and let $\{\epsilon_n\}$ be a null sequence as $n \to \infty$ such that $\|\epsilon_n \mathcal{R}_{\alpha(\epsilon_n)}\| \ge C > 0$. Construct the sequence $g_n = g + \epsilon_n \phi_n$, where $\phi_n \in Y, \|\phi_n\|_Y = 1$. Applying $\mathcal{R}_{\alpha(\epsilon_n)}$ to g_n and subtracting g from it gives

$$\mathcal{R}_{\alpha \epsilon_n}[g + \epsilon_n \phi_n] - g = [\mathcal{R}_{\alpha \epsilon_n} g - g] + \epsilon_n \mathcal{R}_{\alpha \epsilon_n} \phi_n. \tag{3.23}$$

As $n \to \infty$, the first term on the right-hand side of Eq. (3.23) vanishes, whereas the second term does not. Furthermore, as $\epsilon \to 0$, the perturbed data goes over to the exact one and hence $\mathcal{R}_\alpha Af$ should also go over to the exact solution f_e. By Theorem 3.9, this implies that α must vanish. We have, therefore, the following theorem

Theorem 3.9 *Let $\mathcal{R}_\alpha$ be a regularizing operator satisfying the condition of Theorem 3.8. Then $\{\mathcal{R}_\alpha\}$ results in convergent solutions if and only if*

$$\underset{\epsilon \to 0}{\text{limit}} \, \alpha(\epsilon) = 0$$

and

$$\underset{\epsilon \to 0}{\text{limit}} \, \alpha(\epsilon) \|\mathcal{R}_{\alpha(\epsilon)}\| = 0.$$

These conditions will be further encountered later.

As mentioned on p. 71, the regularizing parameter α depends not only on ϵ, but also on g^ϵ and, therefore, indirectly on the exact solution f_e. Thus $\alpha = \alpha(\epsilon, g^\epsilon, f_e, g_e)$. This is one of the reasons why the convergence of $\mathcal{R}_\epsilon$ is not uniform in α. However, suppose that there is no dependence of α on the error ϵ, but only on the perturbed data $\mathbf{g}^\epsilon$. In that case, $\mathcal{R}_\alpha$ can act on any element of Y and result in a solution in X which implies that the inverse operator is defined on all Y and no regularization strategy is necessary. Contrarily, if the inverse operator is discontinuous, then there can be no regularization strategy with a regularization parameter independent of ϵ that would lead to a convergent solution. This is summarized in the theorem below.

Theorem 3.10 (Bakushinsky and Goncharsky, 1994) *Let $A : X \to Y$ be a continuous linear operator and $\mathcal{R}_\alpha$ a corresponding regularization strategy such that the regularization parameter $\alpha = \alpha(g^\epsilon)$ depends only on g^ϵ and not on ϵ. Then for the regularization strategy to yield convergent solutions, the inverse operator, either A^{-1} or $(A^+)^{-1}$ must be bounded in Y and no regularization strategy is thus necessary. On the other hand, if the inverse operator is discontinuous, then there can be no regularization strategy with a regularization parameter independent of ϵ that would lead to a convergent solution.*

In this context the following definition occurs in the literature on regularization.

Definition 3.10. *If a regularization parameter α depends only upon the error level ϵ and not on the perturbed data g^ϵ, then α is said to be determined a priorily and $\alpha = \alpha(\epsilon)$. Otherwise, α is said to be determined a posteriorily.*

Since the regularization parameter depends (indirectly) on the exact solution f_e, the determination of α presupposes the solution of the problem which is what is to be determined to begin with. Therefore, an exact and an in-principle determination of α is not possible. However, its approximate estimation is feasible and will be discussed later in Section 3.8.

3.6.2   The Construction of Regularizers

Having defined a regularizing operator and a regularizing algorithm, the next task is to construct them. This is done through the minimization of Tikhonov's functional (3.17). There are basically three ways to perform this task. These are:

1. Minimize the stabilizer subject to the condition that the residual is minimum.

2. Minimize the residual subject to the condition that the stabilizer is minimum. This can be considered to be the dual of program 1.

3. Minimize the functional directly without any constrained optimization.

Here we will present the construction of a regularizer by appealing to the spectral theory of nonnegative self-adjoint operators.

Consider the functional (3.16). Let ϕ be an element of X and consider the first variation of (3.16). This leads to the variational identity

$$< Af_\alpha, A\phi >_Y + \alpha < f_\alpha, \phi >_X = < \phi, A^*g >$$

from which the minimizer f_α is to be determined. Rewriting, we have

$$< \phi, A^*Af_\alpha > + \alpha < f_\alpha, \phi >_X = < \phi, A^*g >_X .$$

From this the Euler–Lagrange equation follows, namely

$$[A^*A + \alpha I]f_\alpha = A^*g, \tag{3.24}$$

because ϕ is arbitrary. We then have

$$f_\alpha = [A^*A + \alpha I]^{-1}A^*g = \mathcal{R}_\alpha g. \tag{3.25}$$

Since $\alpha > 0, \|(A^*A+\alpha I)f_\alpha\|_X \geq \alpha\|f_\alpha\|_X$ and, consequently, the inverse $\|[A^*A + \alpha I]^{-1}\|$ is less than α^{-1}, implying that $A^*A + \alpha I$ is one-to-one. Hence the functional (3.16) has a unique minimizer.

It is also instructive to look at the problem from another angle. As pointed out in Section 3.3 (see Definition 3.2), in any Hilbert space, the L_2-norm is *rigorously convex* or *rigorously normed*.[5] The functional in (3.16) is, therefore, rigorously normed if X and Y are Hilbert spaces which they are in this discussion and from convex analysis (Rocafeller, 1970) we know that a unique minimum of $T_\alpha(f)$ exists in $\hat{X}$.

Maslov (1968) provided an interesting result. Let A be a closed linear operator from $X \to X, X$ a Hilbert space. Consider a sequence $\{\alpha_n\}$ and a family of sequences $\{g_n\}$ such that $\{g_n\} \to g \in X$ as $n \to \infty$. In other words, $\|g - g_n\|_X \to 0$ as $n \to \infty$. Consider the sequence

$$f_n = A^*[AA^* + \alpha_n]^1 g_n. \tag{3.26}$$

Maslov's theorem tells us that if there exists a solution to the equation

$$Af = g, g \in N(A^*)^\perp, \tag{3.27}$$

then the sequence f_n in Eq. (3.26) converges strongly and the limit is the solution of Eq. (3.27) in $N(A)^\perp$. Then the sequence operators $\mathcal{R}_n = A^*[AA^*+\alpha_n], \alpha_n > 0, n \in Z^+$, is a linear regularizing strategy for the ill-posed equation $Af = g$. A detailed discussion in this line can be found in Domanskii (1987).

From Eq. (3.25), it follows that the operator $\mathcal{R}_\alpha : Y \to X$ is a function of A^*A. This remains unchanged even if $\Omega(f)$ is used in place of $\|f\|_X^2$ (see Chapter 5). Now the normal operators A^*A and AA^* are self-adjoint positive (if zero is not an eigenvalue). Let us consider that A is compact. For a compact self-adjoint operator, the spectral theory (Akhiezer and Glazman, 1961; Liusternik and Sobolev, 1961; Roman, 1975; Stakgold, 1979), gives us the following result

$$A = \sum_{k=1}^{\infty} \lambda_k P_k, \tag{3.28}$$

where P_k is the operator of projection onto the finite-dimensional λ_k eigenspace corresponding to the k-th eigenvalue. Recall that for a compact normal operator, the eigenvalues are at most countably infinite of finite multiplicity and the corresponding eigenspaces are pairwise orthogonal (see Roman, 1975). If A is replaced by a function, say, $F(A)$, then

$$F(A) = \sum_{k=1}^{\infty} F(\lambda_k) P_k. \tag{3.29}$$

---

[5]The situation is different if the space is a Banach space. An example is the norm $\|f\|$ where f is a continuous function. If the Banach space is rigorously convex, then $\|f\|$ is not rigorously normed, but $\|f\|^2$ is.

This can be verified readily by considering a polynomial function of A. Equation (3.28) is called the *spectral resolution* or the *resolution of the identity* of the operator A.

In view of Eqs. (3.28) and (3.29), we can express $\mathcal{R}_\alpha = \mathcal{R}_\alpha(A^*A)$ as

$$\mathcal{R}_\alpha g = \sum_{\lambda_k \in \sigma(\lambda), \lambda_k \neq 0} r_\alpha(\lambda_k) < A^*g, u_k > u_k, \tag{3.30}$$

where Eq. (3.5) was used. The problem of finding the regularizing operator $\mathcal{R}_\alpha$ thus reduces to investigating the spectral function $r_\alpha(\lambda_k)$. Let us first show that $\mathcal{R}_\alpha$ is bounded. We obtain

$$\begin{aligned}
\|\mathcal{R}_\alpha g\|_X^2 &= \sum_{\lambda_k \in \sigma(\lambda), \lambda_k \neq 0} |r_\alpha(\lambda_k)|^2| < A^*g, u_k > |^2 \\
&\leq \sum_{\lambda_k \in \sigma(\lambda), \lambda_k \neq 0} |r_\alpha(\lambda_k)|^2| < g, Au_k > |^2 \\
&= \sum_{\lambda_k \in \sigma(\lambda), \lambda_k \neq 0} |r_\alpha(\lambda_k)|^2 \lambda_k| < g, v_k > |^2,
\end{aligned} \tag{3.31}$$

since $Au_k = \mu_k v_k$. If in Eq. (3.31) we impose the condition that

$$\sup_{\lambda_k \in \sigma(\lambda), \alpha > 0} |r_\alpha(\lambda_k)| \mu_k \leq C_\alpha < \infty,$$

where C_α is a constant, then $\mathcal{R}_\alpha$ is bounded. Furthermore, the bound can be calculated exactly. Indeed if g in Eq. (3.30) is replaced by v_m, then

$$R_\alpha v_m = r_\alpha(\lambda_m)\mu_m u_m$$

from which it follows that

$$\|\mathcal{R}_\alpha\| = \sup_{\lambda \in \sigma(\lambda), \lambda \neq 0, \alpha > 0} |r_\alpha(\lambda_m)| \mu_m,$$

which is the constant C_α.

If $\mathcal{R}_\alpha$ is to be a regularizer for $Af = g$, then $\mathcal{R}_\alpha g$ must reduce to the generalized solution f^+ as $\alpha \to 0$. Thus

$$\begin{aligned}
&\sum_{\lambda_k \in \sigma(\lambda), \lambda_k \neq 0, \alpha > 0} r_{\alpha \downarrow 0}(\lambda_k) < A^*g, u_k > u_k \\
&= \sum_{\lambda_k \in \sigma(\lambda), \lambda_k \neq 0, \alpha > 0} r_{\alpha \downarrow 0}(\lambda_k) \mu_k < g, v_k > u_k \\
&= \sum_{\lambda_k \in \sigma(\lambda), \lambda_k \neq 0, \alpha > 0} \{r_{\alpha \downarrow 0}(\lambda_k) \lambda_k\}\{\mu_k^{-1} < g, v_k > u_k\}.
\end{aligned} \tag{3.32}$$

Comparing Eq. (3.32) with Eq. (3.12), we see that in order for $\mathcal{R}_\alpha g$ to reduce to the generalized solution f^+, we must impose the condition

$$\lim_{\alpha \downarrow 0, \lambda \in \sigma(\lambda), \lambda \neq 0, \alpha > 0} r_\alpha(\lambda_k)\lambda_k = 1.$$

Finally, we consider the quantity $\mathcal{R}_\alpha A f - f, f \in X$. Now from Eq. (3.30)

$$\mathcal{R}_\alpha A f = \sum_{\lambda_k \in \sigma(\lambda), \lambda_k \neq 0, \alpha > 0} r_\alpha(\lambda_k) \lambda_k < f, u_k > u_k.$$

Then using the SVE Eq. (3.5) for f, we have

$$\mathcal{R}_\alpha A f - f = \sum_{\lambda_k \in \sigma(\lambda), \lambda_k \neq 0, \alpha > 0} \left([r_\alpha(\lambda_k) \lambda_k - 1]\right) < f, u_k > u_k.$$

Taking the norm on both sides yields

$$\|\mathcal{R}_\alpha A f - f\|_X^2 = \sum_{\lambda_k \in \sigma(\lambda), \lambda_k \neq 0, \alpha > 0} |r_\alpha(\lambda_k) \lambda_k - 1|^2 | < f, u_k > |^2. \qquad (3.33)$$

Let us next impose the third condition

$$\sup_{\lambda \in \sigma(\lambda), \lambda_k \neq 0, \alpha > 0} |r_\alpha(\lambda_k)| \lambda_k \leq 1,$$

and break up the sum over λ_k in Eq. (3.33) into two parts

$$\sum_{\lambda_k \in \sigma(\lambda), \lambda_k \neq 0, \alpha > 0} = \sum_{\lambda_k \in \sigma(\lambda), \lambda_k \geq \lambda_n, \alpha > 0} + \sum_{\lambda_k \in \sigma(\lambda), \lambda_k \leq \lambda_n, \alpha > 0}.$$

The second sum is

$$C \sum_{\lambda_k \in \sigma(\lambda), \lambda_k \lambda_n, \alpha > 0} | < f, u_k >_X |^2,$$

where

$$C = \sup_{\lambda, \alpha} |r_\alpha(\lambda)| |\lambda - 1|^2.$$

We can select an α such that the sum can be made less than some $\delta > 0$. The same can also be done for the first sum (for further details, see Bakushinsky and Goncharsky (1994)). It then follows that

$$\lim_{\alpha \to 0} \|\mathcal{R}_\alpha A f - f\|_X = 0.$$

Therefore,

Theorem 3.11 *The sequence of operators* $\{\mathcal{R}_\alpha\}$ *defined by*

$$\mathcal{R}_\alpha g = \sum_k r_\alpha(\lambda_k) < g, v_k >_Y u_k, \alpha > 0,$$

is a regularization strategy for the ill-posed equation $Af = g$ *provided that* $r_\alpha(\lambda_k)$ *satisfies the following conditions, namely i.*

$$\sup_{\lambda \in \sigma(\lambda), \lambda \neq 0, \alpha > 0} |r_\alpha(\lambda_k)| \mu = C_\alpha < \infty.$$

ii.

$$\lim_{\alpha \downarrow 0, \lambda \in \sigma(\lambda), \lambda \neq 0, \alpha > 0} r_\alpha(\lambda_k)\lambda_k = 1,$$

iii.

$$\sup_{\lambda \in \sigma(\lambda), \lambda_k \neq 0, \alpha > 0} |r_\alpha(\lambda_k)|\lambda_k \leq 1.$$

Note that from the theory of linear self-adjoint operators (Stakgold, 1979) $\lambda \leq \|A^*A\|$. The norm $\|A^*A\|$ is called the *spectral radius*. Instead of the function $r_\alpha(\lambda_k)$, a modified function $s_\alpha(\lambda_k) = r_\alpha(\lambda_k)\lambda_k$ can be defined. In this case, the conditions i–iii become

i.

$$\sup_{\lambda_k \in \sigma(\lambda)} |s_\alpha(\lambda_k)| \leq C_\alpha \mu_k,$$

ii.

$$\lim_{\alpha \downarrow 0, \lambda \in \sigma(\lambda), \lambda \neq 0, \alpha > 0} s_\alpha(\lambda_k) = 1 \ \forall \ \lambda_k \in (0, \|A^*A\|)$$

iii.

$$\sup_{\alpha, \lambda \in \sigma(\lambda), \alpha > 0} |s_\alpha(\lambda_k)| \leq 1.$$

Using s_α instead of r_α essentially amounts to expressing Eq. (3.30) in terms of the object function f instead of the data g.

3.6.3   The Spectral or Filter Functions

The above formulation was in terms of the singular system which is convenient for a compact operator. However, conditions i–iii can also be formulated as a continuous function of $\lambda \in (0, \infty)$. Recall (references on p. 75) that for a self-adjoint, nonnegative operator, the spectral resolution can be expressed as a *Stielz integral* over the Stielz measure. This measure is related to the spectral family E_λ which is defined as a continuous function of the eigenvalue. This function undergoes jumps at the points of the discrete spectrum and is defined to be either right or left continuous at points of discontinuity. For a compact operator, the spectral family correctly picks out the points of the discrete spectrum. Thus the function (called *spectral* or *filter function*) can be written as $s_\alpha(\lambda), \lambda \in (0, \infty)$ and conditions i through v above remain unchanged.

Some important spectral functions which are used in regularization work are:

i. the *rectangular function* $s_\alpha(\lambda) = H(\lambda - \alpha)$, H being the Heaviside function;

ii. The *Tikhonov function* $s_\alpha(\lambda) = \lambda \backslash (\lambda + \alpha)^{-1}$;

iii. the *triangular function* $s_\alpha(|\xi|) = (1 - |\xi|)H(\alpha^{-1} - |\xi|)$;

iv. the *Hanning window* $s_\alpha(|\xi|) = (1\backslash2)[1 + \cos(\pi\alpha\xi)]H(\alpha^{-1} - |\xi|)$;

v. the *Gaussian window* $s_\alpha(|\xi|) = e^{-\alpha\xi^2\backslash2}$

and many others (see Bertero, 1989). These functions all satisfy the conditions for s_α given above and, therefore, lead to regularizing strategies. For example, conditions i and iii are automatically satisfied by the rectangular spectral function for which the norm $C_\alpha = \alpha^{-\frac{1}{2}}$. The Tikhonov function satisfies conditions ii and iii automatically. C_α for this function is $\frac{1}{2}\sqrt{\alpha}$. The primary characteristics of these functions are that they are close to unity if the singular values are large while being close to zero if otherwise. For example, if $\alpha > 0$, then the Tikhonov function is close to one at high singular values, but zero if these values are small. In this way the high-frequency oscillations of arbitrary amplitudes are eliminated in the inverse solution. It was mentioned above (Section 3.2) that the smoother the kernel, the more unstable is the solution. This means that if the *order of regularization* (Davies, 1992), that is, the order of smoothness m of the function space from which the minimizing sequences are to be selected increases, then the singular values tend to zero more rapidly. For $\alpha > 0$, this makes the filter strong. Therefore, a careful choice of both α and the order m is essential.

The Iterative Filters

Let us introduce at this point an interesting regularization strategy of overwhelming importance, namely, the method of iteration. In its general form, an iteration procedure consists of a sequence of solutions $\{f_n\}$ such that

$$f_{n+1} = F(f_n, g^\epsilon), n = 0, 1, \cdots, N,$$

where F is some function of f_n and g^ϵ. In the case that the number n serves as the regularization parameter, an iteration procedure is a regularization strategy for obtaining a stable, approximate solution to the ill-posed problem $Af = g^\epsilon$. Let $F(f, g)$ be a continuous function of (f, g) except perhaps at the true values (f_e, g_e). Moreover, let the function $F(f, g)$ be such that the iteration $f_{n+1} = F(f_n, g_e)$ converges to the true solution f_e. Then it is shown (Glasco, 1984) that there exists a function $N = N(\epsilon)$ such that

$$\underset{\epsilon \to 0}{\text{limit}} \, ||f_{N(\epsilon)} - f_e||_X \to 0.$$

According to Glasco, a number of gradient methods including the conjugate gradient algorithm satisfy the conditions mentioned.

In the rest of this section, we consider the well-known iterative procedure of Landweber (1951). Consider the operator equation $Af = g$ and write it in the form

$$f = (I - A)f + g = f - (Af - g).$$

This suggests the following iterative procedure:

$$f^{(n+1)} = f^{(n)} - (Af^{(n)} - g) = f^{(n)} - r^{(n)},$$

where the superscript denotes the stage of the iteration and $r^{(n)}$ is the residual after the n-th iteration. Usually, for the purposes of convergence, acceleration of iterations, and so on, the residual is multiplied by the so-called *relaxation parameter* and this parameter is denoted by τ. If τ does not change during iterations, the procedure is called *stationary*, otherwise, it is *dynamic*. We consider the stationary process only. Therefore, the iteration stage is written as

$$f^{(n+1)} = f^{(n)} - \tau(Af^{(n)} - g) = f^{(n)} - \tau r^{(n)}.$$

Strand (1974) generalized the above scheme to the form

$$f^{(n+1)} = f^{(n)} - F(A^*A)r^{(n)},$$

$F(A^*A)$ being a rational function of its argument. In the classical Landweber iteration, therefore, $F(A^*A)$ is simply the relaxation parameter τ.

Let us apply the method to the equation $A^*Af^+ = A^*g$ for obtaining the generalized solution f^+. Consider that the initial guess $f(0) = 0$. The $(n+1)$-th stage of the iteration is

$$f^{(n+1)} = f^{(n)} - \tau r^{(n)}.$$

Reducing the above equation recursively, we obtain

$$f^{(n+1)} = \tau \sum_{j=0}^{n} [I - \tau AA^*]^j A^*g.$$

Write the above equation as

$$f^{(n+1)} = R_n g,$$

where

$$R_n = \tau \sum_{j=0}^{n} [I - \tau AA^*]^j A^*.$$

The corresponding spectral function $s_n(\lambda)$ is given by

$$s_n(\lambda) = \gamma \sum_{j=0}^{n} (1 - \gamma)^j = 1 - (1 - \gamma)^{(n+1)},$$

where $\gamma = \tau\lambda$. Therefore, if we select the relaxation parameter τ in the range $0 < \tau < 2||A||^{-2}$, (remember that $\lambda \in [0, ||A||^2]$), then $0 < 1-\gamma < 1$. Conditions i through iii are satisfied if the regularization parameter α is taken to be the reciprocal of the number of iterations n.

It is clearly seen that in the initial stages of iteration, the algorithm picks up the larger singular values (picking up the largest in the very first iteration) and, consequently, results in stable and characteristic features of the object function. The initial stages then result in an overall good estimate of the solution. However, as the iterations proceed, the algorithm starts entering into the region of instability corresponding to the smaller singular values. As a result, the solution becomes increasingly worse. It is, therefore, necessary that some sort of a *stopping criterion* is designed, that is, to determine the number of iterations n, which depends upon the error level in the data. An *a posteriori* rule was suggested by Vainniko (1982) according to which the iterations must be terminated when

$$||A(f^\epsilon)^{n+1} - g^\epsilon||_Y \leq C\epsilon < ||A(f^\epsilon)^n - g^\epsilon||_Y.$$

This is related to the so-called *discrepancy principle* of Morozov (to be discussed shortly) which is used in determining the regularization parameter. For the generalized iteration scheme, the filter is $[1 - (1 - \lambda_k F(\lambda_k)]_n$. There have been attempts (Strand, 1974; Graves and Prenter, 1978; see also Hansen, 1998) to design the function $F(\eta)$ such that it realizes a step function:

$$
\begin{aligned}
F(\eta) &= 1,\ \eta \leq \lambda, \\
&= 0, 0 \leq \lambda < \eta.
\end{aligned}
$$

Thus during the iterations where the singular values $\mu_j \geq \eta$, the filter will be close to unity, and *vice versa*.

A major disadvantage of Landweber iteration is that its convergence rate is often slow compared to faster methods such as that of the *conjugate gradient*. A most exhaustive discussion of the iterative methods is the text by Bakushinsky and Goncharsky (1994). For a rigorous account of the Landweber iteration, the reader should consult Kirsch (1996)

Consider the rectangular function and the approximate data g^ϵ. From Eq. (3.22)

$$||f_\alpha^\epsilon - f_e||_X = \epsilon||\mathcal{R}_\alpha(\epsilon)|| + ||\mathcal{R}_\alpha(\epsilon)Af_e - f_e||_X.$$

Since $||\mathcal{R}_\alpha(\epsilon)|| = \alpha^{-\frac{1}{2}}$, the above equation becomes

$$||f_\alpha^\epsilon - f_e||_X = \epsilon\alpha^{-\frac{1}{2}} + ||\mathcal{R}_\alpha(\epsilon)Af_e - f_e||_X.$$

The second term goes to zero as $\epsilon \to 0$ by virtue of Theorem 3.9. However, the first term becomes unbounded since the problem is ill-posed. Therefore, in order for $\mathcal{R}_\alpha(\epsilon)$ with the rectangular spectral function to define a regularization algorithm, we must have $\epsilon\alpha^{-\frac{1}{2}} \to 0$ as $\epsilon \to 0$. This is in complete agreement with Theorem 3.10. For the Tikhonov function, this condition is

$$\lim_{\epsilon \to 0} \epsilon \backslash \alpha(\epsilon) \to 0,$$

whereas, for the function $s_n(\lambda) = 1 - (1 - \gamma)^{(n)}$, the corresponding condition is $\lim_{\epsilon \to 0} n\epsilon \to 0$. A regularization strategy satisfying Theorem 3.10 is called *admissible*. We would like to point out that if the operator A happens to be self-adjoint itself, then the spectral function can be derived on the basis of the resolution of the identity of this operator only and the resulting functions are called *Lavrentiev regularizer* (see Bakushinsky and Goncharsky, 1994).

3.6.4   First-order regularization

In this section we consider Tikhonov's regularization of the first-order. By this
we mean that the functional to be minimized is of the form of Eq. (3.16). This
form of the functional is used most often. The only *a priori* knowledge about the
solution that is embodied in Eq. (3.16) is the fact that the solution is bounded.
Depending upon the estimated smoothness of the solution, more complicated
norms (Sobolev, weighted Sobolev, etc.) may appear in the stabilizer with
correspondingly more complex Euler–Lagrange equations. The essentials of the
theory, however, remain the same. A second-order stabilizer will be considered
in Chapter 5. The objectives here are: (a) to show that if the generalized inverse
exists, then the regularized solution reduces to the generalized solution as the
regularization parameter $\alpha \to 0$; and (b) to estimate the error between the
noise-free regularized solution and the corresponding result for noisy data.

Our starting point is Eq. (3.24). First we note that it is not immediately
clear from this equation that f_α belongs to $N(A)^\perp$. However, it can be made
explicit if we consider Eq. (3.25) and use the eigenvalue problem (1) of Section
3.5 and expansions (3.5) and (3.6). A straightforward algebra shows that

$$\begin{aligned}
f_\alpha &= \sum_{j=1}^{\infty} f_j u_j = \sum_{j=1}^{\infty} \frac{g_j}{\alpha + \lambda_j} \{\mu_j u_j\} & (3.34)\\
&= \sum_{j=1}^{\infty} A^* \left(\frac{v_j}{\alpha + \lambda_j} \right) g_j \\
&= A^* \left[AA^* + \alpha I \right]^{-1} g.
\end{aligned}$$

The above representation shows that $f_\alpha \in N(A)^\perp)$.

We demonstrate next that if A^+ exists, that is, if Picard's conditions (2)
(Section 3.5) hold, then the regularized solution f_α goes over to the generalized
solution f^+ as $\alpha \to 0$. Let us first show that f_α remains bounded as $\alpha \to 0$.
Assuming Picard's conditions and taking norms taking norms on both sides of
the first line of Eq. (3.34), squaring and then taking the limit $\alpha \to 0$ gives

$$\begin{aligned}
\lim_{\alpha \to 0} \| f_\alpha \|_X^2 &= \sum_{j=1}^{\infty} \left(\frac{\mu_j}{\alpha + \lambda_j} \right)^2 |<g, v_j>_Y|^2 \\
&\to \sum_{j=1}^{\infty} \lambda_j^{-1} |<g, v_j>_Y|^2 < \infty,
\end{aligned}$$

and f_α, therefore, remains bounded as $\alpha \to 0$

The boundedness can also be seen as follows (Lebedev *et al.*, 1996). Choose
a sufficiently large value of j, say, N. Then

$$\lim_{\alpha \to 0} \left(\frac{\mu_j}{\alpha + \lambda_j} \right) < \mu_j^{-1}$$

for $j \leq N$ and

$$\lim_{\alpha \to 0} \left(\frac{\mu_j}{\alpha + \lambda_j} \right) \to \frac{\mu_j}{\alpha}$$

for $j > N$. If we define by β the maximum of the two, i.e.,

$$\beta = \max \left(\frac{1}{\mu_j}, \frac{\mu_j}{\alpha} \right)$$

then

$$\|f_\alpha\|_X^2 \leq \beta^2 \|g\|^2,$$

showing that f_α is bounded.

In order to see how f_α converges to f^+ as $\alpha \to 0$, we consider the quantity $A^+ g - f_\alpha$ and write (upon using Eqs. (3.12) and (3.34))

$$A^+ g - f_\alpha = \sum_{j=1}^\infty \left(\frac{1}{\mu_j} - \frac{\mu_j}{\alpha + \lambda_j} \right) < g, v_j >_Y u_j$$

from which

$$\|A^+ g - f_\alpha\|_X^2 = \sum_{j=1}^\infty \left(\frac{\alpha}{\mu_j(\alpha + \lambda_j)} \right)^2 | < g, v_j >_Y |^2. \tag{3.35}$$

As $j \to \infty$, the summand in Eq. (3.35) approaches $\lambda_j^{-1} | < g, v_j >_Y |^2$ and by Picard's conditions

$$\|A^+ g - f_\alpha\|_X^2 \leq \sum_{j=1}^\infty \lambda_j^{-1} | < g, v_j >_Y |^2 < \infty. \tag{3.36}$$

Let us now rewrite the series in Eq. (3.35) as

$$
\begin{aligned}
\|A^+ g - f_\alpha\|_X^2 &= \sum_{j=1}^N \left(\frac{\alpha}{\mu_j(\alpha + \lambda_j)} \right)^2 | < g, v_j >_Y |^2 \\
&+ \sum_{j=N+1}^\infty \left(\frac{\alpha}{\mu_j(\alpha + \lambda_j)} \right)^2 | < g, v_j >_Y |^2.
\end{aligned}
$$

Since the series converges as shown in Eq. (3.36), the second sum in the above equation can be made as small as one wishes by taking N as large as is necessary. Therefore, for $\alpha > 0$,

$$
\begin{aligned}
\|A^+ g - f_\alpha\|_X^2 &= \sum_{j=1}^N \left(\frac{\alpha}{\mu_j(\alpha + \lambda_j)} \right)^2 | < g, v_j >_Y |^2 + \frac{\delta}{2} \\
&\leq \alpha^2 \sum_{j=1}^N \lambda_j^{-3} | < g, v_j >_Y |^2 + \frac{\delta}{2} \\
&\leq \alpha^2 S_N + \frac{\delta}{2}.
\end{aligned}
$$

Choosing $\alpha < \delta^{1/2}(2S_N)^{-1/2}$ leads to

$$\|A^+ g - f_\alpha\|_X^2 \ < \ \delta$$
$$\underset{\alpha \to 0}{\text{limit}} \|A^+ g - f_\alpha\|_X \to 0.$$

Let us next consider the physical data g^ϵ and the corresponding regularized solution f_α^ϵ for a certain value of $\alpha > 0$. We would like to estimate the error between f_α and f_α^ϵ and also between f_α^ϵ and the generalized solution f^+. For the first, we write

$$f_\alpha - f_\alpha^\epsilon = \sum_{j=1}^{\infty} \left(\frac{\mu_j}{\alpha + \lambda_j} \right) < g - g^\epsilon, v_j >_Y u_j$$

and therefore,

$$\|f_\alpha - f_\alpha^\epsilon\|_X^2 = \sum_{j=1}^{\infty} \left(\frac{\mu_j}{\alpha + \lambda_j} \right)^2 | < g - g^\epsilon, v_j >_Y |^2 \le \beta^2 \|g - g^\epsilon\|_Y^2,$$

where $\beta = \max(1, \alpha^{-1})$. From this we see that for small values of α,

$$\|f_\alpha - f_\alpha^\epsilon\|_X \le \epsilon \alpha^{-1/2}. \tag{3.37}$$

As to the second estimate

$$\begin{aligned}
\|f_\alpha^\epsilon - f^+\|_X &\le \ \|f_\alpha^\epsilon - f_\alpha\|_X + \|f_\alpha - f^+\|_X \\
&\le \ \epsilon \alpha^{-1/2} + \|f_\alpha - f^+\|_X.
\end{aligned}$$

Again we arrive at the same contradiction that as $\alpha \to 0$, the second term in (3.37) goes to zero, whereas for a fixed ϵ, the first term grows without bound. Therefore, α should be so chosen as to make the ratio $\epsilon^2 \alpha^{-1}$ vanish as $\alpha \to 0$. However, α may not vanish faster than ϵ. Moreover, the fact that $\epsilon^2 \alpha^{-1}$ must go to zero as $\alpha \to 0$ cannot be modified. A regularization scheme is sometimes called *regular* if $f_{\alpha(\epsilon)}^\epsilon \to f^+$ as $\epsilon \to 0$. If the operator A is injective, then a regular scheme implies that $\mathcal{R}_\alpha(\epsilon)g^\epsilon \to A^{-1}$ as $\epsilon \to 0$.

Note that the first term $\|f_\alpha^\epsilon - f_\alpha\|_X$ is nothing but $\epsilon \|\mathcal{R}\|$. It is worth pointing out that the norm of the regularizing operator $\|\mathcal{R}\|$ varies as $\alpha^{-1\backslash 2}$ in general. This can be seen as follows. Let B denote the operator $AA^* + \alpha I$. Clearly, $\|Bf\|_X \ge \alpha \|f\|_X$ yielding $\|B\| \ge \alpha$. Consequently, $\|B^{-1}\| \le \alpha^{-1}$. Obviously, $\|AA^* B^{-1}\| < 1$. Therefore,

$$\|\mathcal{R}g\|_X^2 = < AA^* B^{-1}g, \ B^{-1}g > \le \|g\|_Y^2 \alpha^{-1}$$

from which it follows at once that $\|\mathcal{R}_\alpha\| \le \alpha^{-1/2}$. In Section 3.8 we will demonstrate that for any $\alpha > 0$ there exists an optimum α and a unique α-regularized solution of the first-order for the ill-posed problem $Af = g$.

3.7   Convergence, Stability and Optimality

The discussions thus far have made it clear that stable, approximate solutions to an ill-posed problem (due to noisy data, discontinuous inverse operator, and so on) can be obtained by designing a regularization strategy $\{R_{\alpha(\epsilon)}\}$, $\alpha > 0$. Thus, given a regularization strategy and a method to determine the regularization parameter α, one has a recipe for obtaining approximate solutions to an ill-posed problem. It is natural to expect that as the error in the data goes to zero, the approximate regularized solution should converge to the true solution. This is again the principle of regularization of Glasco mentioned in Section 3.6.1. We, therefore, impose the conditions

$$\alpha(\epsilon) \to 0 \text{ and } \sup\{\|R_{\alpha(\epsilon)}g^{\epsilon} - f\|_X, \|Af - g^{\epsilon}\|_Y \le \epsilon\} \to 0 \text{ as } \epsilon \to 0.$$

Moreover, this should hold for every $f \in X$. A regularization strategy satisfying these requirements is called *admissible* and so is the corresponding pair $\{\alpha, \epsilon\}$.

Several questions arise at this point regarding the ambiguity of the approximate solution and its convergence. These are interrelated. Estimates of the ambiguity or uncertainty in the solution goes under the generic name *stability estimates*. The convergence questions are as follows. First, even though convergence is guaranteed, its rate may be arbitrarily slow. One must, therefore, restrict the solution space on the basis of the available *a priori* information and the elements of the space thus restricted are called *admissible approximate solutions*. Second, one would like the convergence to be of the order of the error level ϵ. If the operator A is boundedly invertible, then it is certainly so because $\|f\|_X \le \|A^{-1}\|\|Af\|_Y$. Therefore, if $\|Af\|_Y \le \epsilon$, then the convergence is of the order of ϵ. One would like to generalize this to the approximate solutions. A regularization strategy for which the convergence is of the order of ϵ is known as *optimal*. These issues are discussed in this section.

3.7.1   Convergence and Stability Estimates

Let us first discuss the restriction of the solution space in order to have a rate of convergence. This can be achieved on the basis of *a priorily* prescribed bounds on the solution and/or the data. The problem in the first case (Problem 1) can be posed as:

find a solution $\tilde{f}_E$ to the ill-posed problem $Af = g$

such that $\|A\tilde{f}_E - g\|_Y = \inf\{\|Af - g\|_Y \|\|f\|_X \le E\}$.

This is essentially the problem of the quasisolution of Section 3.3. When the bound on the data is used, the program (Problem 2) is:

find a solution $\tilde{f}_\epsilon$ to the ill-posed problem $Af = g$

such that $\|\tilde{f}_\epsilon\|_X = \inf\{\|f\|_X \|\|Af - g\|_Y \le \epsilon\}$.

This is the method of Phillips (1962). As noted by Ivanov (1966), the above two problems are dual of each other (also see Glasco, 1984). In the third case, both E and ϵ are prescribed and Problem 3 consists of finding all solutions f to the ill-posed problem $Af = g$ such that

$$||Af - g||_Y \leq \epsilon \text{ and } ||f||_X \leq E.$$

If X_E and X_ϵ denote the solution spaces for Problem 1 and 2, respectively, then the space of solutions for Problem 3 is the intersection $X_{E,\epsilon} = X_E \cap X_\epsilon$. It can be shown (see Bertero, 1989) that this intersection is nonempty if and only if both $\tilde{f}_E$ and $\tilde{f}_\epsilon$ belong to $X_{E,\epsilon}$. The solution of the third problem, namely, to single out a specific element of $X_{E,\epsilon}$ as the solution of the ill-posed problem is not discussed here. The details, however, can be found in Miller (1970), Franklin (1974) and Bertero (1989).

A generalization of the set $X_{E,\epsilon}$, denoted by $\tilde{X}$, can be defined as:

$$\tilde{X} = \{f \in X \big| f \in \hat{X} \subset X, ||Af - g||_Y \leq \epsilon\},$$

where $\hat{X}$ is a subset of X. $\hat{X}$ is X_E, X_ϵ and $\overline{X}$ in Problem 1,2 and 3, respectively. The set $\hat{X}$ contains *a priori* information about the solution, and, therefore, all acceptable, approximate solutions of the ill-posed problem that are compatible with *a priori* knowledge of the solution and the error level in the data.

Let us next replace $||Af - g||_Y \leq \epsilon$ by $||Af||_Y \leq \epsilon$ and define a set $\omega_{\hat{X},\epsilon}$ as

$$\omega_{\hat{X},\epsilon} = \sup\{||f||_X \big| f \in \hat{X} \subset X, ||Af||_Y \leq \epsilon\}.$$

This set is the *modulus of continuity* of the inverse operator A^{-1} defined on $A\hat{X}$. The letter ω was used to indicate this fact. We would like to point out that the null element is included in $\hat{X}$.

What we are actually looking for is a rate of convergence as the noise or the error level goes to zero. Now the regularization parameter α also goes to zero in this limit. Since the regularization strategy is parameterized by α, it is customary to define the rate of convergence in terms of α. In other words, we define a quantity $\sigma(\alpha, \epsilon)$ as:

$$\sigma(\alpha, \epsilon) = \sup\{||R_\alpha g^\epsilon - f||_X \big| f \in \hat{X}, ||Af - g^\epsilon||_Y \leq \epsilon\}$$

and call this the *modulus of convergence* (Franklin, 1974). $\sigma(\alpha, \epsilon)$ is bounded from above and from below. It is bounded above by (cf. Eq. (3.22))

$$\sigma(\alpha, \epsilon) \leq \sup\{||R_\alpha Af - f||_X \big| f \in \hat{X}\} + \epsilon ||R_\alpha||.$$

The lower bound is furnished by the *modulus of continuity* $\omega_{\hat{X},\epsilon}$ itself (Franklin, 1974; see also De Mol, 1992). It then follows that

$$\omega_{\hat{X},\epsilon} \leq \sigma(\alpha, \epsilon).$$

The conclusion is this. Given the set $\hat{X}$ of the solutions, the best possible rate of convergence that a regularizing strategy can attain is $\omega_{\hat{X},\epsilon}$. Again, the null element is included in $\hat{X}$. Moreover, in this case, $\omega_{\hat{X},\epsilon}$ is a continuously increasing function of α. In addition, if $\hat{X}$ is compact, then Tikhonov's theorem (Theorem 3.1) applies implying that the inverse operator A^{-1} defined on $A\hat{X}$ is continuous. Therefore, $\omega_{\hat{X},\epsilon} \to 0$ as $\epsilon \to 0$.

We conclude this subsection by characterizing the set $\tilde{X}$. Specifically, we would like to estimate the *diameter* of this set which will serve as a measure of stability of a regularizing algorithm. Toward this consider two elements $f_1, f_2 \in \hat{X}$. Then $f = (f_1 - f_2)/2$ also belongs to $\hat{X}$. Upon taking the norm of the quantity Af yields

$$\|Af\|_Y \le \frac{1}{2}\left[\|Af_1 - g\|_Y + \|Af_2 - g\|_Y\right] \le \epsilon.$$

Therefore,

diameter$(\hat{X}) = \sup\{\|f_1 - f_2\|_X \big| f_1, f_2 \in \hat{X}\} \le 2\epsilon.$

If the zero element is included in $\hat{X}$, then the above statement can also be written as

diameter$(\hat{X}) = \sup\{\|f_1 - f_2\|_X \big| f_1, f_2 \in \hat{X}\} \le 2\omega_{\hat{X},\epsilon}.$

The robustness or uncertainty or stability of a regularizing algorithm is, therefore, formulated (independent of the data) in terms of the modulus of continuity alone. It shows clearly that as $\omega_{\hat{X},\epsilon} \to 0$ as $\epsilon \to 0$, so does the uncertainty of the solution.

We next discuss the question of optimality of a regularizing algorithm.

3.7.2 The Optimality of a Regularization Strategy

Let us consider the subspace $\hat{X}$ to be such that the norm in this space is *stronger* than that on X. In other words,

$$\|f\|_X \le c\|f\|_{\hat{X}}, \ c > 0, \ \forall \ f \in \hat{X}.$$

Define a set $\hat{X}_E$ as

$$\hat{X}_E = \{f \in \hat{X}_E \big| \|f\|_{\hat{X}_E} \le E\},$$

replace $\hat{X}$ by $\hat{X}_E$ and introduce a modified modulus of continuity

$$\omega_{\hat{x}_E}(\epsilon) = \sup\{\|f\| : f \in \hat{X}_E, \|Af\|_Y \le \epsilon, \|f\|_{\hat{X}_E} \le E\}.$$

Note that this newly defined quantity $\omega_{\hat{x}_E}(\epsilon)$ still is a measure of the uncertainty of the solution. Since the latter has $\omega_{\hat{x}_E}(\epsilon)$ as an upper bound, it is customary to call $\omega_{\hat{x}_E}(\epsilon)$ as the *maximal error* (Vainikko, 1987; Davies, 1992; Natterer 1984. Kirsch (1996) terms it the *worst-case error*) compatible with the bounds E on the solution and ϵ on the data.

It was mentioned at the beginning that we would like $\omega_{\hat{X}_E}(\epsilon)$ not only to go to zero as $\epsilon \to 0$, but to be of the order of ϵ, if possible. The vanishing of $\omega_{\hat{X}}(\epsilon)$ with ϵ is, of course, true if the operator A is boundedly invertible as was shown earlier. However, it turns out (Kirsch, 1996) that this does not hold if the operator A is compact. Thus given a $\epsilon_0 > 0$, there exists a $c > 0$ such that

$$\omega_{\hat{x}_E}(\epsilon) \geq c \; \forall \epsilon \in (0, \epsilon_0].$$

The invocation of a stronger norm $\hat{X}_E$ is to ensure that the vanishing of $\omega_{\hat{x}_E}(\epsilon)$ is guaranteed for a compact operator. If an approximate solution f_a is such that

$$||f_a - f||_\infty \leq c\omega_{\hat{x}_E}(\epsilon), \; f \in \hat{X}_E, \; ||f||_{\hat{X}_E} \leq E, \; ||Af - g||_Y \leq \epsilon,$$

then f_a is considered to be *asymptotically optimal*. The asymptotic normality was illustrated earlier *via* numerical differentiation in Example 4 of Chapter 2.

Let us return to the general case. From the definition of a regularizing strategy (Section 3.6.1), $R_\alpha A f \to f, \; \forall f \in X$ as $\alpha \to 0$. However, what one is interested in is an optimal strategy, i.e., a strategy the rate of convergence of which is of the same order as the maximal error. This depends upon the smoothness assumption on the solution. Consider, for example, the Tikhonov functional defined by Eq. (3.17) in which such assumptions are embodied in the stabilizer $\Omega(f)$. Usually, this functional is considered to be of the form

$$T_{\{\alpha, m\}}(f) = ||Af - g||_Y^2 + \alpha||f||_m^2,$$

where $|| \cdot ||_m$ indicates norm in some Sobolev space $H^m, \; m \geq 0$. The strategy is then characterized by the pair $\{\alpha, m\}$, the number m being the *order of regularization* which was already mentioned in Section 3.6.3. Different degree of smoothness also result if the solution is assumed to belong to the ranges of the operators A^* and $(A^*A)^m$. The higher the order m, the smoother is the solution. As demonstrated in Example 4 of Chapter 2, $f \in H^1$ resulted when f belonged to $R(A^*)$, whereas f was found to be in H^2 when it was in the range of A^*A.

The basic equation which serves as the point of departure here is Eq. (3.22). Let $f \in A^*y$ with $||y||_Y \leq E$. Then Eq. (3.22) reduces to

$$||f_\alpha^\epsilon - f||_X \leq \frac{\epsilon}{\sqrt{\alpha}} + ||R_\alpha A f - f||_X$$

taking into account that $||R_\alpha|| \leq 1\backslash\sqrt{\alpha}$. Using Eq. (3.30) and the spectral function $s_\alpha(\lambda_j)$ of Section 3.6.2, we have

$$||R_\alpha A f - f||_X^2 = \sum_{j=1}^\infty [s_\alpha(\lambda_j) - 1]^2 \lambda_j |<y, v_j>|^2.$$

At this point, impose the condition that

$$|s_\alpha(\lambda_j) - 1| \leq C_1 \frac{\sqrt{\alpha}}{\mu}, \forall \alpha > 0, 0 < \mu \leq ||A||.$$

Then we obtain the estimate

$$||R_\alpha A f - f||_X^2 \leq C_1^2 \alpha ||y||_Y^2$$

yielding

$$||R_\alpha A f - f||_X \le C_1 \sqrt{\alpha} ||y||_Y.$$

This leads to

$$||f_\alpha^\epsilon - f||_X \le \frac{\epsilon}{\sqrt{\alpha}} + \sqrt{\alpha} ||y||_Y.$$

If we now set $\alpha(\epsilon) = c\epsilon/E$, then an optimal convergence is obtained, namely

$$||f_\alpha^\epsilon - f||_X \le (C^{-1} + C) \sqrt{E\epsilon},$$

where $C = \sqrt{c}$.

In an analogous manner, if $f = A^* A z$ with $||z||_X \le E$, we replace Eq. () by

$$|s_\alpha(\lambda_j) - 1| \le C_2 \frac{\alpha}{\lambda}, \forall \alpha > 0, 0 < \mu \le ||A||,$$

assume that $\alpha(\epsilon) = c(\epsilon/E)^{\frac{2}{3}}$, and the estimate corresponding to Eq. () is found to be

$$||f_\alpha^\epsilon - f||_X \le (C^{-1} + C^2) E^{\frac{1}{3}} \epsilon^{\frac{2}{3}}.$$

Similar estimates hold for Tikhonov regularized approximate solutions as well. However, it turns out (Davies, 1992; Kirsch, 1996) that for the order of regularization $m \ge 2$, optimality is not obtained for this regularization algorithm. This is however, not the case for the regularization strategies of Landweber iteration and conjugate gradient method. Space does not allow us to go into the details of these two important methods both of which are treated rigorously in Kirsch (1996).

In Section 3.5, the notions of severe and mild ill-posedness were introduced. Similar notions also occur in regularization theory depending upon the behavior of the modulus of continuity $\sigma_S(\epsilon)$. If it so happens that $\lim_{\epsilon \to 0} \sigma_S(\epsilon)$ behaves as some power of ϵ), say, $\sim \epsilon^p, 0 < p \le 1$, then one speaks of the *Hölder type of continuity* (John, 1960) and the problem is considered to be well-behaved. In this case, the accuracy in the reconstructed solution bears a well-defined relation to the accuracy of the data. More specifically, the number of accurate digits in the solution is a fixed percentage of that in the data (De Mol, 1992). If the behavior of $\sigma_S(\epsilon)$ is logarithmic: $\sigma_S(\epsilon) \sim |\log \epsilon|^{-p}, p > 0$, then the problem is considered to be *severely ill-posed*. The ill-posedness in this case is severe enough that no enhancement in the accuracy of the solution is obtained (De Mol, 1992) even if the noise level in the data is reduced significantly. When Hölder type of continuity can be restored by incorporating *a priori* information about the solution into the regularization algorithm, one speaks of a *mildly ill-posed* problem. We saw in Chapter 2 how numerical differentiations were stabilized by knowing *a priorily* the prescribed bounds on a number of higher-order derivatives. Also in Example 1 of Chapter 2, it was seen how Hölder's continuity was introduced in an otherwise ill-posed backward heat-conduction problem. This also happens in the problem of full-angle computed tomography as was mentioned at the end of Section 3.5. These are then examples of mild ill-posedness. On the other hand, for analytic and bandlimited kernels, for example, the smoothing by the integral operator is too severe to restore mild ill-posedness. The singular values simply decay exponentially to zero and the problem remains severely ill-posed. For further details, see Natterer (1986), Louis (1989) and Bertero (1989).

3.8   The Determination of α

We present a brief discussion on the determination of the regularization parameter α. The problem is important because α controls the smoothness of the solution by preventing the occurrence of arbitrarily large amplitude oscillations in it for only a slight variation in the data. If α is too small, the solution may become unstable because of the ill- posedness of the original problem. On the other hand, if α is unwontedly large, the solution may be too smooth to be a good approximation of the true solution. The fundamental problem in determining the regularization parameter is embedded in Eq. (3.22). It can be seen from this equation that the desired value of α should be such that the L.H.S. attains its minimum. It is, therefore, necessary to establish a bound for the R.H.S. Now the R.H.S. contains the true solution which is unknown. Hence the determination of an optimal α cannot be done in practice. However, approximate methods for its determination can be devised of which there are basically two types: one in which it is assumed that certain bounds can be prescribed for both the solution and the data and, the other which does not require such bounds, but lets the data itself decide the value of α. The second method applies only to discrete data. The most well-known of the methods of the first type is the *discrepancy principle* of Morozov (1984) which is discussed below. The second technique of letting the discrete set of data decide the value of α is called *cross-validation* and will be introduced only briefly at the end.

We first demonstrate that an optimal value of α exists even though its true determination may not be possible in practice. Next is discussed how this optimal α is determined by applying the *discrepancy principle* of Morozov (1984). Finally, it is shown that a regularization algorithm using α so determined is admissible and optimal.

3.8.1   The Existence of an Optimal Value of α

Let g_e be the exact data and $f_\alpha = \mathcal{R}_\alpha g_e$ the corresponding regularized solution. We show that the norm functional $\|f_\alpha\|_X$ is a nonincreasing function of α. From Eq. (3.25) we have

$$\|f_\alpha\|_X^2 = \|B^{-1}A^*g\|_X^2,$$

where $B = A^*A + \alpha I$. Since $\|B\| \leq \alpha^{-1}$ (see below Eq. (3.25)), it follows that $\|f_\alpha\|_X \to 0$ as $\alpha \to \infty$. If we set $\alpha = 0$ in Eq. (3.25), then $\|f_\alpha\|_X = \|f^+\|_X$ (cf. Eq. (3.3)). If a generalized solution exists, then $\|f_\alpha\|_X$ is finite at $\alpha = 0$. Otherwise, $\|f_\alpha\|_X$ is unbounded there. For a fixed data, $\|f_\alpha\|_X$ is a monotonically nonincreasing function of α.

We can also consider the problem from another angle. Consider two values for α, namely, $\alpha_1 > 0$ and $\alpha_2 > 0$. Denote the corresponding solutions by f_{α_1} and f_{α_2}, respectively. Subtracting the Euler–Lagrange equations (3.24) for f_{α_1} and f_{α_2} results in (A^*g is the same for both)

$$0 = (\alpha_1 + A^*A)f_{\alpha_1} - (\alpha_2 + A^*A)f_{\alpha_2}$$

$$= (\alpha_1 + A^*A)(f_\alpha)_{12} - \alpha_{12}f_{\alpha_2}, \tag{3.38}$$

where $(f_\alpha)_{12} = f_{\alpha_1} - f_{\alpha_2}$ and $\alpha_{12} = \alpha_2 - \alpha_1$. We can also derive the following from Eq. (3.42):

$$\alpha_1||(f_\alpha)_{12}||^2 + ||A(f_\alpha)_{12}||^2 \leq |\alpha_{12}|| < (f_\alpha)_{12}, f_{\alpha_2} > |$$
$$\leq |\alpha_{12}|||f_{\alpha_2}||||(f_\alpha)_{12}||.$$

Two things follow from this:
i

$$\alpha_1||(f_\alpha)_{12}|| \leq |\alpha_{12}|||f_{\alpha_2}|| \leq |\alpha_{12}|\frac{||g||}{\sqrt{\alpha_2}} \tag{3.39}$$

since $f_{\alpha_2} = R_{\alpha_2}g$ and the norm of the regularizing operator varies as $\alpha^{-1\backslash2}$.
ii. If $\alpha_2 > \alpha_1 > 0$, then

$$< f_{\alpha_2}, (f_\alpha)_{12} > \geq 0 \longrightarrow ||f_{\alpha_2}|| \leq ||f_{\alpha_1}||. \tag{3.40}$$

Equation (3.43) shows that the map $\alpha \to f_\alpha$ is continuous, whereas Eq. (3.44) tells us that the norm functional $||f_\alpha||_X$ is monotonically nonincreasing with α. In order for the monotonicity to be strict, the null solution $||f_\alpha||_X = 0$ must be avoided and A^*g must be nonzero. This can be seen as follows. If $A^*g \neq 0$, the using the Euler–Lagrange equations corresponding to α and β, subtracting and taking norm results in

$$\beta||f_\beta|| - \alpha||f_\alpha|| \leq ||A^*A||(||f_\alpha|| - ||f_\beta||).$$

The quantity $||A^*A||$ being positive, the above identity can be satisfied if and only if $||f_\alpha|| \neq ||f_\beta||$. Since for $\beta > \alpha$, we have $||f_\beta|| \leq ||f_\alpha||$, the strict monotonicity follows.

We next demonstrate a similar, but contrary monotonicity for the norm functional $||Af_\alpha - g||_Y$. Let us call this the *discrepancy function* and denote it by the symbol ϵ_α in order to emphasize that the function depends on α. Noting that

$$R_\alpha = A^*(AA^* + \alpha I)^{-1}$$

and $g = Pg + Qg$, where $Q = I - P$, we have, in a general situation,

$$
\begin{aligned}
Af_\alpha - g &= AR_\alpha - g = AR_\alpha - Pg - Qg \\
&= A \sum_{j=1}^{\infty} \frac{\mu_j}{\alpha + \lambda_j} < g, v_j > u_j - \sum_{j=1}^{\infty} < g, v_j > v_j - Qg \\
&= \sum_{j=1}^{\infty} \left[\frac{\lambda_j}{\alpha + \lambda_j} - 1 \right] < g, v_j > u_j - Qg \\
&= \alpha \sum_{j=1}^{\infty} \frac{1}{\alpha + \lambda_j} - < g, v_j > u_j - Qg \\
&= \alpha \left[(AA^* + \alpha I)^{-1} Pg \right] - Qg.
\end{aligned}
$$

It then follows that

$$
\|Af_\alpha - g\|_Y^2 = \alpha^2 \left[(AA^* + \alpha I)^{-1} Pg \right]_Y^2 + \|Qg\|_Y^2.
$$

Therefore, as $\alpha \to 0$, $\|Af_\alpha - g\|_Y \to \|Qg\|_Y$, whereas in the limit $\alpha \to \infty$, $\|Af_\alpha - g\|_Y \to \|g\|_Y$. Moreover, $\|Af_\alpha - g\|_Y$ clearly varies continuously with α and the variation is monotonically increasing starting from $\|Qg\|_Y$ at $\alpha = 0$ increasing all the way to $\|g\|_Y$ as $\alpha \to \infty$. If $A(X)$ is dense in Y, then $\|Af_\alpha - g\|_Y \to 0$ at $\alpha = 0$.

The strict monotonicity with α of the norm functionals $\|f_\alpha\|_X$ (decreasing) and $\|Af_\alpha - g\|_Y$ (increasing) shows that an optimal value of the regularization parameter α exists. All this is shown in Figure 3.4.

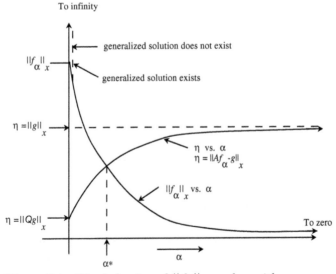

Figure 3.4. The behavior of $\|f_\alpha\|_X$ and η with α. $\eta = \|Af_\alpha - g\|_Y$. α^* indicates the choice of the regularizing parameter α.

Suppose that *a priori* information is available about the bound of the solution f. In other words, $f \in X_E$ and $||f||_X \leq E$ for some constant E. Let us also assume that a generalized solution exists and the bound E is less than $||f^+||$. Then form Figure 3.4, there exists a unique value of α, say α_E corresponding to E. The strictly monotonic decrease of $||f_\alpha||_X$ in α, which is depicted in Figure 3.4, indicates that all values of $\alpha > \alpha_E$ are allowed. Referring to Figure 3.3, this means that all solutions within the ball can be accepted. However, for any $\alpha > \alpha_E$, the discrepancy parameter ϵ_α would also increase. This then shows that a constrained optimization solution subject to satisfying the bound on the solution leads to smaller values of α. Therefore, $\alpha = \alpha_E$ gives a regularized solution which is compatible with the prescribed bound on the solution and which at the same time minimizes the discrepancy between the actual data and the data computed on the basis of the regularized solution.

3.8.2 The Discrepancy Principle

In this section, we discuss the other situation, namely, determining α when the error bound on the data is prescribed – the so-called discrepancy principle. Using this principle, the optimal value of the regularization parameter is to be determined by replacing the discrepancy function ϵ_α by the prescribed data error ϵ. In other words, the constraint is now

$$||Af_\alpha - g||_Y = \epsilon. \tag{3.41}$$

In a well-posed problem, the error in the solution bears a direct relation to that in the data. Equation (3.45) is essentially a reiteration of this fact for an ill-posed problem. Note that the discrepancy relation does not hold if α is too small thereby implying stability. Again, if we follow the same line of reasoning as in the case of the prescribed bound on the solution presented above, then a unique value $\alpha = \alpha_\epsilon$ exists corresponding to a prescribed error level ϵ. Moreover, it gives an approximate solution which is compatible with the prescribed data error and results in the minimum norm solution.

It is important to realize that the norm $||g^\epsilon||$ must exceed the error level ϵ, that is, $||Af - g||_Y \leq \epsilon \leq ||g||_Y$. Otherwise, the continuity of the discrepancy functional will result in the trivial solution $f_\alpha = 0$. From a practical standpoint, this simply means that there is no signal to work with and whatever information there may be is indistinguishable from the noise.

Before proceeding further, let us demonstrate that if the discrepancy principle applies, then $||f_\alpha||_X \leq ||f||_X, \forall f \in X$. Now f_α is a minimizer of Tikhonov's functional. In other words,

$$||Af_\alpha - g||_Y^2 + \alpha||f_\alpha||_X^2 \leq ||Af - g||_Y^2 + \alpha||f||_X^2.$$

The application of the discrepancy principle then yields

$$\alpha||Af_\alpha - g||_Y^2 + \epsilon^2 \leq \alpha||Af_\alpha - g||_Y^2 + \epsilon^2$$

proving the assertion.

We demonstrate next that a regularization strategy based on the discrepancy principle is admissible (see Section 3.7 for definition). Let us assume that A is injective and its domain is dense in Y (Kirsch, 1996). There exists, therefore, an element $z \in D(A^*)$ such that $||f - A^*z||_X$ can be made as small as is desired. Now consider

$$
\begin{aligned}
||f_{\alpha(\epsilon)}^\epsilon - f_e||_X^2 &= \; < f_{\alpha(\epsilon)}^\epsilon - f_e, \; f_{\alpha(\epsilon)}^\epsilon - f_e >_X \leq \\
&= \; ||f_{\alpha(\epsilon)}^\epsilon||_X^2 + ||f_e||_X^2 - 2\mathrm{Re} < f_{\alpha(\epsilon)}^\epsilon, \; f_e >_X \\
&\leq \; 2\mathrm{Re}[||f_e||_X^2 - < f_{\alpha(\epsilon)}^\epsilon, \; f_e >_X] \\
&\leq \; 2\mathrm{Re} < f_e - f_{\alpha(\epsilon)}^\epsilon, \; f_e >_X \\
&\leq \; 2\mathrm{Re} < f_e - f_{\alpha(\epsilon)}^\epsilon, f_e - A^*z >_X + 2\mathrm{Re} < f_e - f_{\alpha(\epsilon)}^\epsilon, A^*z >_X \\
&\leq \; 2C_1 ||f_e - f_{\alpha(\epsilon)}^\epsilon||_X + 2\mathrm{Re} < g_e - Af_\alpha^\epsilon, \; z >_Y \\
&= \; 2C_1 ||f_e - f_{\alpha(\epsilon)}^\epsilon||_X + 4\epsilon ||z||_Y, \qquad\qquad (3.42)
\end{aligned}
$$

$C_1 \leq ||f_e - A^*z||_X$ being a constant. The third line in the derivation of Eq. (3.46) was written using the fact (see above) that $||f_\alpha||_X \leq ||f||_X, \forall f \in X$. Now the both terms on the R.H.S. of Eq. (3.46) can be made as small as desired by choosing ϵ as small as possible. It, therefore, follows that a regularization strategy obtained on the discrepancy principle is also admissible, that is,

$$
\underset{\epsilon \to 0}{\mathrm{limit}} \, ||f_e - f_{\alpha(\epsilon)}^\epsilon||_X \to 0.
$$

The discrepancy principle also leads to a regularization strategy which is optimal. We assume that $f = A^*z$ and $||z||_Y \leq E$. Using Eq. (3.46), we obtain

$$
\begin{aligned}
||f_{\alpha(\epsilon)}^\epsilon - f_e||_X^2 &= \; 2\mathrm{Re} < f_e - f_{\alpha(\epsilon)}^\epsilon, \; A^z >_X \\
&= \; 2\mathrm{Re} < A(f_e - f_{\alpha(\epsilon)}^\epsilon), \; z >_Y \\
&= \; 2\mathrm{Re} < g_e - g^\epsilon + (g^\epsilon - Af_{\alpha(\epsilon)}), \; z >_Y \\
&= \; 4\epsilon ||z||_Y \leq 4\epsilon E.
\end{aligned}
$$

It follows immediately that

$$
||f_{\alpha(\epsilon)}^\epsilon - f_e||_X^2 \leq 2\sqrt{\epsilon} \omega_{\tilde{x}}(\epsilon). \qquad\qquad (3.43)
$$

Equation (3.47) proves the optimality of a regularization strategy based on the discrepancy principle.

The discrepancy principle is implemented in practice by applying Newton's method to the root-finding of the equation

$$
||Af_{\alpha(\epsilon)}^\epsilon - g^\epsilon||_Y - \epsilon = 0.
$$

The convex nature of the functional (with α replaced by its reciprocal) was established by Morozov (1984). Thus, given any initial approximation, Newton's method converges. The derivatives of the functional with respect to α follows the same Euler–Lagrange equation as for the solution except the right hand side. More specifically, if y represents the derivative, that is, if

$$\frac{\mathrm{d}}{\mathrm{d}\alpha} f^\epsilon_{\alpha(\epsilon)} = y,$$

then

$$(AA^* + \alpha I)y = -f^\epsilon_{\alpha(\epsilon)}.$$

It was suggested by Kirsch (1996) that

$$\alpha(\epsilon) = \epsilon \frac{||A||}{(||f^\epsilon_{\alpha(\epsilon)}||x - \epsilon)}$$

be used as an initial starting point.

In summary, the discrepancy principle for the Tikhonov functional (3.16) for a compact, injective operator provides an optimal regularization strategy with admissible regularized solutions. Note that all solutions for which $\alpha \leq \alpha_d$, where α_d means the parameter obtained by applying the discrepancy principle, are acceptable solutions. However, $\forall \alpha < \alpha_d$, the norm $||f^\epsilon_{\alpha(\epsilon)}||x$ is larger by virtue of the strict monotonic decrease of this norm (Figure 3.4). Therefore, the solution given by $f^\epsilon_{\alpha_{dp}(\epsilon)}$ is the one that is compatible with the measured data and has minimal norm.

As an aside, it is worth mentioning that if Eq. (3.45) is modified to read

$$||Af^\epsilon_{\alpha(\epsilon)} - g^\epsilon||_Y = \kappa\epsilon$$

instead of only ϵ on the R.H.S., then other regularization strategies such as spectral windowing and iterative methods can also be made to converge strongly to the generalized solution. For details, see Vainikko (1982), Defrise and De Mol, (1987) and Bertero (1989).

For an extensive, in-depth discussion of various methods of determining α within the framework of prescribed bounds on the solution and the data, see Glasco (1984).

Let us next introduce the method of cross-validation (Wahba, 1977; Wahba and Wold, 1975; Golub et al., 1979). We merely present the basic idea leaving the details to be found in the references just cited. The method is designed for discrete data only and the idea here is to let the data themselves choose the regularization parameter. In this method, the Tikhonov functional that is considered is

$$T_{\alpha,k} = N^{-1} \sum_{n \neq k} ||(Af)_n - g_n||^2 + \alpha||f||^2_X,$$

where N is the total number of data points, $g_n = g_{en} + \epsilon_n$ and the subscripts (α, k) indicate that the functional is the same as the actual functional T_α, but with the k-th data point missing. Let $f_{\alpha,k}$ be the minimizing solution of $T_{\alpha,k}$. The expectation is that if α is really a good choice of the regularization parameter, then the data calculated with the solution $f_{\alpha,k}$ will be closer to the actual data on average. That is, $Af_{\alpha,k}$ will be closer to the actual k-th data point g_k on average than $Af_{\tilde{\alpha},k}$, where $f_{\tilde{\alpha},k}$ is the solution calculated by any other $\tilde{\alpha}$ which is different from α. On the basis of this a *cross-validation* function is defined by

$$V_\alpha = N^{-1} \sum_{n=1}^{N} ||(Af)_n - g_n||^2 w_n(\alpha),$$

where $w_n(\alpha)$ are weight functions. The value of α which minimizes the functional V_α is considered to be the optimal value of the regularization parameter.

Let $\tilde{f}_{\alpha,k}$ be the solution which minimizes T_α with α as the regularization parameter. Then

$$T_\alpha = N^{-1} \sum_{n=1}^{N} ||A\tilde{f}_{\alpha,k} - g_n||^2.$$

If α is a good choice, then it is expected that T_α and V_α will be close to each other. The following is shown by Wahba (1977). Assume that the data is smooth. For example, the data could be the result of applying the kernel $K_I(x,y)$ of the iterated operator $A_I(x,y) = AA$. Then if the kernel $K(x,y)$ of the operator A is smooth, the kernel $K_I(x,y)$ of the iterated operator $A_I(x,y)$ is even smoother. Let $\{\lambda_i\}$ be the eigenvalues of the operator $A_I(x,y)$ which decay as Constant i^{-2m}, $m \geq 1$. Under these conditions, it is shown that if α_T and α_V minimize the expectation values $\overline{T}$ and $\overline{V}$, respectively, then their difference is $\sim o(1)$, the parameters depending on the constant m and the variance σ of the data. For further details regarding the numerics of the minimization processes, the reader is referred to the references already cited above.

3.9   An Application

As an example of Tikhonov regularization, consider the processing of a blurred and noisy image by *total variation minimization* or TVM in short (Rudin *et al.*, 1992; Rudin and Osher, 1994; Szymczak *et al.*, 1998). The intensity of the image f^ϵ is written as

$$g^\epsilon = Af_e + \epsilon, \tag{3.44}$$

where A is a blurring operator (usually convolutional) and ϵ is the noise of zero mean and variance $\sigma_\epsilon^{1/2}$. The Tikhonov functional in TVM is

$$T_\alpha(f) = ||Af - g^\epsilon||^2_{L_2(\Omega)} + \alpha ||\nabla f||_{L_1(\Omega)}, \tag{3.45}$$

where Ω is the region of the image. The TV-stabilizer uses the L_1-norm of the derivative instead of the usual L_2-norm. The motivation behind using the L_1-norm is to preserve the edges in the image. This can be appreciated if we note that a step discontinuity (as occurs at an edge) yields a finite value if Eq. (3.50) is used, but becomes infinite in the L_2-norm. It has been demonstrated (Alvarez *et al.*, 1993) that TVM using a L_1-norm diffuses the solution anisotropically, smoothing the solution only orthogonally to the edges, whereas in a L_2-norm, the diffusion is isotropic and, therefore, results in the smoothing of the edges. A regularized solution is obtained by minimizing $T_\alpha(f)$. The parameter $\alpha = \alpha(\sigma^2)$, that is, α depends upon noise. If $\sigma^2 = 0$, then $f_\alpha^\epsilon \to f_e$. On the other hand, if $\sigma^2 \to \sigma_\epsilon^{1/2}, \alpha \to \infty$ and the solution $f_\alpha^\epsilon \to \overline{f}$, where $\overline{f}$ is the constant mean value of f_α^ϵ and an image of uniform intensity results.

The solution to the minimization problem for the Tikhonov functional (3.50) can be obtained in a number of ways. However, the more efficient way (particularly for large images) is to consider the Euler-Lagrange equation which, for the functional of Eq. (3.50) takes the form

$$\alpha \nabla \cdot \left[\frac{\nabla f}{|\nabla f|_\delta} \right] = f_\alpha^\epsilon - f_e. \tag{3.46}$$

In Eq. (3.51), $|\nabla f|_\delta = \sqrt{|\nabla f|^2 + \delta}$, where a sufficiently small $\delta > 0$ is required in practice for stability purposes (see Szymczak *et al.*, 1998). The magnitude of δ depends upon the intensity resolution of the image.

Figure 3.5 shows the processing of a blurred and noisy image using Tikhonov's regularization with TV-norm. Figure 3.5 (a) shows the original image which consists of two objects lying on a randomly rough surface. The bright parts in the figure are the objects, whereas the associated dark parts correspond to the shadows cast by them. Figures 3.5 (b) and (c) show the processing for various values of α indicated in the figures. The matrix A in Eq. (3.49) was the identity matrix in the processing of this figure.

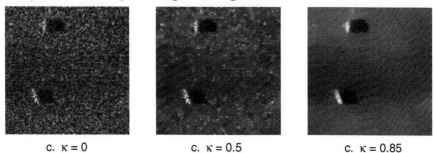

c. $\kappa = 0$ c. $\kappa = 0.5$ c. $\kappa = 0.85$

Figure 3.5 – An application of Tikhonov regularization using a TV-L_1-norm stabilizer. The constraint is described by $\|f - f^e\|_\Gamma = \kappa\sigma^2$. κ is the regularization parameter and σ^2 the variance of the image intensity. The original image (a) corresponds to $\kappa = 0$. Reconstructions are shown for three values of κ. (From Szymczak *et al.*, 1998).

3.10 The Method of Mollification

3.10.1 The Method

The mollification method of solving an ill-posed problem $Af = g$ is essentially a projection method (Louis and Maass, 1991; Murio, 1993; Duc and Hao, 1994).

The inexact data g^ϵ in Y (which may not be in the range of A) is *mollified* so that the mollified data belongs to a subspace of the image space Y in which the problem may be well-posed. The data is mollified by applying a sequence of *mollifying operators* $\{M_\rho\}, \rho \in (0, \infty)$ being the *mollification parameter*, which characterizes the subspace into which the data is mapped by the operator M_ρ. The error is then minimized with respect to the mollification parameter for optimality. M_ρ is analogous to $\mathcal{R}_\alpha$, the mollification parameter ρ playing the role of the regularization parameter α.

Let us first introduce the concept, properties and some examples of mollification. We begin by defining the operation of convolution.

Definition 3.11 (Convolution) *The convolution $f * g$ of two functions f and g is the integral*

$$f * g = \int_{R^n} f(x - y)g(y)\mathrm{d}y, \ x \in R^n,$$

where f and g are interchangeable.

The convolution can be viewed as the action of a linear operator f^* on elements g. The quantity $f * g$ is a sort of a product of two functions f and g and furnishes an alternative to the ubiquitous pointwise product fg. The convolutional product is useful in smoothing or mollifying. For example, a function which is only integrable (and may thus be quite rough) can be mollified to be infinitely differentiable.

The Mollifier

Next consider an infinitely differentiable function $w(x)$ which is compactly supported in a ball $B(0, 1)$ of unit radius centered at the origin $x = 0$ such that

$$\int_{R^n} w(x)\mathrm{d}x = 1. \tag{3.47}$$

Thus $w(x)$ is a $C_0^\infty(B(0, 1))$ function which is normalized to unity according to Eq. (3.52). Associated with $w(x)$ is a family of functions $w_\rho(x), \rho \in (0, \infty)$ defined by

$$w_\rho(x) = \rho^{-n}w(\rho^{-1}x)$$

such that

$$\int_{R^n} w_\rho(x)\mathrm{d}x = 1. \tag{3.48}$$

Thus $w_\rho(x) \in C_0^\infty(B(0, \rho))$ and is normalized to unity according to Eq. (3.53).

Let ϕ be a function in $\Omega \subset R^n$ and $\mathcal{E}$ a linear operator of extension from $L_2(\Omega) \to L_2(R^n)$ such that

$$
\begin{aligned}
(\mathcal{E}\phi)(x) &= \phi(x), \forall x \in \Omega \\
&= 0, x \in R^n \setminus \Omega.
\end{aligned}
$$

Finally, consider the convolution

$$J_\rho \phi = w_\rho * \mathcal{E}\phi, \rho \in (0, \infty).$$

$J_\rho \in \mathcal{L}(L_2(\Omega), L_2(R^n))$, where $\mathcal{L}(X, Y)$ is the usual Banach space of linear continuous operators between X and Y. The function J_ρ is called a *mollifier*. Furthermore, since $\|w_\rho\| = 1$, $\|J_\rho\| \leq 1$. The mollified function $J_\rho\phi$ can be made to be arbitrarily close to the original function ϕ if the parameter ρ is taken to be sufficiently small. This is demonstrated below.

We use *Young's inequality* (see Marti, 1986) stated below.

Young's Inequality. $\forall \phi \in L_1(R^n)$ *and* $g \in L_2(R^n)$, *the convolution* $f * g \in L_2(R^n)$. *Moreover,* $\|f * g\| \leq \|f\|_1 \|g\|_2$, $\|f\|_1$ *being the upper bound of the norm of the linear operator of convolution* $f*$.

Let us calculate $\|J_\rho\phi - \mathcal{E}\phi\|_2^2$. Using the normalization property of the mollifier we obtain,

$$\int_{R^n} \left[\int_R^n w_\rho(x - y) \left((\epsilon\phi)(y) - (\mathcal{E}\phi)(x) \right) dy \right]^2 dx$$

$$= \int_{R^n} \left[\int_{R^n} w\rho(z) \left((\epsilon\phi)(x - z) - (\mathcal{E}\phi)(x) \right) dz \right]^2 dx$$

$$= \int_{R^n} \left[\int_{R^n} w\rho(z) \left((T_z - I)\mathcal{E}\phi \right)(x) dz \right]^2 dx$$

$$\leq \int_{R^n} \left[\int_{R^n} w\rho(z) \left((T_z - I)\mathcal{E}\phi \right)^2 (x) dz \right] dx$$

$$= \int_{R^n} w\rho(z) \|(T_z - I)\mathcal{E}\phi\|_2^2 dz \leq \epsilon^2. \tag{3.49}$$

Now let $(T_z \psi)(x) = \psi(x - z)$ be a translation operator. Then $\forall \psi \in L_2(R^n)$, $\text{limit}_{z \to 0}(T_z \psi = \psi \in L_2(R^n))$.

Rewriting the inequality (3.54) in terms of the translation operator T_z, using Young's inequality and the fact that $\|w_\rho\| = 1$, we obtain

$$\|J_\rho\phi - \mathcal{E}\phi\|_2^2 \leq \int_{R^n} \left[\int_{R^n} w_\rho(z)\{(T_z\mathcal{E}\phi - \mathcal{E}\phi)(x)\}^2 dz \right] dx$$

$$= \int_{R^n} w_\rho(z)\|T_z\mathcal{E}\phi - \mathcal{E}\phi\|^2 dz \to 0$$

as $z \to 0$. This leads to the following theorem.

Theorem 3.12 *Let* $\Omega \subset R^n$ *be a domain and* $\rho \in (0, \infty)$ *a real parameter. Then*

$$\underset{\rho \to 0}{\text{limit}} \|J_\rho\phi - \mathcal{E}\phi\| = 0, \forall \phi \in L_2(\Omega).$$

The mollifier $J_\rho \in L(L_2(\Omega), L_2(R^n))$ *and* $\|J_\rho\| = 1$.

Examples of Mollifiers

The following are a few examples of the functions $w_\rho(x)$.

1. $w_\rho(x) = \text{const.}H(|x - y| - \rho)$, H being Heaviside's function.
2. $w_\rho(x) = \text{const. sinc } \rho(x - y)$.
3. The rapidly decaying *heat* or *Gaussian kernel*:

$$G_\rho(x) = \frac{1}{\rho\pi^{1/2}}e^{-\frac{x^2}{\rho^2}}, \ x \in R^1.$$

$$\int_{R^1} G_\rho(x)dx = 1$$

4. *Test Functions*: A typical test function $\phi(x)$ is given by (See Jones, 1966; Kanwal, 1997; Stakegold, 1979)

$$\phi_\rho(x) \quad = \quad \rho_0 e^{\frac{\rho^2}{|x|^2 - \rho^2}}, \ |x| \leq \rho$$
$$= \quad 0, |x| > \rho.$$

ρ_0 can be selected so that the integral is unity. For example, if $\rho = 1$, then

$$\rho_0 = \left(\int_{w_n} e^{(|x|^2 - 1)^{-1}} dx \right)^{-1}.$$

5. The *Dirichlet kernel* D_ν^* (Nikolskii, 1975). By definition, for $x = \{x_j\}_1^n \in R^n$,

$$D_\nu^*(x) = \Pi_1^n \frac{\sin(\nu x_j)}{x_j} = \Pi_1^n D_\nu(x_j).$$
$$\pi^{-n} \int_{R^n} D_\nu^*(x) \ dx = 1.$$
$$\left(\tfrac{2}{\pi} \right)^{\frac{n}{2}} \hat{D}_\nu^* = (1)_{\Delta_\nu},$$
$$\Delta_\nu = \{x | |x_j| < \nu\}.$$

$\hat{D}_\nu^*$ is the Fourier transform of the Dirichlet kernel. The quantity $(1)_{\Delta_\nu}$ is unity on Δ_ν and zero otherwise.

6. The kernel of *de la Vallee Poussin* (Nikolskii, 1975): V_ν. The de la Vallee Poussin kernel V_ν on R^n is defined as

$$V_\nu(x) = \alpha^n \Pi_1^n x_j^{-2}[\cos(\alpha x_j) - \cos(2\alpha x_j)], \ \alpha > 0. \qquad (3.50)$$
$$\pi^{-n} \int_{R^n} V_\nu(x) \ dx = 1.$$
$$\left(\tfrac{2}{\pi} \right)^{\frac{n}{2}} \hat{V}_\nu = (1)_{\Delta_\nu},$$
$$\Delta_\nu = \{x | |x_j| \leq \nu, j = 1, 2, \cdots, n\}.$$

As an aside, the de la Vallee Poussin kernel is an arithmetic mean of the Dirichlet kernels and

$$V_\rho(x) = \frac{1}{N} \sum_{\nu=N+1}^{2N} D_\nu^*(x).$$

In characterizing the kernels such as Dirichlet or de la Vallee Poussin, it is customary to consider the space $M_{\nu,p}$ of entire functions of exponential type ν in $L_p(R^n), 1 \le p \le \infty$. It may be recalled that if f is a function of exponential type ν, then

$$|f(x)| \le C_\delta e^{(\nu+\delta)|x|}, \delta > 0.$$

That is, f grows no more than exponentially with ν at infinity. Note that functions in $M_{\nu,p}$ are L_p in each variable x_j. In terms of these spaces, the Dirichlet kernel $D_\nu^* \in M_{\nu,p}$ and the de la Vallee Poussin kernel $V_\nu \in M_{2\nu,p}$. The Fourier transforms of these kernels which were given above follow from a general result that if $\phi \in M_{\nu,p}(R^n)$, then its Fourier transform $\hat{\phi}$ has support on $\Delta_\nu = \{x \mid |x_j| \le \nu, j = 1, 2, \cdots, n\}$. For later use, we will need one more property of the de la Vallee Poussin kernel $V_\nu(R^n)$, namely

$$\pi^{-1} \int_{R^1} |V_\nu(x)| dx < 2\sqrt{3}. \tag{3.51}$$

Moreover, the mollification using the de la Vallee Poussin kernel as the mollifier, that is

$$J_\rho \phi = V_\rho * \phi = \pi^{-1} \int_{R^n} V_\rho(x - y)\phi(y) dy$$

satisfies the bound

$$\|V_\rho * \phi - \phi\|_p \le (1 + C)E_\rho(\phi)_p, C(\text{constant}). \tag{3.52}$$

$E_\rho(\phi)_p$ is the best approximation of ϕ in $M_{\nu,p}$ and is given by

$$E_\rho(\phi)_p = \inf_{\psi_\rho \in M_{\nu,p}} \|\phi - \psi_\rho\|_p.$$

Using the bound in Eq. (3.55) and $E_\rho(\phi)_p$ above, Eq. (3.57) becomes

$$\|V_\rho * \phi - \phi\|_p \le (1 + 2\sqrt{3})E_\rho(\phi)_p. \tag{3.53}$$

Next we discuss the error estimate. Let $g^\epsilon = g_e + \epsilon$ be the inexact data and $f_\rho^\epsilon = A^{-1}M_\rho g^\epsilon$ the corresponding mollified solution. f_ρ^ϵ is a regularized solution with $A^{-1}M_\rho$ acting as a regularizing operator. The error estimate is obtained as follows.

$$
\begin{aligned}
\|f_e - f_\rho^\epsilon\| &\le \|f_e - f_{\rho e}\| + \|f_{\rho e} - f_\rho^\epsilon\| \\
&= \|f_e - f_{\rho e}\| + \|A^{-1}M_\rho(g^\epsilon - g_e)\| \\
&\le \|f_e - f_{\rho e}\| + \epsilon\|A^{-1}M_\rho\|,
\end{aligned}
$$

which is the mollified version of Eq. (3.22). Again the first term goes to zero as the mollification parameter $\rho \to 0$ by Theorem 3.12, whereas the second term becomes unbounded. Moreover, depending upon how the norm $\|A^{-1}M_\rho\|$ grows with ρ, the problem can be either mildly or severely ill-posed. If the growth is algebraic in powers of ρ, the problem is mildly ill-posed, whereas if the growth is exponential in ρ, the ill-posedness is severe. The determination of the optimal mollifier width now takes the place of determining the regularization parameter α and is discussed rigorously in Murio (1993) and Duc and Hao (1994), to which the interested reader is referred to for further details.

We end this brief section by illustrating the method through the example of numerical differentiation which is patterned after Duc and Hao (1994).

3.10.2   An Example: Numerical Differentiation

Consider a function f defined on R^1 and belonging to $L_p, 1 \leq p \leq \infty$. The function is supposed to be known approximately, i.e., $\|f - f^\epsilon\|_p \leq \epsilon > 0$. The problem is to differentiate the inexact function f^ϵ and obtain an error estimate. We mollify the function using the de la Vallee Poussin kernel (3.55) for R^1. This yields

$$f_\alpha^\epsilon(x) = (\alpha\pi)^{-1}(F_1 - F_2) * f^\epsilon, \tag{3.54}$$

where the kernels F_1 and F_2 of the convolutions are given by

$$F_1 = x^{-2}\cos(\alpha x) \text{ and } F_2 = x^{-2}\cos(2\alpha x),$$

respectively. The derivatives of the mollified function f_α^ϵ can be calculated by virtue of the differentiability of the convolution kernels. Hence the problem of differentiating f_α^ϵ is well-posed.

Let $f \in L_p(R^1)$ have derivatives up to and including $(k + 1)$-th order which also belong to $L_p(R^1)$ and $\|f^{(k+1)}\|_p \leq C_1$ and consider the quantity $\|f^{(k)} - (f_\alpha^\epsilon)^{(k)}\|_p$.

$$\|f^{(k)} - (f_\alpha^\epsilon)^{(k)}\|_p = \|f^{(k)} - (f_\alpha)^{(k)}\|_p + \|(f_\alpha)^{(k)} - (f_\alpha^\epsilon)^{(k)}\|_p. \tag{3.55}$$

The mollified function f_α^ϵ in Eq. (3.59) with F_1, F_2 defined above, is known to be an entire function of exponential type 2α (Nikolskii, 1975), and

$$|f_\alpha^\epsilon(x)| \leq C_\delta e^{(2\alpha+\delta)|x|}, \ \delta > 0.$$

Let $W_{2\alpha,p}(R^1)$ denote the class of the entire functions of exponential type 2α which are also in $L_p(R^1)$. We are going to use the following well-established results (Nikolskii, 1975; Duc and Hao, 1994; Meinardas, 1967; Rivlin, 1969):

1. $\pi^{-1}\int_{R^1}|V_\alpha(x)|\mathrm{d}x = 2\sqrt{3}$.

2. $\|V_\alpha * f - f\|_2 \leq (1 + 2\sqrt{3})E_\alpha(f)_p$. This is Eq. (3.58) for $n = 1$.

3. $\|\partial_x f\| \leq \alpha\|f\|_{L_p}$.

The quantity $E_\alpha(f)_p$ is a measure of the best approximation of $f \in W_{2\alpha,p}(R^1)$ by a polynomial from the same space. Thus

$$E_\alpha(f)_p = \|f - g^*\| = \inf_{g_\alpha \in W_{2\alpha,p}(R^1)} \|f - g_\alpha\|.$$

The inequality (3.60) is also known as the *Bernstein–Nikolskii* inequality (Duc and Hao, 1994).

Using these facts in Eq. (3.60) yields

$$\|f^{(k)} - \left(f_\alpha^\epsilon\right)^{(k)}\|_p = \leq (1 + 2\sqrt{3})E_\alpha(f)_p + (2\alpha)^k \|f_\alpha - f_\alpha^\epsilon\|_{L_p}$$
$$\leq (1 + 2\sqrt{3})E_\alpha(f^{(k)})_p + 2\sqrt{3}\epsilon(2\alpha)^k. \tag{3.56}$$

In obtaining Eq. (3.62), use was made of the fact that in differentiating a convolution product, the differentiation can be shifted from the convolver to the convolved function. In other words,

$$D^k(f * g) = f^{(k)} * g = f * g^{(k)},$$

where D^k represents the k-th order differentiation. From the general results in the theory of function approximation, (usually goes under *Jackson's theorems*, see Chapter 4), we have

$$E_n(\phi)_p \leq c(k)n^{-k}\|\phi^{(k)}\|_p. \tag{3.57}$$

Using Eq. (3.63) in Eq. (3.62) yields

$$\|f^{(k)} - \left(f_\alpha^\epsilon\right)^{(k)}\|_p = \leq c(1 + 2\sqrt{3})\alpha^{-k}\|f^{k+1}\|_p + 2\sqrt{3}\epsilon(2\alpha)^k$$
$$\leq c(1 + 2\sqrt{3})E_1\alpha^{-k} + 2\sqrt{3}\epsilon(2\alpha)^k, \tag{3.58}$$

where by assumption $\|f^{k+1}\|_p \leq E_1$. Choosing α to be $\epsilon^{-1/2k}$, Eq. (3.64) reduces to

$$\|f^{(k)} - \left(f_\alpha^\epsilon\right)^{(k)}\|_p \leq \{c(1 + 2\sqrt{3})E_1 + 2^{k+1}\sqrt{3}\}\sqrt{\epsilon}. \tag{3.59}$$

The estimate (3.65) implies Hölder continuity of the derivative in the error ϵ. All this can be summed up in the following theorem

Theorem 3.13 (Duc and Hao) *Let a function f and its derivatives up to and including (k + 1)-th order be in $L_p(R^1)$. Further suppose that $\|f^{k+1}\|_p \leq E_1, E_1 > 0$ is finite. Then $\forall \rho > 0$, the problem of differentiating the mollified function f_ρ^ϵ as given by Eq. (3.58) is well-posed and varies continuously with the error ϵ in the following manner:*

$$\|f^{(k)} - \left(f_\alpha^\epsilon\right)^{(k)}\|_p = \leq c\sqrt{\epsilon},$$

where c is a function only of the bound E_1 and k.

A similar result is also obtained in $L_p(\pi, \pi)$, i.e., for periodic functions in L_p having a period of 2π using the trigonometric form of the de la Vallee Poussin kernel. In $L_p(\pi, \pi)$, it is given by

$$V_\rho(x) = 4k\sin^2(\frac{x}{2})]^{-1}[\cos(k + 1)x - \cos(2k + 1)x].$$

The method of mollification is also applicable when the data is known inexactly and only at discrete points Murio (1993) and Duc and Hao (1994).

Chapter 4

Regularization by Projections

4.1 Introduction

The analysis in Chapter 3 was in the setting of an infinite-dimensional function space X. An element in such a space can be considered as requiring an infinite number of basis functions for its representation. In practice, however, it is not possible to use anything other than a finite basis set and the solution must perforce be obtained in a finite-dimensional space X_n. The problem is thus to obtain an approximate, finite-dimensional solution $f_n \in X_n$ of an otherwise infinite-dimensional solution f of an ill-posed problem $Af = g$. This is the objective of the present chapter. Regularization in a finite-dimensional space X_n can be viewed as being *discrete* in the sense that the space of solutions is spanned by a finitely discrete number of basis functions. These functions themselves are, however, continuous. It turns out that posing the problem in X_n amounts to regularizing the solution. The case when X_n is not a function, but a Euclidean space, will be discussed in the following chapter.

The problem is usually transferred from X to X_n by a *projection operator* $P_n, P_n^2 = P_n P_n = I, ||P_n|| = 1$ if the projection is orthogonal, and the ensuing methods are called the *methods of projections*. Two important examples of P_n are the operators of interpolation and finite Fourier transform. Different projection methods are obtained for different choices of (P_n, X_n). The most well-known are the *least-square*, *collocation* and *Galerkin* techniques and their variants such as *dual least-square* and *Petrov–Galerkin* procedures.

4.2 The Basic Projection Methods

Let us briefly summarize the essence of the methods. The equation $(Af)(x) = g(x), x \in \Omega \subset R^m$, is to be solved for the unknown function f. We seek an approximation of f in the form

$$f_n(x) = \sum_{j=1}^{n} f_j \phi_j(x),$$

where $\{\phi_j(x)\}_1^n$ is some known basis set. Since $Af_n(x)$ cannot equal $g(x)$, there is a *residual* error $r_n(x) = r_n(\mathbf{f_n}, x) = Af_n(x) - g(x)$, where $\mathbf{f_n} = \{f_1, f_2, \cdots, f_n\}^T$ denotes a vector in R^n the components of which are the coefficients f_j in the above series. We would like to determine $\mathbf{f_n}$ such that r_n is minimum and hope that the error $f - f_n(x)$ in the solution is also minimum in some sense. Of course, since the true solution f is not known, neither is the error. However, it is still possible to estimate the problem-dependent norm $||f - f_n||$. Note that the residual r_n depends on x and, for various reasons, it is usual to consider the minimization (in $\mathbf{f_n}$) of r_n not in a pointwise, but in an averaged sense, the averaging being in x over Ω. The minimization problem can then be stated as: *find f_n such that* $\int_\Omega r_n(\mathbf{f_n}, x)w_j(x)\mathrm{d}x = 0$, $j = 1, 2, \cdots n$, w_j *being a weight function*. The method is thus called *the method of weighted residuals*. Moreover, since there are finitely many weight functions, the qualification *discrete* should be added in order to distinguish from the continuous case.

Different choices of weight functions give rise to different methods. If $w_j = 2r_{n,j}$, the subscript j indicating differentiation with respect to f_j, then the problem is one of solving

$$2 \int_\Omega r_n(\mathbf{f_n}, x)r_{n,j}(\mathbf{f_n}, x)\mathrm{d}x = 0, \ j = 1, 2, \cdots n. \tag{4.1}$$

This is the same as minimizing the square of the residual since $2r_n r_{n,j} = (r_n^2)_{,j}$ and the least-square method is obtained as a result. If $w_j = \delta(x - x_j)$ at a discrete set of points $\{x_j\}_1^n$, then

$$\int_\Omega r_n(\mathbf{f_n}, x)w_j(x)\mathrm{d}x = r_n(\mathbf{f_n}, x_j) = 0, \ j = 1, 2, \cdots n. \tag{4.2}$$

The residual is now minimized at a discrete set of the so-called *collocation points* $\{x_j\}_1^n$ and one obtains the collocation method. Finally, if w_j is chosen to be the basis set $\{\phi_j\}_1^n$ itself, then

$$\int_\Omega r_n(\mathbf{f_n}, x)\phi_j(x)\mathrm{d}x = 0, \ j = 1, 2, \cdots n, \tag{4.3}$$

or equivalently, $< Af_n, \phi_n > = < g, \phi_n >$. What ensues in this case is the method of Galerkin also known as *standard Galerkin* or *Bubnov–Galerkin* method. Note that for the Galerkin method of Eq. (4.3), $A : X \to X$ and $X_n = Y_n$. On the other hand, if w_j is other than the basis set $\{\phi_j\}_1^n$ for the solution itself, one obtains the Petrov–Galerkin method.

If A is a bounded linear operator with a bounded inverse, then the approximate finite-dimensional solution f_n computed by minimizing the residual

r_n converges to the exact solution f. Let ϵ be the error in the solution, i.e., $||f_n - f|| = \epsilon$. Then

$$||r_n(\mathbf{f_n})|| = ||A(f_n(\mathbf{f_n}) - f)|| \leq ||A||\epsilon.$$

Let $\tilde{\mathbf{f}}_n$ be the minimizing vector. Then

$$||r_n(\tilde{\mathbf{f}}_n)|| \leq ||r_n(\mathbf{f_n})||, \ \forall \mathbf{f_n}.$$

Thus

$$
\begin{aligned}
||f_n(\tilde{\mathbf{f}}_n) - f|| &\leq ||A^{-1}|| ||r_n(\tilde{\mathbf{f}}_n)|| \\
&\leq ||A^{-1}|| ||r_n(\mathbf{f_n})|| \leq ||A^{-1}|| ||A||\epsilon.
\end{aligned}
$$

Since A is bounded and has a bounded inverse, the convergence follows by letting ϵ go to zero. The result holds even if A is not bounded (Linz, 1978).

The methods described by Eqs. (4.1) through (4.3) can be considered to be the special cases of a more general formulation. Notice that the method of the weighted residuals consists in setting the inner product $< r_n, w_j >$ to zero for each w_j. This gives: $< Af_j, w_j >=< g, w_j >$. Each method is, therefore, of the form

$$\Phi[Af_n] = \Phi[g],$$

where Φ is a linear functional. If $\Phi[\cdot] =< A\phi_n, [\cdot] >$, then the least-square method (Section 4.4) is obtained. If $\Phi[\cdot] =< \phi_n, [\cdot] >$, then the Galerkin method results and, finally, if $\Phi[\cdot] = [\cdot](x_i)$, the collocation method is recovered. In the collocation method, the inner product is with a delta function and hence the matrix elements are simpler compared to the other two. However, there are difficulties with collocation. For example, if $A = I$ and $\phi_i = x^{i-1}$, then the problem is one of approximating a function by polynomial interpolation. Convergence can be problematic in this case (see Section 4.5).

Under special circumstances, the least-square and the Galerkin method may coincide. From Eq. (4.1), the least-square solution can be seen to be characterized by

$$< Af_n, Az_n >=< g, Az_n >, \ \forall z_n \in X_n.$$

This follows from the fact that the linear system associated with Eq. (4.1) can be written as

$$\sum_{j=1}^{n} f_j < A\phi_j, A\phi_i >=< g, A\phi_i > .$$

On the other hand, for the Galerkin method, the corresponding equations are $< Af_n, z_n >=< g, z_n >$ and $< Af_n, \phi_n >=< g, \phi_n >$, respectively. A comparison of the two shows that the methods coincide if the $A : X \to X$ and $X_n = Y_n$.

Furthermore, if the operator A is self-adjoint and positive definite, i.e., $< Af, g >=< g, Af >$ and $< Af, f >> 0, \forall f \in X$, then the Galerkin method is synonymous with the classical *Rayleigh–Ritz* procedure (see, for example,

Collatz, 1966). This can be seen as follows. It is well-known that for a self-adjoint and positive definite operator A, the problem of solving the equation $Af = g$ is equivalent to finding the minimum of the functional Φ defined as:

$$\Phi(z) = <Az, z> - <z, g> - <g, z> = <Az, z> - 2\text{Re} <z, g>.$$

If $z = f$ minimizes this variational equation, then f solves the original problem $Af = g$ uniquely (the uniqueness is a consequence of self-adjointness and positivity). In the discrete case, the functional $\Phi(z)$ is to be minimized over X_n. This procedure is known as the *Rayleigh–Ritz method*. Let x_n be the minimum solution. Then $\Phi(z_n) - \Phi(x_n) \geq 0$. Writing $z_n = x_n \pm \delta w_n$, $w_n \in X_n$, and after some straightforward algebraic manipulation, it is found that

$$\Phi(z_n) - \Phi(x_n) = \pm 2\delta \text{Re} <Ax_n - g, w_n> + \delta^2 <Aw_n, w_n> \geq 0.$$

Dividing by δ and letting $\delta \to 0$, we obtain $<Ax_n - g, w_n> = 0$ which is the Galerkin condition above. Contrarily, if one starts with the Galerkin condition, then the nonnegativity of $\Phi(z_n) - \Phi(x_n)$ can also be demonstrated. Therefore, for a self-adjoint and positive definite operator, the Rayleigh–Ritz and the Galerkin procedures are equivalent.

A few other remarks: i. If A is a differential operator, then the collocation method closely resembles the finite-difference scheme, the collocation nodes acting as the grid points. ii. If the basis happens to be orthogonal, then the Galerkin method amounts to determining the solution by setting each of the n Fourier coefficients of the residual individually to zero. iii. It is worth mentioning that for global basis functions, as in the above, the basis set tends toward linear dependency as its size increases. This can be appreciated by considering the set of functions $x^n(1 - x)$ over a finite interval and gradually increasing the values of n. At large values of n, the functions become almost imperceptible from one another. It is, therefore, desirable to work with a basis the elements of which are either orthogonal or close to being so. iv. Finally, the Galerkin method enjoys a wide popularity and has a broad appeal primarily due to the fact that it constitutes the basis of the much celebrated *finite-element method* or FEM in short. FEM is a widely used numerical method for obtaining weak solutions of numerous physical problems for which classical solutions are at best difficult to obtain. This will be discussed in Section 4.6.

4.3   The Method of Projections: General Framework

The problem is to solve the equation $Af = g$, $f \in X$, $g \in Y$, for f. $A : X \to Y$ is a linear, bounded and injective integral operator between two infinite-dimensional function spaces X and Y. Let us begin by stating the well-known *Geometric Series Theorem* for linear bounded operators (Conway, 1990). The theorem is fundamental in the error analysis of the approximations.

Theorem 4.1 (The Geometric Series Theorem) *Let X be a normed linear space and $A : X \to X, ||A|| < 1$, a bounded linear operator. Assume that either (a) X is complete (a Banach space) or (b) A is compact. Then the inverse operator $(I - A)^{-1} : X \to X$ is bounded. Moreover,*

$$(I - A)^{-1} = \sum_{n=0}^{\infty} A_n$$

and

$$||(I - A)^{-1}|| \leq (1 - ||A||)^{-1}.$$

The series in the theorem is known as the *Neumann Series*. For $||A|| < 1$, $||A(x - y)|| \leq \epsilon||x - y||$, $\epsilon < 1$, and the Geometric Series Theorem is called the *Contractive Mapping Theorem*. It is also known by another name as *Banach's Fixed-Point Theorem*.

Again we seek an approximate, finite-dimensional solution $f_n \in X_n \subset X$ to the infinite-dimensional problem $Af = g$, $A : X \to X$. The approximate problem is

$$Af_n = g, g \in X. \tag{4.4}$$

Applying the projection operator P_n to Eq. (4.4), we obtain

$$P_n A f_n = P_n g, \ f_n \in X_n. \tag{4.5}$$

Let $\{\phi_i\}_1^n$ be a spanning set for X_n. Then

$$f_n(x) = \sum_{j=1}^{n} f_j \phi_j(x) \tag{4.6}$$

and

$$(P_n g)(x) = \sum_{j=1}^{n} g_j \phi_j(x). \tag{4.7}$$

With expansions of Eqs. (4.6) and (4.7), the projection equation (4.5) reduces to a linear system of algebraic equation given by

$$\mathbf{M f_n} = \mathbf{g_n}, \tag{4.8}$$

where the elements M_{ij} of the $n \times n$ matrix $\mathbf{M}$ are given by

$$\mathbf{M}_{ij} = \int K(x_i, z)\phi_j(z)dz, \ i, j = 1, 2, \cdots, n.$$

$K(x, z)$ is the kernel of the integral operator A, $\mathbf{f_n} = \{f_1, f_2, \cdots, f_n\}^T$ and $\mathbf{g_n} = \{g_1, g_2, \cdots, g_n\}^T$ are two n-dimensional vectors. The linear system (Eq. (4.8)) has a unique solution if and only if $\det |\mathbf{M}| \neq 0$, i.e., the matrix $\mathbf{M}$ is nonsingular.

It is hoped that that a solution of Eq. (4.8) can be so constructed that the residual $r_n = Af_n - g$ will be zero and the resulting function f_n will prove to be a good approximation to the true solution in some well-defined sense.

The formal solution of Eq. (4.5) is

$$f_n = (P_n A)^{-1} P_n Af = R_n Af, \tag{4.9}$$

where $R_n = (P_n A)^{-1} P_n : X \to X$ is a linear operator. The method is considered to be *convergent* if $R_n Af \to f$, $\forall f \in X$, as $n \to \infty$. Notice that if n^{-1} is called α, then this definition of convergence coincides with that of a regularization strategy defined in Theorem 3.9 of Chapter 3. Therefore, the projection solution converges to the actual solution if and only if R_n defined by Eq. (4.9) is a regularization strategy. Let it be assumed that the union $\bigcup_{n=1}^{\infty} X_n = \overline{X}_n$ is dense in X. Then from Eq. (4.9) $R_n A z_n = z_n, \forall z_n \in X_n$ and

$$
\begin{aligned}
\|f - f_n\| &= \|(I - R_n A)f\| = \|(I - R_n A)(f - z_n)\| \\
&\leq \|I - R_n A\| \, \|f - z_n\|, \forall z_n \in X_n. \tag{4.10}
\end{aligned}
$$

Since $\overline{X}_n$ is dense in X, $\|f - z_n\|$ can be made as small as desired by taking n as large as desired. Hence if $\|R_n A\|$ is bounded for all $n \in Z^+$, the projection method converges. In addition, from Eq. (4.10), the error bound is

$$\|f - f_n\| \leq (1 + C) \min_{z_n \in X_n} \|f - z_n\|, C \leq \infty.$$

Hence the following theorem.

Theorem 4.2 *Let the approximate solution of $Af = g, f, g \in X$, be given by Eq. (4.9). Then $f_n = R_n g \in X_n$ converges to the exact solution f, $\forall f \in X$, i.e. for $\forall g = Af$, provided that*

$$\|R_n A\| \leq C \text{(constant)}, \ \forall n \in Z^+.$$

It is assumed that $U_{n=1}^{\infty} X_n$ is dense in X. Moreover, the following error estimate holds:

$$\|f - f_n\| \leq (1 + C) \min_{z_n \in X_n} \|f - z_n\|.$$

R_n given by Eq. (4.9) is a regularizing operator as defined in Theorem 3.9.

In the rest of this section we illustrate the essentials of the methods of projections for an integral equation of the second kind. Consider, therefore, the equation

$$(I - \lambda A)f = g, \tag{4.11}$$

and its projection form

$$(I - \lambda P_n A)f_n = g_n, \ f_n, \ g_n \in X_n \subset X. \tag{4.12}$$

It is assumed that $A : X \to X$ is compact and λ is a regular value of the equation (Cochran, 1972). Furthermore, assume that the formal solution

$$f = (I - \lambda A)^{-1} g$$

exists and is unique. The inverse operator $(I - \lambda A)^{-1}$ is, therefore, bounded. From Eq. (4.12), the approximate solution is

$$f_n = (I - \lambda P_n A)^{-1} g_n. \tag{4.13}$$

The moot point is the behavior of the inverse operator $(I - \lambda P_n A)^{-1}$ as $n \to \infty$. Now

$$
\begin{aligned}
(I - \lambda P_n A) &= (I - \lambda A) + \lambda(A - P_n A) \\
&= (I - \lambda P_n A)\left[I + \lambda(I - \lambda A)^{-1}(A - P_n A)\right].
\end{aligned}
$$

Therefore,

$$(I - \lambda A)^{-1} = \left[I + \lambda(I - \lambda A)^{-1}(A - P_n A)\right]^{-1}(I - \lambda A)^{-1}$$

and hence

$$\|(I - \lambda P_n A)^{-1}\| \le \|(I - \lambda A)^{-1}\| \|[I + \lambda(I - \lambda A)^{-1}(A - P_n A)]^{-1}\|.$$

Now $(I - \lambda A)^{-1}$ is bounded by assumption. If it is further assumed that

$$|\lambda| \, \|A - P_n A\| \, \|(I - \lambda A)^{-1}\| < 1, \tag{4.14}$$

then by the Geometric Series Theorem 4.1, it follows that

$$\|(I - \lambda P_n A)^{-1}\| \le M(\text{constant}). \tag{4.15}$$

Next consider Eqs. (4.11) and (4.12). Operating on both sides of Eq. (4.11) by P_n, subtracting from Eq. (4.12) and after relatively minor algebra, we obtain

$$f - f_n = (I - \lambda P_n A)^{-1}(f - P_n f).$$

This leads to the error estimate (using Eq. (4.15))

$$\|f - f_n\|_X \le M\|f - P_n f\|_X. \tag{4.16}$$

An alternative error estimate can be obtained using Eq. (4.14) and Theorem 4.1. Some fairly straightforward manipulation provides the bound (Baker, 1977)

$$
\begin{aligned}
\|f - f_n\| \le \quad & \|[(I - \lambda A)^{-1}\| \, [1 - |\lambda| \\
& \cdot \quad \|(I - \lambda A)^{-1}\| \, \|A - P_n A\|]^{-1} \\
& \cdot \quad [\|g - g_n\| + |\lambda| \, \|A - P_n A\| \, \|f\|]
\end{aligned} \tag{4.17}
$$

It is worth mentioning that the roles of A and $P_n A$ can be interchanged in these estimates. A number of other bounds can also be obtained for the error.

However, only those that lead to their computability in practice are of interest. In this regard, an interesting bound is given in Kress (1989).

The convergence results based on Eqs. (4.16) and (4.17) require that $||A - P_n A|| \to 0$ as $n \to \infty$. If A is compact, then it can indeed be shown that $||A - P_n A|| \to 0$ as $n \to \infty$ provided that $P_n x \to x$, $\forall x \in X, n \to \infty$.[1] Assume that $\{P_n\}$ is a family of bounded projection operators such that

$$\underset{n\to\infty}{\text{limit}} \, ||P_n x - x|| \to 0 \,\forall\, x \in X. \tag{4.18}$$

Let $A : X \to X$ be compact and consider the norm $||P_n A - A||$. By definition:

$$||P_n A - A|| = \sup_{||x|| \le 1} ||(P_n A)x - Ax|| = \sup_{z \in D} ||P_n z - z||,$$

where $D = \{Ax|\ ||x|| \le 1,\ x \in X\}$. Now D is the image of a bounded subset under a compact operator A is thus compact itself. By the assumption (4.18) above, $||P_n A - A|| \to 0$ as $n \to \infty$ on any compact subset of X. The convergence of $f_n \to f$ in the estimates in Eqs. (4.16) and (4.17) is thus established. More importantly, the convergence is of the same order as $||f - P_n f||$. In other words, f_n converges to f exactly as fast as $P_n f$ converges to f as $n \to \infty$. A reader interested in a more rigorous proof using the *Banach–Steinhaus* theorem should consult Atkinson (1997).

We collect these results into the theorem below.

Theorem 4.3 *Let X be a normed linear space, $A : X \to X$ a bounded linear operator and $\{P_n\}$ a family of linear, bounded projection operators from $X \to X_n$, $X_n \subset X$. Let*

$$(I - \lambda A)f = g, \ f, g \in X.$$

be a Fredholm integral equation of the second kind with λ as its regular value. Assume that

$$|\lambda|\ ||(I - \lambda A)^{-1}||\ ||A - P_n A|| < 1$$

and either (a) X is complete, i.e., a Banach space or (b) A is compact. If f_n is a solution of the approximate problem

$$(I - \lambda P_n A)f_n = P_n g,$$

then the inverse operator $(I - \lambda P_n A)^{-1}$ exists as a bounded operator from X to X. Moreover, the following estimate holds for the error $f - f_n$:

$$\begin{aligned} ||f - f_n|| \ &\le\ ||(I - \lambda A)^{-1}||\left[1 - |\lambda|\ ||(I - \lambda A)^{-1}||\ ||A - P_n A||\right]^{-1} \\ &\cdot\ \{||g - g_n|| + |\lambda|\ ||A - P_n A||\ ||f||\}. \end{aligned}$$

Alternatively,

$$||f - f_n|| \le M||f - P_n f||,$$

----

[1] It is important to point out that there are situations where $P_n x \to x, \forall x \in X$, and yet $||P_n A - A||$ does not go to zero with n tending to infinity. We will come across this in the sequel.

where

$$M = ||(I - \lambda P_n A)^{-1}|| < \infty.$$

Moreover, $||A - P_n A|| \to 0$ *as* $n \to \infty$, *if* $P_n f \to f, \forall f \in X$, *and* f_n *converges to* f *at the same rate at which* $P_n f$ *converges to* f *in the limit* $n \to \infty$. *Also,* A *and* $P_n A$ *can be interchanged in the error estimates.*

4.4 The Method of Least-Square

The basic minimum residual problem for the least-square method is expressed by Eq. (4.1). If $\tilde{f}_n \in X_n$ is a least-square solution of $Af = g, f \in X, g \in Y$, then

$$||A\tilde{f}_n - g|| \leq ||Az_n - g||, \ \forall z_n \in X_n \tag{4.19}$$

in the L_2-norm. Now

$$\begin{aligned}
||r_n(\tilde{\mathbf{f}}_\mathbf{n})||^2 &= <A\tilde{f}_n - g, A\tilde{f}_n - g> \\
&= \sum_{i=1}^{n}\sum_{j=1}^{n} \tilde{f}_i \tilde{f}_j < A\phi_i, A\phi_j > -2\sum_{i=1}^{n} \tilde{f}_i < A\phi_i, g > + < g, g >.
\end{aligned}$$

Upon differentiating the above relation with respect to $\tilde{f}_i, i = 1, 2, \cdots, n$, and setting the results to zero, the linear system for Eq. (4.1) is obtained, namely

$$\sum_{j=1}^{n} \tilde{f}_j < A\phi_i, A\phi_j >=< g, A\phi_i >. \tag{4.20}$$

From Eq. (4.20), it is seen that a least-square solution can be characterized by the identity

$$< A\tilde{f}_n, Az_n >=< g, Az_n >, \ \forall z_n \in X_n. \tag{4.21}$$

Moreover, Eq. (4.19) shows that the least-square solution is also the *best approximation* of g in Y_n. At this point, let us recall the definition of the best approximation (Rivlin, 1969; Davies, 1965).

Definition 4.1. (Best Approximation) *Let* V *be a linear space and* $W \subset V$ *a subset of* V. *For* $v \in V$, *let* $w^* \in W$ *be such that* $||v - w^*|| \leq ||v - w||, \forall w \in W$. *Then* w^* *is called the best approximation of* v *in* W.

The least-square solution has a geometric interpretation. If Eq. (4.21) is written in the form $< A\tilde{f}_n - g, Az_n >= 0, \forall z_n \in X_n$, then the best approximation, least-square solution $\tilde{f}_n$ is that solution for which the difference $g - A\tilde{f}_n \perp Y_n$. In other words, $A\tilde{f}_n$ is the normal projection of g on Y_n. From the theory of function approximation in Hilbert spaces, the best approximation exists and is unique. This is a consequence of the fact that for Hilbert spaces, the norm is strictly convex (see Chapter 3, Definition 3.2). This immediately implies that the solution $\tilde{f}_n$ is unique if A is one-to-one. Now the matrix $\mathbf{M}$ of Eq. (4.20), the elements of which are $< A\phi_i, \phi_j >$, is symmetric over a real field and Hermitian over a complex field. The above uniqueness result proves that $\mathbf{M}$ is also positive-definite.

Thus far it was established that the least-square problem has a unique solution. It turns out that $\tilde{f}_n$ does not necessarily converge to the actual solution f as n approaches infinity. One can so construct an operator A and a function f such that no convergence can be obtained. An example was provided by Seidman (1980). In this example, f was defined as

$$f = \sum_{j=1}^{\infty} j^{-1}\phi_j$$

and the operator A was constructed so that

$$A \sum_{j=1}^{n} u_j\phi_j = \sum_{j=1}^{n} [\alpha_j u_j + \beta_j u_1]\phi_j,$$

where $\beta_1 = 0, \beta_j = j^{-1}, \alpha_j = j^{-1}$ if j is odd and $\alpha_j = j^{-2.5}$ if j is even. From this, a variational problem for the least-square representation $\mathbf{f_n}$ can be formulated as the minimization of the functional F given by

$$F = \sum_{j=1}^{n} [\alpha_j(\tilde{f}_j - j^{-1}) + \beta_j(\tilde{f}_j - 1)]^2 + \sum_{j=n+1}^{\infty} [(\alpha_j j^{-1})(1 + \alpha_j - \tilde{f}_j)]^2.$$

Differentiating F in $\tilde{f}_j$ and then setting the resulting equations to zero, one obtains

$$\tilde{f}_1 = 1 + \Big[\sum_{j=n+1}^{\infty} \alpha_j j^{-2} \Big]\Big[1 + \sum_{j=n+1}^{\infty} \alpha_j j^{-2}\Big]$$

and·

$$\tilde{f}_j = j^{-1} + (\alpha_j j)^{-1}(\tilde{f}_j - 1), \ 2 \leq j \leq n.$$

It is then obtained that

$$\|\tilde{f}_n - P_n f\| = \left[(\alpha_j j)^{-2} \left(\sum_{j=n+1}^{\infty} \alpha_j j^{-2} \right)^{-2} \left(1 + \sum_{j=n+1}^{\infty} j^{-2} \right)^{\frac{1}{2}} \right],$$

from which it follows that $\|\tilde{f}_n - P_n f\| \sim n$ and $\tilde{f}_n$, therefore, does not converge.

Notice that in the above example, the sequence $\{\tilde{f}_n\}$ does not remain bounded and since by Eq. (4.9), $\tilde{f}_n = R_n A f$, this implies that $R_n A$ is not bounded, thereby violating the condition of Theorem 4.2 for convergence. Additional conditions are to be imposed on the solution in order for convergence to occur.

One can still estimate the norm of R_n as follows. Define

$$V_n = \{u | u_n \in X_n; \|Au_n\| = 1\}$$

and let $U_n = \max ||u_n||$, $u_n \in V_n$. Then for $u_n \in V_n$,

$$
\begin{aligned}
||\tilde{f}_n|| & = & ||Au_n||||\tilde{f}_n|| \le ||A||||u_n||||\tilde{f}_n|| \\
& \le & ||u_n||||A||||\tilde{f}_n|| \le ||u_n||||A\tilde{f}_n|| \le U_n||A\tilde{f}_n||.
\end{aligned}
\tag{4.22}
$$

From Eq. (4.21), on the other hand,

$$
\begin{aligned}
< A\tilde{f}_n, A\tilde{f}_n > & = & ||A\tilde{f}_n||^2 \\
& = & || < g, A\tilde{f}_n > || \le ||g||||A\tilde{f}_n||,
\end{aligned}
\tag{4.23}
$$

implying that $||A\tilde{f}_n|| \le ||g||$. Combining Eqs. (4.22) and (4.23) results in $||\tilde{f}_n|| \le U_n||g||$ showing that $||R_n|| \le U_n$.

An error estimate can also be determined. Following the same arguments leading to Eq. (4.22), we have

$$
||\tilde{f}_n - z_n|| \le U_n||A(\tilde{f}_n - z_n)||, \quad z_n \in X_n.
\tag{4.24}
$$

Furthermore, in view of Eq. (4.21), it follows that

$$
||A(\tilde{f}_n - z_n)||^2 \le ||A(f - z_n)||||A(\tilde{f}_n - z_n)||
$$

resulting in $||A(\tilde{f}_n - z_n)|| \le ||A(f - z_n)||$. Using this in Eq. (4.24), we have

$$
||\tilde{f}_n - z_n|| \le U_n||A(f - z_n)||, \quad z_n \in X_n.
\tag{4.25}
$$

Now using the inequalities in Eqs. (4.24) and (4.25) it is seen that

$$
||f - \tilde{f}_n|| \le ||f - z_n|| + U_n||A(f - z_n)||.
$$

Since this is valid for all $z_n \in X_n$, we can also write

$$
||f - \tilde{f}_n|| \le \min_{z_n \in X_n} [||f - z_n|| + U_n||A(f - z_n)||].
\tag{4.26}
$$

Now

$$
||f - \tilde{f}_n|| \le (1 + ||R_n A||)||f - z_n||.
\tag{4.27}
$$

$R_n = (\tilde{P}_n A)^{-1} \tilde{P}_n : Y \to X_n$, where $\tilde{P}_n : Y \to Y_n$ is the operator of projection from Y to Y_n. Also, $R_n A z_n = z_n$. Comparing Eqs. (4.26) and (4.27), it follows that $||R_n A||$ is bounded. Thus, under the condition that Eq. (4.26) is satisfied, we have convergence according to Theorem 4.2, assuming that $\overline{X}_n$ is dense in X.

Let us next consider data with perturbation g^ϵ where $|g^\epsilon - g| \le \epsilon > 0$. Again by Eq. (4.26) and the bound $||R_n|| \le U_n$,

$$
||f - \tilde{f}_n^\epsilon|| \le U_n \epsilon + ||R_n A f - f||.
$$

Let us combine all this succinctly into the theorem below.

Theorem 4.4 *Let $A : X \to Y$ be a linear, bounded, one-to-one operator, X and Y being Hilbert spaces. Consider the equation $Af = g$, $f \in X$, $g \in Y$. Let $X_n \subset X$ be an n-dimensional subset of X and it is assumed that the union $\cup_n X_n$ is dense in X. Let $\tilde{f}_n$ be the least-square solution of $Af = g$. Then $\tilde{f}_n$ converges to f provided that*

$$\min_{z_n \in X_n} \left[||f - z_n|| + U_n ||A(f - z_n)|| \right]$$

is bounded for all $f \in X$ where

$$U_n = \max\{ ||u_n|| \, | \, u_n \in X_n; \, ||Au_n|| = 1 \}.$$

The error in the solution is bounded by

$$||f - \tilde{f}_n|| \leq \min_{z_n \in X_n} \left[||f - z_n|| + U_n ||A(f - z_n)|| \right].$$

The least-square solution is given by $\tilde{f}_n = R_n Af$, $R_n = (\tilde{P}_n A)^{-1} \tilde{P}_n : Y \to X_n$ is the regularization strategy with $\tilde{P}_n : Y \to Y_n$ being the operator of projection from Y to $Y_n \subset Y$. The norm $||R_n|| \leq U_n$ and the norm of the combined operator $||R_n A||$ is bounded. Furthermore, if g in the equation $Af = g$ is replaced by the perturbed data g^ϵ, $|g^\epsilon - g| \leq \epsilon > 0$, then the solution error is bounded by

$$||f - \tilde{f}_n^\epsilon|| \leq U_n \epsilon + ||R_n Af - f||.$$

A variant of the least-square method is the so-called *dual least-square method*. The problem is the dual of the above problem in the sense that the least-square method is now applied to the adjoint equation which is $A^* g = f$. Everything remains the same if X_n is replaced by Y_n and A by its adjoint A^*. The problem is to solve $g_n = P_n g$, for some g_n such that

$$< AA^* g_n, z_n > = < g, z_n >, \forall z_n \in Y_n.$$

The orthogonality condition is replaced by

$$< A^* g_n, A^* z_n > = < g, A^* z_n >, \forall z_n \in Y_n.$$

We now write for g_n the following expansion

$$g_n = \sum_{j=1}^{n} g_j \hat{y}_j,$$

where $\{\hat{y}_j\}_i^n$ is the span$\{Y_n\}$. The linear system to be solved is now given by

$$\sum_{j=1}^{n} g_j < A^* \hat{y}_j, A^* \hat{y}_i > = < g, \hat{y}_i > .$$

The solution is given by

$$f_n = A^* g_n = \sum_{j=1}^{n} g_j A^* \hat{y}_j.$$

Theorem 4.4 remains unchanged if X_n is replaced by Y_n and A by A^*. This concludes our discussion about the method of least-square.

4.5 The Method of Collocation

Next we consider projection methods of *collocation* type and describe their essentials by again considering the integral equation (4.11). If f_n is the approximate solution, then the residual error $r_n(x)$ at x is

$$r_n(x) = \{(I - \lambda A)f_n\}(x) - g(x).$$

Referring to Eq. (4.2), the collocation equations in this case can be written as

$$\{(I - \lambda P_n A)f_n\}(x_i) = g_n(x_i), \ i = 1, 2, \cdots, n.$$

Writing out fully

$$\sum_{j=1}^{n} f_j \left\{ \phi_j(x_i) - \lambda \int_D K(x_i, y)\phi_j(y)dy \right\} = g_n(x_i). \tag{4.28}$$

The matrix elements $\mathbf{M}_{ij}$ of the corresponding linear system are given by

$$\mathbf{M}_{ij} = \phi_j(x_i) - \lambda \int_D K(x_i, y)\phi_j(y)dy. \tag{4.29}$$

If $f \in C(D)$, then the error is

$$\|f - f_n\|_\infty = \inf_{z_n \in X_n} \|f - z_n\|_\infty. \tag{4.30}$$

Now by definition

$$\inf_{z_n \in X_n} \|f - z_n\|_\infty \le \omega(f; h) \sim O(h),$$

where $\omega(f; h)$ is the modulus of continuity

$$\omega(f, h) = \max_{\substack{x,y \in D \\ |x-y| \le h}} |f(x) - f(y)|.$$

If $f \in C^2(D)$, then $\|f - f_n\|_\infty \sim 0(h^2)$. In general, for a piecewise polynomial interpolation, $\|f - f_n\|_\infty \sim 0(h^{m+1})$ provided that $f \in C^{m+1}(D)$ (Atkinson, 1997).

Let us next consider the perturbed data $g^\epsilon = g + \epsilon$. The corresponding collocation solution is f_n^ϵ. We would like to estimate the error

$$\|f_n^\epsilon - f\|_\infty \le \|f_n^\epsilon - f_n\|_\infty + \|f_n - f\|_\infty. \tag{4.31}$$

The second term in Eq. (4.31) was already given above in Eq. (4.30). The first term can be estimated as follows. For the perturbed data, the solution of Eq. (4.8) for the collocation method becomes

$$\mathbf{f_n^\epsilon} = \mathbf{M}^{-1}\mathbf{g_n^\epsilon},$$

with $\mathbf{M}$ given by Eq. (4.29). Similarly, for the unperturbed approximate solution $\mathbf{f_n}$ with $\mathbf{g_n}$ replacing $\mathbf{g_n^\epsilon}$. Let Φ be an operator that transforms a vector, say $\mathbf{u_n}$, to a function u_n by

$$\Phi\mathbf{u_n}(x) = \sum_{j=1}^{n} u_j \phi_j(x).$$

Subtract $\mathbf{f_n}$ and $\mathbf{f_n^\epsilon}$ from each other and obtain

$$
\begin{aligned}
||f_n^\epsilon - f_n||_\infty &= ||\Phi||\{\mathbf{M}^{-1}(\mathbf{g_n^\epsilon} - \mathbf{g_n})\}||_\infty \\
&\leq ||\Phi||\,||\mathbf{M}^{-1}||\,||\mathbf{g_n^\epsilon} - \mathbf{g_n}||_\infty \\
&\leq ||\Phi||\,||\mathbf{M}^{-1}||\epsilon.
\end{aligned}
\tag{4.32}
$$

From the definition of the operator norm

$$||\Phi|| = \left\{ \max \left| \sum_{j=1}^{n} u_j \phi_j \right| : ||\mathbf{u}|| = 1|| \right\}. \tag{4.33}$$

Inserting Eqs. (4.30) and (4.32) into Eq. (4.31) with $||\Phi||$ given by Eq. (4.33), we arrive at the error estimate

$$
\begin{aligned}
||f_n^\epsilon - f||_\infty &\leq \left\{ \max \left| \sum_{j=1}^{n} u_j \phi_j \right| : ||\mathbf{u}|| = 1 \right\} ||\mathbf{M}^{-1}||\epsilon \\
&\quad + \text{const.} \min_{z_n \in X_n} ||f - z_n||_\infty.
\end{aligned}
\tag{4.34}
$$

From the spectral theory (Collatz, 1966. See also Appendix A.5.1), the matrix norm $||\mathbf{M}^{-1}||$ can be expressed in terms of the spectral radius which is the lowest singular value of $\lambda_{\min}$ of $\mathbf{M}$. In terms of $\lambda_{\min}$, the estimate (4.34) becomes

$$
\begin{aligned}
||f_n^\epsilon - f||_\infty &\leq \frac{\epsilon}{\lambda_{\min}} \left\{ \max \left| \sum_{j=1}^{n} u_j \phi_j \right| \, ||\mathbf{u}|| = 1 \right\} \\
&\quad + \text{const.} \min_{z_n \in X_n} ||f - z_n||_\infty.
\end{aligned}
$$

Note that if $\{\phi_j\}_1^n$ are orthonormal, the $||\Phi|| = 1$. We collect all this into the following theorem.

Theorem 4.5 *Let $A: X \to X$ be bounded, X being a Banach space and f solves the integral equation of the second kind:*

$$(I - \lambda A)f = g, \quad f, g \in X,$$

λ being a regular value of the equation. Let $P_n: X \to X$ be an interpolatory projection onto the subspace $X_n \subset X$. Assume that the collocation method

$$(I - \lambda P_n A)f_n = P_n g$$

converges for a set of collocation nodes $\{x_i\}_1^n$. Let f_n^ϵ be the collocation solution corresponding to the perturbed data $g^\epsilon = g + \epsilon$, $\epsilon \geq 0$ being the data error. Then the following error estimate holds, namely

$$||f_n^\epsilon - f_n||_\infty \leq ||\Phi|| \frac{\epsilon}{\lambda_{\min}} + \text{const.} \min_{z_n \in X_n} ||f - z_n||_\infty,$$

where various quantities in the estimate were as defined above.

Denoting the inverse operator $(I - \lambda A)$ by K, and noting that $P_n P_n = P_n$, we again obtain the approximate solution f_n in terms of the regularizing operator R_n, namely

$$f_n = (P_n K)^{-1} P_n K f = R_n K f.$$

Nothing was said so far about the evaluation of the integrals

$$\int_D K(x_i, y) \phi_j(y) \mathrm{d}y, \ \forall \ x_i \in \{x_k\}_1^n \tag{4.35}$$

in the collocation Eq. (4.28). In practice the integrals are evaluated numerically by applying a suitable quadrature scheme. Let

$$\int_D \psi(z) \mathrm{d}z = \sum_{k=1}^n \omega_k \psi(z_k) \tag{4.36}$$

be a quadrature rule with weights ω_k and *quadrature nodes* $\{z_k\}_1^n$ which may or may not coincide with the collocation nodes $\{x_i\}_1^n$. When the integrals are evaluated numerically, the collocation procedure is called *fully discrete*. Furthermore, the fact that the fully discrete collocation solution may differ from the solution f_n when the integrals are evaluated exactly, will be emphasized by writing $\tilde{f}_n$ for the former solution. The collocation equation (4.28), with the integral in Eq. (4.35) discretized according Eq. (4.36), becomes

$$\sum_{j=1}^n \tilde{f}_j \left\{ \phi_j(x_i) - \lambda \sum_{k=1}^n \omega_k K(x_i, y_k) \phi_j(y_k) \right\}. \tag{4.37}$$

At this point it is convenient to introduce a numerical operator A_n in order to describe the quadrature sum in Eq. (4.37) which then reduces to

$$\sum_{j=1}^n \tilde{f}_j \{ \phi_j(x_i) - \lambda (A_n \phi_j)(x_i) \}.$$

It is also assumed that the numerical operator is pointwise convergent, *i.e.*, $||A\phi - A_n \phi|| \to 0$ as $n \to \infty$, $\forall \phi \in X$ and that $P_n x \to x$, $n \to \infty$, $\forall x \in X$, $P_n : C(D) \to C(D)$. Since $\{A_n\}$ is pointwise convergent in $C(D)$ and $P_n f \to f$, $n \to \infty$, $\forall f \in C(D)$, it follows (from the *principle of uniform boundedness*) that

$$\sup ||P_n|| < \infty = C_P$$

and

$$\sup ||A_n|| < \infty \ = C_A.$$

Also define by C_n the norm $||(I - \lambda A_n P_n)^{-1}||_\infty$. In terms of the constants C_P, C_A and C_n, Atkinson (1997) shows that the error between the true and the approximate solution $\tilde{f}_n$ can be given by

$$||f - \tilde{f}_n||_\infty \le C_n[(1 + C_P C_A)||f - P_n f||_\infty + C_P||(A - A_n)f||_\infty].$$

4.6   The Standard Galerkin Method

The basic idea behind the Galerkin approximation and its importance in computational physics and mathematics were outlined in Section 4.1. In view of its wide popularity, a somewhat detailed discussion of the method here was deemed worth while, especially, in reference to the important finite-element method (FEM).[2] The primary focus is on the standard Galerkin technique (Section 4.1), the Petrov–Galerkin variation being discussed only briefly at the end. From Eq. (4.3), the Galerkin approximation consists in projecting the residual error $r_n(x)$ on the subspace X_n of X spanned by a basis set $\{\phi_i\}_1^n$, setting each component $< r_n, \phi_j >$ individually to zero and then solving the resultant linear system of equations. In terms of the basis functions ϕ_i, we have

$$< r_n, \phi_j > = \sum_{j=1}^{n} [f_j < A\phi_j, \phi_i > -g_j \phi_j], \ j = 1, 2, \cdots, n,$$

where $f_k = < f_n, \phi_k >$, $g_k = < g, \phi_k >$, and

$$A\phi_j = \int_D K(x, y)\phi_j(y)\mathrm{d}y.$$

Let us begin by considering Galerkin's approximation in one dimension, but in various function spaces. This will allow us to better appreciate the importance of FEM.

4.6.1   The Galerkin Approximation in one Dimension

Let us first consider $f \in L_2(D)$ and show that $P_n f \to f, \forall f \in L_2(D), n \ge N, N$ sufficiently large. Let $\{f_m\}$ be a set of continuous functions converging to f in $L_2(D)$. Recall that the continuous functions are dense in L_2. Then, noting that

---

[2]Most recently, the finite-element techniques are being increasingly applied to problems of atoms and molecules. A major application in this area is to solve the many-particle Schrödinger equation *via ab initio* Hartree–Fock or density functional calculations including the presence of intense electro-magnetic fields of lasers. See, for example, Schweizer *et al.*, 1999; Murakami *et al.*, 1992; Gusev *et al.*, 1992. Another highly interesting application of FEM is in the so-called across-the-length calculation (Tabbar, 2000; Abraham *et al.*, 1998).

$\|P_n\| = 1$, the projection being orthogonal, we have

$$
\begin{aligned}
\|f - P_n f\|_2 &\leq \|f - f_m\|_2 + \|f_m - P_n f_m\|_2 + \|P_n(f_m - f)\|_2 \\
&\leq 2\|f - f_m\|_2 + \|f_m - P_n f_m\|_2 \\
&\leq \frac{\epsilon}{2} + \|(I - P_n)f_m\|_2
\end{aligned}
\tag{4.38}
$$

for a particular value of m and $\epsilon > 0$. Moreover, since P_n is orthogonal, it follows that for any $x, z_n \in C(D)$,

$$
\begin{aligned}
\|x - P_n x\|_2 &\leq \min_{z_n \in X_n} \|x - z_n\|_2 \\
&\leq \sqrt{l}\|x - z_n\|_\infty,
\end{aligned}
\tag{4.39}
$$

where l is the length of the integration interval. Thus $\|f - P_n f\|_2 \to 0, \forall f \in C(D)$. Using this result in Eq. (4.37) yields

$$
\|f - P_n f\|_2 \leq \epsilon, \forall f \in L_2(D), n \geq N,
$$

N sufficiently large. Therefore, $P_n f \to f, \forall f \in L_2(D)$. From Eqs. (4.38) and (4.39) we have the error estimate

$$
\begin{aligned}
\|f - f_n\|_2 &\leq \text{const.} \min_{z_n \in X_n} \|f - z_n\|_2 \\
&\leq \text{const.} \|f - z_n\|_\infty, \ f \in L_2(D).
\end{aligned}
$$

Moreover, the following two results can also be derived

$$
\begin{aligned}
\|f - f_n\|_2 &\leq \text{const.} \ \omega(x; h), f \in C(D), \\
&\leq \text{const.} \ \frac{h^2}{8}\|f''\|_\infty, f \in C_2(D).
\end{aligned}
$$

Next consider the space $L_2(0, 2\pi)$ of complex-valued square-integrable functions periodic in the interval $(0, 2\pi)$. The functions $\phi_j = (2\pi)^{-1\backslash 2} \exp(ik_j x), j \in Z^+$, constitute a complete orthonormal basis in this space. Therefore,

$$
f(x) = \sum_{j=-\infty}^{\infty} < f, \phi_j > \phi_j, \ f \in L_2(0, 2\pi).
$$

From the theory of Fourier series in L_2, it is known (Folland, 1992) that the partial sums $S_n f$ converge to f in the mean-square sense as $n \to \infty$. Therefore, $P_n f \to f, \forall f \in L_2(0, 2\pi)$, as $n \to \infty$. In addition,

$$
\|f - P_n f\|_2 = \left[\sum_{|j|>n} | < f, \phi_j > |^2 \right]^{\frac{1}{2}},
\tag{4.40}
$$

from which it follows that $\|f - P_n f\|_2 \leq \|f\|_2$ by Parseval's theorem.

However, more informative error estimates can be derived in the Sobolev space $H^r(0, 2\pi), r \geq 0$, with norm

$$\|\psi\|_r = \Big[\sum_{j=-\infty}^{\infty} (1+j^2)^r |< \psi, \phi_j >|^2 \Big]^{\frac{1}{2}}, \ \psi \in L_2(0, 2\pi)$$

and inner product

$$
\begin{aligned}
< \psi_1, \psi_2 >_r \ &= \ < \psi_1, \phi_0 > + < \psi_2, \phi_0 > \\
&= \ \sum_{j=-\infty, j\neq 0}^{\infty} (1+j^2)^r |< \psi_1, \phi_j >< \psi_2, \phi_j >^*,
\end{aligned}
$$

where the superscript * denotes complex conjugation. Error estimates in these spaces make clear the role of smoothness of the functions in their finite-dimensional approximations.

Let $s \leq r$. Then from Eq. (4.40)

$$
\begin{aligned}
\|f - P_n f\|_s^2 \ &\leq \ \sum_{|j|\geq n} (1+j^2)^s |< f, \phi_j >|^2 \\
&= \ \sum_{|j|\geq n} (1+j^2)^s |f_j|^2 \\
&\leq \ \sum_{|j|\geq n} (1+j^2)^{s-r} \Big[(1+j^2)^r |f_j|^2 \Big] \\
&\leq \ (1+n^2)^{s-r} \|f\|_r^2 \\
&\leq \ n^{2(s-r)} \|f\|_r^2 .
\end{aligned}
$$

If $s = 0$, then we recover the L_2-norm and

$$\|f - P_n f\|_2 \leq n^{-r} \|f\|_r .$$

It is clear that as the smoothness (i.e., r) increases, $P_n f \to f$ faster as $n \to \infty$. Since $P_n f \to f, \forall f \in L_2(0, 2\pi)$, then $\|A - P_n A\| \to 0, n \to \infty$, Theorem 4.3 applies and we have the error estimate

$$\|f - P_n f\|_{L_2(0,2\pi)} \leq \text{const. } n^{-r} \|f\|_r, n \geq 0, f \in H^r(0, 2\pi).$$

Next we consider $f \in C_p(0, 2\pi)$, the space of continuous functions which are periodic in $(0, 2\pi)$. Some problems arise in this case. But before proceeding, let us state a general theorem in function approximation (Powell, 1997).

Theorem 4.6 *Let X be a normed linear space and X_n a finite-dimensional subspace of X. Let $P_n : X \to X_n$ be a linear projection operator. For any $f \in X$, let $E_n(f)$ define the minimum distance of f from the subspace X_n. That is,*

$$E_n(f) = \min_{g_n \in X_n} \|f - g_n\|_X .$$

Then the error of the approximation is bounded by

$$\|f - P_n f\|_X \leq [1 + \|P_n\|] E_n(f).$$

We encountered similar results on various occasions. The important point is that the norm of the projection operator $||P_n||$ enters the equation.[3]

Now let $f \in C_p(0, 2\pi)$ be approximated by trigonometric polynomials of degree $\leq n$. Then

$$S_n f(x) = \frac{f_0}{2} + \sum_{j=1}^{n} [f_{cj} \cos(jx) + f_{sj} \sin(jx)]. \tag{4.41}$$

The Fourier coefficients are given by

$$f_{cj} = (2\pi)^{-1} < \cos(jx), f >$$

and

$$f_{sj} = (2\pi)^{-1} < \sin(jx), f > .$$

Inserting these coefficients into the Fourier series (4.41) and after some straightforward reduction, we obtain

$$S_n f(x) = (2\pi)^{-1} \int_0^{2\pi} \frac{\sin(n + \frac{1}{2})\theta}{2 \sin \frac{1}{2}\theta} f(x + \theta) d\theta.$$

A further calculation shows that

$$||S_n|| > \text{const.} \ln(n + 1). \tag{4.42}$$

Equation (4.42) shows that $||S_n||$ is unbounded with n and the fact that $P_n f \to f, n \to \infty$, fails to hold $\forall f \in C_p(0, 2\pi)$. However, being $\sim O(\ln n)$, the growth is not catastrophic, especially, in view of the fact that in an actual implementation, n is not expected to be large.

Fortunately, only mild restrictions are needed for convergence and these can be obtained from *Jackson's theorems* (Meinardus, 1967; Rivlin, 1969; Powell, 1997). The results of these theorems are summarized below.

Jackson's II-Theorem: $E_n(f) \leq \pi L[2(n+1)]^{-1}, f \in C_p(0, 2\pi)$ is Lipschitz continuous with L as the Lipschitz constant: $|f(x) - f(y)| \leq L|x - y|$.

Jackson's III-Theorem: $E_n(f) \leq \frac{3}{2}\omega(\frac{\pi}{2n+1}), f \in C_p(0, 2\pi)$, and ω is the modulus of continuity.

Jackson's IV-Theorem: $E_n(f) \leq \left[\pi[2(n+1)]^{-1}\right]^k ||f^{(k)}||_\infty, f \in C_p^{(k)}(0, 2\pi)$, the space of k-times differentiable 2π periodic functions.

In view of Theorem 4.6, the error estimate can be written as

$$||f - P_n f||_\infty \leq (1 + ||P_n||) E_n(f).$$

Upon using the above theorems of Jackson, the following estimates are obtained.

$$||f - P_n f||_\infty \sim O \left| (\ln n)\omega \left(\frac{\pi}{2n + 1} \right) \right|, \quad f \in C_p(0, 2\pi).$$

---

[3]The norm $||P_n||$ is sometimes called the *Lebesgue constant* (see Gautschi, 1997; Baker, 1977).

$$\|f - P_n f\|_\infty \sim O\left|(\ln n)\pi L[2(n+1)]^{-1}\right|, \; f \in C_p(0, 2\pi)$$

is Lipschitz continuous with the Lipschitz constant L. If f is Hölder continuous with the Hölder exponent $\alpha, 0 \le \alpha < 1$, then the error estimate is

$$\|f - P_n f\|_\infty \sim O\left|(\ln n)n^{-\alpha}\right|.$$

Finally,

$$\|f - P_n f\|_\infty \sim O\left|(\ln n)\pi[2(n+1)]^{-1}\right|^k \|f^{(k)}\|_\infty, f \in C_p^{(k)}(0, 2\pi).$$

It is seen that $P_n f \to f, n \to \infty$, in all the above cases. In the first case where $f \in C_p(0, 2\pi)$, the convergence result is

$$\lim_{n\to\infty}\left|(\ln n)\omega\left(\frac{\pi}{2n+1}\right)\right| \to 0.$$

Fortunately, this is true for almost all continuous functions (see Powell, 1997). In the case of the Hölder continuous functions, it is clear that the rate of convergence increases with the Hölder exponent α, i.e., with smoothness. The same is also true for functions in $C_p^{(k)}(0, 2\pi)$.

The above discussion is closely related to a result of Faber (Hämmerlin and Hoffmann, 1991; Kress, 1998) which is stated in the theorem below.

Theorem 4.7 (Faber) *Let $f \in C(a, b)$ and P_k the space of the polynomials of degree $\le k$. Then given any sequence of interpolation nodes $\{x_i\}_1^k, \exists$ a function f such that $P_n f \in P_k$ does not converge to f uniformly on $C(a, b)$.*

Faber's theorem draws attention to a drawback of the spaces of algebraic or trigonometric polynomials as the subspaces for approximation. For continuous functions, the convergence of the approximation cannot be guaranteed as $n \to \infty$.[4]

Three points are to be noted in the analysis so far. These are: (1) The problems were all in one space dimension; (2) the approximations were global, i.e., the functions were all approximated over the entire domain of their definitions; and (3) The basis functions spanning the subspace X_n were orthonormal. The third point leads to a significant simplification since the linear system, Eq. (4.8), then becomes diagonal. Moreover, as was already pointed out in Section 4.1, orthogonal or near orthogonal bases are desirable from the standpoint of stability. However, although orthonormal bases for spaces such as $C_p(0, 2\pi)$ or L_2 are well-known, at least for functions on the real line or in two- and three-dimensional regions of simple shapes (e.g., the spherical harmonics for spherical geometries), finding such bases for arbitrary function spaces and domains is an entirely nontrivial task. In addition, it must be shown that as the dimension of the subspace increases toward infinity, the original infinite-dimensional space is recovered (in the sense that $\|f - P_n f\| \to 0$). In order to deal with this situation, it is necessary to consider the Galerkin method in a more general form. It is in this context that the Galerkin method is intimately associated with FEM.

---

[4]There is also a theorem due to Marcinkiewicz which seems to directly contradict Faber's result. For more on these two theorems and an explanation of the apparent contradiction between the two, see Hämmerlin and Hoffmann (1991).

4.6.2   The General Case

Let ℓ be a *bounded linear functional* on a complex Hilbert space X. Recall that a linear functional on X is an operator $X \to \mathcal{K}$, $\mathcal{K}$ being either a real or a complex field. Now the functional $\ell \in \mathcal{L}(X, \mathcal{K})$. $\mathcal{L}(X, \mathcal{K})$ is a normed linear space which is termed the *dual* space of X and is usually designated by X' (also by X^*). For a Hilbert space, it turns out that there exists a unique element $\eta \in X$ such that if $\xi \in X$ is any element of X, then $\ell[\xi] = <\xi, \eta>_X$. This is the celebrated *Riesz representation theorem*, which we state below.

Theorem 4.8 (The Riesz Representation Theorem) *Let X be a Hilbert space and ℓ be a bounded linear functional on X. Then there exists a unique $\eta \in X$ such that $\ell[\xi] = <\xi, \eta>_X$, $\forall \eta \in X$. Moreover, $||\ell[\xi]|| = ||\eta||_X$.*

In order to see the boundedness of ℓ, simply note that $\ell[\eta] = <\eta, \eta>_X = ||\eta||_X^2$. Then from the definition of the operator norm it follows at once that $||\ell|| = ||\eta||_X$. For a Hilbert space, X and X' are essentially the same space. In the parlance of functional analysis, these two spaces are said to be *isometrically isomorphic* to each other.

The functional ℓ maps a single element $\xi \to \mathcal{K}$. A functional can also be defined so as to map a pair of elements, say, $(u, v) \in X \times X \to \mathcal{K}$. Writing a for ℓ, the linear operator $a(u, v) \in X \times X \to \mathcal{K}$ is called a *sesquilinear form* defined on the product space $X \times X$. This means that $\forall \alpha, \beta \in C$ and $\forall u, v \in X$,

i. $a(u, v) = b \in C, u, v \in X$: boundedness.

ii. $a(\alpha u_1 + \beta u_2, v) = \alpha a(u_1, v) + \beta a(u_2, v)$: 4 linearity.

iii. $a(u, \alpha v_1 + \beta v_2) = \overline{\alpha} a(u, v_1) + \overline{\beta} a(u, v_2)$: conjugate linearity.

iv. $a(u, v) = \overline{a(v, u)}$.

v. $|a(u, v)| \le C_a ||u|| ||v||, C_a > 0$ constant: continuity, and finally,

vi. $|a(u, u)| > C_e ||u||_X^2, C_e > 0$: coercivity (strong ellipticity).

The overlines in (ii) and (iv) indicate complex conjugation. If the field is over the reals instead of C, then $a(u, v)$ is called *bilinear* and condition (iv) reduces to the condition of symmetry. The final condition (vi) of coercivity is analogous to positive definiteness of matrix operators.

It is known from functional analysis (see Debnath and Mikusinski, 1998) that there exists a unique operator $Q : X \to X$, such that $a(u, v) = <u, Qv>_X, \forall u \in X$. If v solves the equation $Qv = w$, then we have the identity $a(u, v) = l[u]$. In terms of the problem under discussion, namely, $Af = g$, this means that

$$a(u, f) = l[u], \quad l[u] = <u, g>_X . \tag{4.43}$$

In an infinite-dimensional function space, Galerkin's problem can be posed as: *find a solution $u \in X$ so that $a(u, v) = \ell[v], \forall v \in X$.* Note that Eq. (4.43) is really the Galerkin equation $<Af_n, z_n> = <g, z_n>$. That there exists a unique solution of the variational problem (4.43) follows from the well-known *Lax–Milgram theorem* (see references on FEM later) given below.

Theorem 4.9 (Lax–Milgram) *Let a be a bounded, coercive, bilinear form on a Hilbert space X. Then for every bounded linear functional l on X, there exists a unique solution $f \in X$ such that $a(u, f) = l[u], \forall u \in X$ satisfying the bound*

$$\|f\| \le \frac{1}{C_e}\|l\|.$$

The theorem does not require property iv. This is, however, crucial in an alternate *variational* formulation which runs thus:

find $u \in X$ such that $J(u) \le J(v)$, $\forall v \in X : J(v) = \frac{1}{2}a(v,v) - \ell[v]$.

Property iv is necessary in establishing the first part of the statement $J(u) \le J(v)$.

Thus far everything was in the continuous regime. Let us now return to the discrete level and consider the Galerkin solution $f_n \in X_n$ of $f \in X$. The discrete Galerkin method consists in solving the variational problem $a(u_n, v_n) = \ell[u_n]$, $\forall u_n \in X_n$. The corresponding linear system is $\mathbf{M f_n} = \mathbf{g_n}$, where $\mathbf{M}_{ij} = a(\phi_i, \phi_j)$. Let us show that the system has a unique solution. If otherwise, then $\exists v_n = \sum v_j \phi_j$ such that $\mathbf{M f_n} = 0$ implying that $v_n = 0$ by the coercivity of a. In other words, the matrix $\mathbf{M}$ is positive-definite, besides being symmetric. This not only shows that the infinite-dimensional system can be discretized almost trivially, but the uniqueness property of the former is transferable to the finite-dimensional system as well.

Let us first note that the error in the solution $e = f - f_n$ is orthogonal to X_n. Indeed from the Galerkin equation and Eq. (4.43), we have

$$< z_n, A f_n >= a(z_n, f_n) =< z_n, g > . \tag{4.44}$$

Moreover,

$$< z_n, A f >= a(z_n, f) =< z_n, g > . \tag{4.45}$$

Subtracting Eqs. (4.44) and (4.45) from one another, the result follows. Therefore, $a(e, v_n) = 0$, $\forall v_n \in X_n$. Thus

$$
\begin{aligned}
a(f - f_n, f - f_n) &= a(f - f_n, f - z_n - f_n + z_n) \\
&= a(f - f_n, f - z_n) - a(e, f_n - z_n) = a(f - f_n, f - z_n).
\end{aligned}
$$

From conditions (v) and (vi) above, we obtain

$$C_e\|f - f_n\|_X^2 < a(f - f_n, f - z_n) \le C_a\|f - f_n\|_X\|f - z_n\|_X,$$

from which it follows at once that

$$\|f - f_n\|_X \le \frac{C_a}{C_e}\|f - z_n\|_X, \forall z_n \in X_n.$$

These results constitute what is called *Cea's lemma* which is introduced below.

Cea's Lemma *Let a be a bilinear form satisfying conditions (i)–(vi), and l a bounded linear functional on X. Then the variational problem $a(u, f) =$*

$l[u], u, f \in X$ has a unique finite-dimensional solution $f_n \in X_n$ which satisfies the bound

$$\|f_n\| \leq \frac{1}{C_e}\|l\|.$$

Moreover, the approximation error is given by

$$\|f - f_n\|_X \leq \frac{C_a}{C_e} \inf_{z_n \in X_n} \|f - z_n\|_X. \tag{4.46}$$

If X_n is a closed subspace, then the infimum in Eq. (4.46) can be replaced by the minimum.

Consequently, for the Galerkin approximation $f_n \in X_n$ to converge to the solution f of the original problem $Af = g, f, g \in X$, it is sufficient that there exists a family of subspaces $\{X_n\}$ such that

$$\lim_{n \to \infty} \inf_{z_n \in X_n} \|f - z_n\|_X \to 0.$$

The uniqueness result in Cea's Lemma also follows from the fact that the matrix $\mathbf{M}$ of Eq. (4.8) is positive-definite in the Galerkin case. The linear system, Eq. (4.8), is, therefore, uniquely solvable. Now, $\forall z_n \in X_n$,

$$\begin{aligned} a(f - z_n, f - z_n) &= a(f - f_n + f_n - z_n, f - f_n + f_n - z_n) \\ &= a(e + w, e + w) = a(e, e) + a(w, w) + 2a(e, w) \\ &= a(e, e) + a(w, w) \geq a(e, e). \end{aligned}$$

Therefore,

$$\|e\|_a \leq \|f - z_n\|_a \leq \inf_{z_n \in X_n} \|f - z_n\|_a.$$

Since $\exists f_n$ for which the infimum is attained, the infimum can be replaced by the minimum and we obtain

$$\|e\|_a = \min_{z_n \in X_n} \|f - z_n\|_a.$$

This shows that when measured in the so-called *energy norm* $\|\cdot\|_a = \{a(v, v)\}^{\frac{1}{2}}$, the Galerkin solution is the *best* among all solutions of the form

$$v_n = \sum_{j=1}^{n} v_j \phi_j.$$

In X-norm, Cea's lemma tells us that the Galerkin solution is quasi-optimal, being proportional to $\inf_{z_n \in X_n} \|f - z_n\|_a$. Furthermore, the bound given in Eq. (4.46) can be sharpened by using the fact that the norms $\|\cdot\|_X$ and $\|\cdot\|_a$ are *equivalent*. This means that

$$C_1\|\cdot\|_X \leq \|\cdot\|_a \leq C_2\|\cdot\|_X,$$

C_1 and C_2 being constants. Using properties v and vi of $a(u, v)$ above, it follows that $C_1 = \sqrt{C_e}$ and $C_2 = \sqrt{C_a}$. The estimate (4.46) then becomes

$$\|f - f_n\|_X \leq \left\{ \frac{C_a}{C_e} \right\}^{\frac{1}{2}} \inf_{z_n \in X_n} \|f - z_n\|_X. \tag{4.47}$$

Equation (4.47) gives the best quasi-optimal error estimate for the Galerkin solution in the X-norm. The bound (4.47) is important since $C_a > C_e$.

4.6.3   The Galerkin Method and FEM

On p. 118 we discussed some of the problems involved in approximating a function in an arbitrary space and domain. Of these, the most important difficulty was to find a suitable basis set for use in Galerkin's solution for an arbitrarily shaped domain Ω. Moreover, the basis set must also be dense in X for reasons of convergence. This is reflected in the requirement that $X_n \rightarrow X$ as $n \rightarrow \infty$. Alternately, the index n can be replaced by a parameter $h \in (0, 1)$ in which case the equivalent statement is: $X_h \rightarrow X$ as $h \rightarrow 0$. If $h = n^{-1}$, then, of course, cannot take any value but the reciprocals of n. In order to alleviate the problem of the basis set, Ω is decomposed into subdomains ω_j so that $\Omega = \cup_{j=1}^{N} \omega_j$, and the function is approximated in each subdomain ω_j by a piecewise polynomial less than or equal to some degree k. Each polynomial is then nonzero only over a relatively small part of Ω, being zero everywhere else. Moreover, the denseness of X_h is obtained in some sense in this technique (see p. 123). All results are first obtained locally, that is, for the individual elements ω_j. These are then summed to yield the global approximation over the entire Ω. The Galerkin approximation with piecewise polynomial bases is synonymous with the finite-element approximation (Babuska and Aziz, 1972; Strang and Fix, 1973; Oden and Carey, 1982; Wait and Mitchell, 1985; Dautray and Lions, 1990; Qarteroni and Valli, 1994; Reddy, 1998; Langtangen, 1999). The subdomains ω_j are called the *finite element meshes* which must be *admissible* in a certain differential geometric sense.[5] Furthermore, interelemental continuity must be assumed as the minimum requirement for the existence of the nodes.

Now $\inf_{z_n \in X_n} \|f - z_n\|_X \leq \|f - \tilde{z}_n\|_X$ for any $\tilde{z}_n \in X_n$. Let $\tilde{z}_n$ be an *interpolate* of f in X_n. Then from Cea's lemma $\|f - f_n\|_X \leq \text{const.}\|f - \tilde{z}_n\|_X$. The Galerkin error $\|f - f_n\|_X$ can then be estimated via the estimation of the interpolation error $\|f - \tilde{z}_n\|_X$. The problem thus reduces to that of minimizing the interpolation error by which the Galerkin error is bounded above. This error is again first determined locally over each individual mesh and then summed to yield the error in the global approximation. Usually, the error estimates are given in some Sobolev space norm (and for good reason). It is entirely possible that the Galerkin method can lead to a null solution in the interior of the elements as, for example, if $\Delta u = \lambda u$ is approximated by linear functions. The

---

[5]For example, if the region is polygonal and the meshes triangular, then the triangulation is admissible if the sides of a triangle are either parts of the boundary or sides of another element.

gradients ∇u across the elements are, however, nonvanishing and are actually delta functions which lead to the equations $< \nabla u, \nabla \phi_i > = \lambda < u, \phi_i >$. It, therefore, makes sense to use a Sobolev space norm. A detailed error analysis is outside the scope of this book. However, for the sake of completeness of the discussion, we summarize the results into the following two theorems. The contents of the theorems are derived on the basis of two well-known lemmas, namely, those due to *Bramble and Hilbert* and *Deny and Lions* (see Qarteroni and Valli, 1994).

Let us introduce the following notations. Let $X_h \subset X$ be the finite-dimensional approximating subspace and $P_h^k : X \to X_h$ be the operator of projection from X to X_h. The superscript k indicates that this global projection operator uses polynomials of degree $\leq k$, $k \geq 1$. A local projection operator is denoted by $^j P_h^k$, the subscript j indicating that the projection is restricted to the j-th element. Also $^j P_h^k = P_h^k|_{\omega_j}$, that is, the local projection operator $^j P_h^k$ is the restriction of the global projection operator P_h^k to the j-th element ω_j. It is also necessary to characterize the mesh size. Locally, the mesh size h_j is the *diameter* of ω_j. That is,

$$h_j = \max |x - y|, \ \forall x, y \in \omega_j.$$

Globally, the mesh size is $h = \max h_j$, $\forall \omega_j \subset \overline{\Omega}$. The parameter h essentially determines the size of the subspace X_h. In other words, dim $X_h = h$. In terms of h, the convergence can be defined as $||u - P_h^k u|| \to 0$, as $h \to 0$. This is also the sense in which $X_h \to X$ as $h \to 0$.

The local interpolation error is given by the following theorem.

Theorem 4.10 (Local Interpolation Error) *Let* $0 \leq m \leq k + 1, k \geq 1$. *Then* $\exists$ *a constant* $C = C(\omega_j, {}^j P_h^k, k, m)$ *such that*

$$||f - {}^j P_h^k f|_m \leq Ch_j^{k+1-m}|f|_{k+1}, \forall f \in H^{(k+1)}(\Omega_j).$$

The quantity $|f|_{k+1}$ is the Sobolev space *seminorm* given by $||\nabla f||_{L_2(\Omega)}$. Recall that a seminorm, unlike a norm, can be zero even when the element (of which it is the seminorm) is nonzero (for example, if $u = 1$). Globally, a function u, defined over Ω, is represented in terms of the FEM basis functions as

$$(P_h^k u)(x) = \sum_{i=1}^{N_n} u(x_i)\phi_i(x) = \sum_1^N u(x_i)\phi_i(x). \tag{4.48}$$

In Eq. (4.48), N_n is the total number of nodes and N the total number of elements. Note that the number of basis functions at a particular node equals the number of meshes connected to that node. Furthermore, the global projection

$$(P_h^k u)(x) = \sum_{\text{all elements } j} ({}^j P_h^k u)(x),$$

which is the sum of the local projections. The global problem is, therefore, to estimate the error $u - P_h^k u$. This is given by the following theorem.

Theorem 4.11 (Global or the Galerkin Error Estimate) *Let $\{\Omega_s\}$ be a regular family of triangulation of a polygonal domain Ω. Let $m = 0, 1$, and $k \geq 1$. The $\exists$ a constant C, independent of h, such that*

$$|f - P_h^k f|_m \leq Ch^{l+1-m}|f|_{l+1}, \forall f \in H^{(l+1)}(\Omega), 1 \leq l \leq k.$$

Before leaving this section, let us point out a heuristic interpretation of the error estimates. Note the occurrence of the factor $(k+1)$ in these estimates. This is a reflection of the following. Suppose that a function can be exactly interpolated by polynomials of degree k. That is, the function and its first k-derivatives are exactly matched by the polynomials. The interpolation error, therefore, must be of the order of its $(k+1)$-th derivative as in the estimates above.

4.6.4   The Perturbed Data

Let us next consider the Galerkin solution for the perturbed data g^ϵ with error level ϵ. As usual, f_n, f_n^ϵ are the solutions corresponding to the actual and the perturbed data, respectively. The error in the solution is then given by

$$
\begin{aligned}
\|f - f_n^\epsilon\| &\leq \|f - f_n\| + \|f_n - f_n^\epsilon\| \tag{4.49}\\
&= \|R_n(g - g^\epsilon)\| + \|f - f_n\| \leq \|R_n\|\epsilon + \|R_n A f - f_n\|.
\end{aligned}
$$

Just as in the collocation method, we introduce the operator Φ and use the linear system (4.8) in order to obtain a similar error estimate for the Galerkin solution which turns out to be

$$\|f - f_n^\epsilon\| \leq \frac{\|\Phi\|}{\lambda_{\min}}\epsilon + \|R_n A f - f_n\|,$$

where $\|\Phi\|$ and $\lambda_{\min}$ are as defined in the collocation case.

Let the operator A be coercive so that Cea's lemma holds good. Then the results of Cea's lemma can be used to obtain

$$\|f - f_n\| \leq \frac{C_a}{C_e} \inf_{z_n \in X_n} \|f - z_n\|. \tag{4.50}$$

Furthermore,

$$R_n g = \sum_{j=1}^{n} f_j \phi_j.$$

Therefore,

$$\|R_n\| = \sup\|f_n\|, \quad \|g\| = 1.$$

From the Lax–Milgram Theorem 4.8

$$\|f_n - f_n^\epsilon\| \leq \frac{1}{C_e}\|\ell\| = \frac{1}{C_e}\|g - g^\epsilon\|. \tag{4.51}$$

Combining Eqs. (4.50) and (4.51) into Eq. (4.49) yields the error estimate

$$\|f - f_n^\epsilon\| \leq \frac{1}{C_e}\left[\epsilon + C_a \inf_{z_n \in X_n} \|f - z_n\|\right].$$

The various error estimates in Galerkin's solution are summarized in the theorem below.

Theorem 4.12 *Let the Galerkin solution of $a(f, v) = \ell[v], \forall f, v \in X$, be given by the solution of the approximate, finite-dimensional problem $a(f_n, v_n) = \ell[v_n]$, $\forall f_n, v_n \in X_n$. Then among all solutions $v_n \in X_n$, the Galerkin solution f_n is the best approximation when measured in the energy norm $\| \cdot \|_a$ and*

$$\|f - f_n\|_a \leq \|f - v_n\|_a, \ \forall v_n \in X_n.$$

It is also quasi-optimal in the X-norm and the error is bounded by

$$\|f - f_n\|_X \leq \left\{\frac{C_a}{C_e}\right\}^{\frac{1}{2}} \inf_{z_n \in X_n} \|f - z_n\|_X.$$

If the data is perturbed and $|g - g^\epsilon| \leq \epsilon > 0$, then the following estimate applies in general, namely

$$\|f - f_n^\epsilon\| \leq \frac{\|\Phi\|}{\lambda_{\min}}\epsilon + \|R_n A f - f_n\|,$$

where $\|\Phi\|$ and $\lambda_{\min}$ are as defined above. If, in addition, the operator A is coercive or strongly elliptic, then the error bound is given by

$$\|f - f_n^\epsilon\| \leq \frac{1}{C_e}\left[\epsilon + C_a \inf_{z_n \in X_n} \|f - z_n\|\right].$$

4.6.5 The Petrov–Galerkin Method

Finally, before concluding this chapter, let us briefly mention the Petrov–Galerkin variant of the standard Galerkin method. For more details, the reader is referred to Quarteroni and Valli (1994), Kress (1989), Johnson (1987) and Langtangen (1999). As discussed in Section 4.1, in Petrov–Galerkin variation, the residual error r_n and the approximate solution f_n are projected on two different subspaces spanned by two different basis sets ϕ_i and ψ_j. The Lax–Milgram Theorem 4.8 needs to be modified since the subspaces for the solution and the test function are different from each other. There exists such a modification due to Necas (see Qarteroni and Valli, 1994). The bilinear form is now $a(u, v) :\ X \times Y \rightarrow \mathcal{K}$, $u \in X$, $v \in Y$. The bounded condition is: $\|a(u, v)\| \leq$ const. $\|u\|_X \|v\|_Y$ and the coercivity takes the form $\sup_{u \in X, u \neq 0} a(u, v) \geq C_e \|u\|_X \|v\|_Y$, $\forall y \in Y$. The variational problem is to *find* $u \in X$ such that $a(u, v) = \ell[v]$, $\forall y \in Y$. There exists a unique solution which is bounded by $\|u\|_X \leq C_e^{-1}\ell[v]$.

The use of two different basis sets lends more flexibility to the problem. There are situations where the Petrov–Galerkin method results in approximations which are superior in quality to the standard Galerkin solution. This is especially the case for *reactive-diffusive* and *convective-diffusive* problems. For example, in convective-diffusive problems, it is found (Reddy, 1998) that when the convection term dominates, the standard Galerkin solution becomes oscillatory bearing little relation to the actual solution. But the Petrov–Galerkin method results in approximations which are of superior quality. It has also been applied to beam problems in elasticity (Loula *et al.*, 1987).

Chapter 5

Discrete Ill-posed Problems

5.1 Introduction

In Chapter 3 we formulated the theory of regularization in infinite-dimensional function spaces which was subsequently reduced to finite-dimensions in Chapter 4. The results obtained in these two chapters were meant to hold pointwise. Such analyses assume that the data are known everywhere in some domain of measurement and the solutions over the entire domain of its definition. In a scattering problem, for example, the scattered fields are assumed to be known over the entire (or a patch) of the unit sphere continuously, without interruptions. Similarly, the scattering solutions are also determined continuously, without interruptions, over the boundary and/or the domain of the scatterer and, therefore, could be considered as elements of a suitable function space such as C^k or H^m. Analogously, if the problem involves a Fourier transform, then it is assumed that the spectrum is known for all frequencies in a given interval continuously, without interruptions.

In practice, data cannot be collected continuously, without interruptions, and neither can the solution be obtained at each and every point of its domain of definition. Except perhaps in very specialized circumstances, the data must *perforce* exist only at a finite number of discrete points represented by a finite number of sensors. Similarly, the solution can only be given discretely. One must, therefore, deal with a situation where both data and solution are elements in Euclidean spaces. In other words, the finite-dimensional function space X_n of Chapter 4 is to be replaced by a finite-dimensional Euclidean space R^n. We describe the corresponding regularization as *fully discrete*. It is also necessary to clarify the relation between a continuous, function space and a discrete, Euclidean space solution of the same problem. An example is the celebrated *Nyström method* (see Appendix A.5.2 at the end of this chapter) of solving an integral equation of the second kind. In Nyström's method, the equation is essentially solved in a Euclidean space, whereas the mathematical analysis is carried out within the framework of functional analysis (for a clear exposition of this point, see Linz, 1978).

In this chapter, we present the essentials of regularization theory for fully discrete (in the sense just described) ill-posed problems. The subject matter is substantially complex and full of subtleties. In addition, because of its practical importance, numerous algorithmic considerations abound. Clearly, it cannot be treated in any depth in the span of a single short chapter and we must perforce be content with the essentials only. We begin with discrete decompositions.

5.2   Discrete Decompositions

Let $\{g_n\}_1^N$ be a data set (also denoted by $\mathbf{g}$). In general

$$g_n = g(x_n) = \int K(x_n, y) f(y) \mathrm{d}y,$$

where K is a smoothing kernel. $\{g_n\}$ is a set of N continuous linear functionals $\{\ell_n(f)\}$ of f in a Hilbert space. By the Riesz representation theorem, $g_n(x) = < f, \phi_n >_X, n = 1, 2, \cdots, N, \phi_n$ being a function in a Hilbert space. In the absence of errors, the linear inverse problem for the discrete data $\mathbf{g}$ means finding a function f in an object space such that $g_n(x) = < f, \phi_n >_X, n = 1, 2, \cdots, N$. If the set $\{\phi_n\}$ is linearly independent, we have the following semi-discrete finite-dimensional version of the continuous problem

$$f(x) = \sum_{i=1}^{N} \sum_{j=1}^{N} (\mathbf{G})_{ij} \phi_j(x) g_i = \sum_{i=1}^{N} g_i \phi_i^*(x), \qquad (5.1)$$

where $\mathbf{G}$ is the *Gram matrix* with matrix elements $(\mathbf{G})_{ij} = < \phi_i, \phi_j >_X$, and

$$\phi_i^*(x) = \sum_{j=1}^{N} (\mathbf{G})_{ij} \phi_j(x)$$

is the *dual basis*. For linearly independent vectors, the Gram determinant is always positive and is zero otherwise (Akhiezer and Glatzman, 1961). It is clear that the solution f given by Eq. (5.1) depends continuously on the data. This follows from the fact that an operator in a finite-dimensional space (i.e., a matrix) is continuous (Naylor and Sell, 1982).

It would, therefore, seem that any discrete inverse problem is well-posed. Indeed if we truncate the SVE of Eq. (3.9), Picard's condition, Eq. (3.12), is satisfied and no regularization is necessary. From a purely mathematical standpoint, some regularizing effect is always present when the dimension of the space is reduced as in the projection methods of the previous chapter. However, a discrete formulation is only a finite-dimensional version of an otherwise ill-posed problem and the ill-posedness must somehow be reflected in the finite-dimensional version also. In other words, if some information about the solution is missing to begin with, discretization cannot make it reappear. The fact of the matter is that ill-posedness creeps in through the ill-conditioning (see Appendix A.5.1 for definition) of the matrix operator of the associated linear system (the Gram matrix $\mathbf{G}$ in Eq. (5.1) or $\mathbf{M}$ in Eq. (4.8)) instead of the nonclosedness of the operator's range as in the continuous case since the range of a matrix operator is necessarily closed, the dimension of the space being finite. Moreover, as the dimension N of the data vector $\mathbf{g}$ is increased, the ill-conditioning should also increase. It, therefore, stands to reason that we use the term *ill-posed* when referring to a continuous problem while reserving the term *ill-conditioned* for a discrete situation.

Let $\mathbf{f}$ and $\mathbf{g}$ be finite-dimensional vectors in a Euclidean space and consider the SV-expansion of Eq. (3.7). This expansion can be considered to represent the action of an *impulse response function*, say, I_A, on f. The representation of I_A is *via* the sum over j in that equation and its action on f takes the form of an inner product. A discrete analog of the continuous I_A can be obtained by truncating the infinite sum in Eq. (3.7) at, say, $j = n$, and then considering the resulting finite sum as a matrix product. The continuous impulse response function I_A now becomes a matrix operator $\mathbf{A}$ which can be expressed as

$$\mathbf{A} = \mathbf{U\Sigma V^T} = \sum_{j=1}^{n} \mathbf{u}_j \sigma_j \mathbf{v}_j^T, \qquad (5.2)$$

where $\mathbf{A}$ is a real $m \times n$, $m \geq n$, matrix, and $\mathbf{U}, \mathbf{V}, \Sigma$ are $m \times n$, $n \times n$ and $n \times n$ matrices, respectively. $\mathbf{u}_j$ is a $m \times 1$ column matrix and $\mathbf{v}_j$ a $n \times 1$ row matrix. The superscript T in definition (5.2) indicates transpose operation which replaces the adjoint symbol $*$ for a continuous operator.[1] This is a matrix version of the diagonalizability of any linear mapping between two spaces X and Y of different dimensions, i.e., $A : X \to Y$, and the matrix of the mapping is rectangular of dimension $m \times n$.

Written explicitly,

$$\mathbf{U} = \begin{pmatrix} u_{11} & \cdots & u_{1n} \\ \cdot & & \cdot \\ \cdot & & \cdot \\ u_{m1} & \cdots & u_{mn} \end{pmatrix}.$$

$$\mathbf{V} = \begin{pmatrix} v_{11} & \cdots & v_{1n} \\ \cdot & & \cdot \\ \cdot & & \cdot \\ v_{n1} & \cdots & v_{nn} \end{pmatrix}$$

---

[1] Note that we should really write A^H instead of A^T in order to be more general so that complex matrices can be allowed. In other words, A^H could stand either for the frequently used notation A^T for a real matrix or $A^\dagger$ for the complex conjugate transpose if the matrix is complex. However, we will continue to write A^T for all these cases inspite of its nonconformity.

and

$$\Sigma = \begin{pmatrix} \sigma_1 & \cdots & 0 \\ & \cdot & \\ & \cdot & \\ 0 & \cdots & \sigma_n \end{pmatrix}.$$

This is, however, not the only way to write Eq. (5.2). One could as well assume that $\mathbf{U}$ is $m \times m$, $\mathbf{V}$ is $n \times n$ and Σ is $n \times m$, in which case, we would have

$$\mathbf{U} = \begin{pmatrix} u_{11} & \cdots & u_{1m} \\ \cdot & & \cdot \\ \cdot & & \cdot \\ u_{m1} & \cdots & u_{mm}u_n \end{pmatrix}$$

$$\mathbf{V} = \begin{pmatrix} v_{11} & \cdots & v_{1n} \\ \cdot & & \cdot \\ \cdot & & \cdot \\ v_{n1} & \cdots & v_{nn} \end{pmatrix}$$

and

$$\Sigma = \begin{pmatrix} \sigma_1 & \cdots & 0 \\ \cdot & & \cdot \\ \cdot & & \cdot \\ \cdot & & \cdot \\ 0 & \cdots & \sigma_n \\ 0 & \cdots & 0 \\ \cdot & & \cdot \\ \cdot & & \cdot \\ \cdot & & \cdot \\ 0 & \cdots & 0 \end{pmatrix}.$$

These are simply the matrix versions of a linear mapping.

$\mathbf{U}, \mathbf{V}$ are unitary matrices with orthonormal columns. Thus $\mathbf{U}^T\mathbf{U} = \mathbf{V}^T\mathbf{V} = \mathbf{I}_n$, the $n \times n$ identity matrix. Σ is a diagonal matrix having nonzero nonnegative real numbers $\{\sigma_j\}$ on the diagonal which are arranged in a nonincreasing order:

$$\sigma_1 \geq \sigma_2 \geq \sigma_3 \cdots.$$

The representation (5.2) which was written under the assumption that $m \geq n$, is known as the *singular value decomposition* or SVD in short (Atkinson, 1989; Stewart, 1973; Golub and Van Loan, 1989; Hansen, 1998) of the matrix $\mathbf{A}$.[2] $\{\sigma_j\}$ denote the singular values and $\mathbf{u_j}, \mathbf{v_j}$ the *left* and *right singular vectors*,

---

[2]If $n > m$, then matrices $\mathbf{U}$ and $\mathbf{V}$ are to be interchanged and the matrix $\mathbf{A}$ to be replaced by its transpose $\mathbf{A}^T$. The SVD of $\mathbf{A}^T$ is $\mathbf{V}\Sigma^T\mathbf{U}^T$.

respectively. The singular system $\{\mathbf{U}, \Sigma, \mathbf{V}\}$ is essentially unique[3] for a given matrix. The matrix Σ has n number of diagonal values all of which may not be nonzero. In other words, we may have $\sigma_1 \geq \sigma_2 \geq \cdots \sigma_r, \sigma_{r+1} = \sigma_{r+2} = \cdots \sigma_n = 0$. In that case, the number r is called the *rank* of the matrix $\mathbf{A}$. Moreover, $\mathbf{A}$ is diagonal in the bases provided by $\mathbf{U}$ and $\mathbf{V}$ since their columns are orthonormal. SVD is also defined if $\mathbf{A} \in C^{m \times n}$ is a $m \times n$ complex matrix (see Stewart, 1973).

At this point, let us illustrate the SVD procedure by a simple example (Cantrell, 2001). Consider the matrix

$$\mathbf{A} = \begin{pmatrix} 0 & 2 \\ 0 & 0 \\ 3 & 0 \end{pmatrix}.$$

Its transpose is

$$\mathbf{A}^T = \begin{pmatrix} 0 & 0 & 3 \\ 2 & 0 & 0 \end{pmatrix}$$

giving

$$\mathbf{A}^T \mathbf{A} = \begin{pmatrix} 9 & 0 \\ 0 & 4 \end{pmatrix}.$$

Clearly, the eigenvalues of $\mathbf{A}^T \mathbf{A}$ are $\lambda_1 = 9$ and $\lambda_2 = 4$. The singular values are, therefore, $\sigma_1 = 3$ and $\sigma_2 = 2$. The eigenfunctions of $\mathbf{A}^T \mathbf{A}$ corresponding to the eigenvalues 9 and 4 can be determined as

$$\mathbf{v}_1 = (1, 0)^T \quad \text{and} \quad \mathbf{v}_2 = (0, 1)^T,$$

respectively. Since $\mathbf{A}\mathbf{v}_i = \sigma_i \mathbf{u}_i$, we see that

$$\mathbf{u}_1 = (0, 0, 1)^T \quad \text{and} \quad \mathbf{u}_2 = (1, 0, 0)^T.$$

Moreover, the singular value matrix Σ is given by

$$\Sigma = \begin{pmatrix} 3 & 0 \\ 0 & 2 \\ 0 & 0 \end{pmatrix}.$$

Now choosing $\mathbf{u}_3 = (0, 1, 0)^T$, it is easily seen that the original matrix $\mathbf{A}$ is indeed the product $\mathbf{U}\Sigma\mathbf{V}^T$. Note that both $\mathbf{U}$ and $\mathbf{V}$ have columns which are orthogonal. Moreover, the rank r of $\mathbf{A}$ is 2.

Interestingly, the SVD gives us the ranges and null spaces of both $\mathbf{A}$ and $\mathbf{A}^T$. These spaces are spanned by the column partitions (assuming that the

---

[3]This is because the matrices $\mathbf{U}$, $\mathbf{V}$ are less well-determined. For example, if $\mathbf{A}^T \mathbf{A}$ has multiple eigenvalues, then the columns of $\mathbf{U}$ can be chosen to be any orthonormal basis spanning the corresponding eigenspace (see Stewart, 1973).

rank is r)

$$\mathbf{U} = \begin{pmatrix} \underbrace{\mathbf{u_1} \dots \mathbf{u_r}}_{\text{basis of range}[\mathbf{A}]} & \underbrace{\mathbf{u_{r+1}} \dots \mathbf{u_m}}_{\text{basis of null}[\mathbf{A}^T]} \end{pmatrix}.$$

Similarly,

$$\mathbf{V} = \begin{pmatrix} \underbrace{\mathbf{v_1} \dots \mathbf{v_r}}_{\text{basis of range}[\mathbf{A}^T]} & \underbrace{\mathbf{v_{r+1}} \dots \mathbf{v_n}}_{\text{basis of null }[\mathbf{A}]} \end{pmatrix}.$$

Therefore,

$$\begin{aligned} \text{range}[\mathbf{A}] &= \text{span}[\mathbf{u_1} \dots \mathbf{u_r}], \\ \text{null}[\mathbf{A}] &= \text{span}[\mathbf{v_{r+1}} \dots \mathbf{v_n}], \\ \text{range}[\mathbf{A}^T] &= \text{span}[\mathbf{v_1} \dots \mathbf{v_r}], \\ \text{null}[\mathbf{A}^T] &= \text{span}[\mathbf{u_{r+1}} \dots \mathbf{u_m}]. \end{aligned}$$

Moreover,

$$\dim[\text{range}[\mathbf{A}]] = r.$$

$$\dim[\text{null}[\mathbf{A}]] = n - r.$$

$$\dim[\text{range}[\mathbf{A}^T]] = r.$$

and

$$\dim[\text{null}[\mathbf{A}^T]] = m - r.$$

These are the four so-called *fundamental subspaces*. An excellent discussion of the fundamental subspaces in linear algebra is given in Strang (1988).

It is also not difficult to see that

$$\begin{aligned} \text{null}[\mathbf{A}] &= \text{range}[\mathbf{A^T}]^{\perp} \\ \text{null}[\mathbf{A}^T] &= \text{range}[\mathbf{A}]^{\perp} \\ \text{range}[\mathbf{A}^T] &= \text{null}[\mathbf{A}]^{\perp} \\ \text{range}[\mathbf{A}] &= \text{null}[\mathbf{A^T}]^{\perp} \end{aligned}$$

It immediately follows that

$$\text{rank} + \text{nullity} = \text{dimension of the domain}$$

which is well-known in matrix theory. The various relations between the fundamental subspaces mentioned above are described in Figure 5.1 which is a discrete, matrix version of Figure 3.2 of Chapter 3 for linear continuous operators.

In sum, the continuous singular functions are replaced by the finite-dimensional singular vectors. The SVD is essentially a generalization of classical diagonalizability (valid for any matrix with distinct eigenvalues, especially the Hermitian matrices) and provides a discrete analogue of singular function expansion. The singular values have close analogy with the eigenvalues of Hermitian matrices. In fact, if $\mathbf{A}$ is Hermitian, then the singular values are the same as the absolute values of the eigenvalues of $\mathbf{A}$. The analogy beside, the SVD provides a stable numerical process for determining the rank of a matrix (which is an ill-posed problem in itself) and for constructing least-square solutions of a linear system. The rank-revealing property of the SVD can be appreciated from the fact that it is easier to determine the rank of a diagonal matrix.

It is interesting to point out the connection between the SVD and the least-square solution of a rectangular matrix $\mathbf{A}^{m \times n}$. A solution to the linear system $\mathbf{Af} = \mathbf{g}$, $\mathbf{A}$ being rectangular, is obtained as a least-square solution. That is, if $\tilde{\mathbf{f}}$ is a solution, then it is a minimizer of the norm $\|\mathbf{g} - \mathbf{Af}\|_2$. The solution is unique if and only if the null-space of $\mathbf{A}$ is empty. Thus we can expect a unique least-square solution only when the n columns of the matrix are linearly independent (assuming that $m > n$). Since $\tilde{\mathbf{f}}$ is a least-square solution, we must have $\mathbf{A}^T \mathbf{A} \tilde{\mathbf{f}} = \mathbf{A}^T \mathbf{g}$. By our above assumption of n linearly independent columns, the normal matrix $\mathbf{A}^T \mathbf{A}$ is of full rank n. Thus $\tilde{\mathbf{f}}$ is the unique solution of the equation

$$\tilde{\mathbf{f}} = \mathbf{A}^T \mathbf{A}^{-1} \mathbf{A}^T \mathbf{g}.$$

The matrix $\mathbf{A}^T \mathbf{A}^{-1} \mathbf{A}^T$ is the *Moore–Penrose generalized inverse* $\mathbf{A}^+$ of $\mathbf{A}$. Just as $\mathbf{f} = \mathbf{A}^{-1} \mathbf{g}$ tells us that $\mathbf{f}$ is a solution of the equation $\mathbf{Af} = \mathbf{g}$, so is the equation $\tilde{\mathbf{f}} = \mathbf{A}^+ \mathbf{g}$ is a statement that $\tilde{\mathbf{f}}$ solves the least-square problem of finding the minimum of $\|\mathbf{g} - \mathbf{Af}\|_2$. The SVD solution, given by

$$\mathbf{f}_{\text{SVD}} = \mathbf{V} \Sigma^{-1} \mathbf{U}^T \mathbf{g}$$

generalizes the least-square solution

$$\tilde{\mathbf{f}} = \mathbf{A}^T \mathbf{A}^{-1} \mathbf{A}^T \mathbf{g}$$

in the case when the null-space of $\mathbf{A}$ is nonempty. The solution $\mathbf{f}_{\text{SVD}}$ is the unique least-square solution in the sense that of all the other such solutions, it has the minimum 2-norm.

At this point, we would like to mention the following. It turns out that in general, the singular values of the linear system of a discretized ill-posed problem decay gradually to zero without any well-delineated gap in the spectrum. This is unlike a rank-deficient matrix in which case a gap in the spectrum is normal (Hansen, 1998). This means that the notion of a *numerical rank* (see below) does not generally exist for a discrete ill-posed problem and, consequently, it is difficult to approximate it by a well-determined problem having the numerical rank as its full rank. The method of solving ill-conditioned problems is thus generally variational. In a rank-deficient problem with a well-determined gap in the spectrum, the numerical rank sets the limit, in a sense, as to what information can be extracted regarding the unknown exact system. In a discretized ill-posed

problem, on the other hand, this depends upon the norms of the solution as well
as the residual (Section 5.2). In general, these problems are underdetermined.

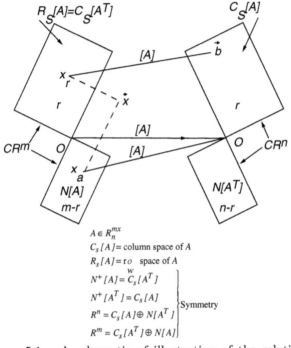

$$A \in R_n^{mx}$$
$C_s[A]$ = column space of A
$R_s[A]$ = row space of A
$$N^+[A] = C_s[A^T]$$
$$N^+[A^T] = C_s[A]$$
$$R^n = C_s[A] \oplus N[A^T]$$ $\Big\}$Symmetry
$$R^m = C_s[A^T] \oplus N[A]$$

Figure 5.1 – A schematic of illustration of the relations
between various subspaces of $a(m \times n)$ matrix **A** and its
$(n \times m)$ transpose $\mathbf{A}^T$.

As already mentioned, the SVD expression (5.2) is a discrete analog of the
SVE representation of the continuous Eq. (3.7) (with the minor notational exception that u_i, v_i of (3.7) now correspond to $\mathbf{v_i, u_i}$ of (5.2), respectively). The
action of the impulse response $I_{\mathbf{A}}$ on f in Eq. (3.7) now becomes

$$\mathbf{Af} = \sum_{j=1}^{n} \sigma_j \mathbf{u}_j (\mathbf{v}_j^T \mathbf{f}) \tag{5.3}$$

and

$$\mathbf{f} = \sum_{j=1}^{n} (\mathbf{v}_j^T \mathbf{f}) \mathbf{v}_j. \tag{5.4}$$

The scalar product of vectors in the discrete case is denoted by $\mathbf{a}^T\mathbf{b}$ as opposed
to the continuous inner product $< a, b >$. Equation (5.4) is the discrete analog
of the continuous SVE of Eq. (3.5). Moreover, the decay of the singular values,
increasing oscillations in the singular vectors with the increase in the index of
the singular values found in the continuous case, also carry over to the discrete

situation. In addition, Eq. (5.3) exhibits the same smoothing effect of the operator as was found in the continuous case, namely, that the high-frequency information of the object is damped in the data leading to instability in inversion. This is because the singular values σ_i decrease with the increase in the index i. The ill-posedness of the original problem is, therefore, carried over to the discrete case through the ill-conditioning of the matrix, the measure of which is provided by the singular value matrix Σ.

From definition (5.2), it follows that

$$\mathbf{A}\mathbf{A}^T = (\mathbf{U}\Sigma\mathbf{V}^T)(\mathbf{V}\Sigma^T\mathbf{U}^T) = \mathbf{U}(\Sigma\Sigma^T)\mathbf{U}^T,$$

and similarly,

$$\mathbf{A}^T\mathbf{A} = (\mathbf{V}\Sigma^T\mathbf{U}^T)(\mathbf{U}\Sigma\mathbf{V}^T) = \mathbf{V}(\Sigma^T\Sigma)\mathbf{V}^T.$$

From the orthonormality of $\mathbf{U}$ and $\mathbf{V}$, it is readily seen that $\mathbf{U}$ is the eigenvector matrix for $\mathbf{A}\mathbf{A}^T$ with the $n \times n$ eigenvalue matrix $\Sigma\Sigma^T$ with diagonal elements $\sigma_i^2, i = 1, 2, \cdots r$, whereas $\mathbf{V}$ is the eigenvector matrix for $\mathbf{A}^T\mathbf{A}$ with the $n \times n$ eigenvalue matrix $\Sigma^T\Sigma$ with diagonal elements $\sigma_i^2, i = 1, 2, \cdots r$. It also follows that

$$\mathbf{A}\mathbf{v}_i = \sigma_i\mathbf{u}_i \longrightarrow ||\mathbf{A}\mathbf{v}_i||_2 = \sigma_i$$

and similarly,

$$\mathbf{A}^T\mathbf{u}_i = \sigma_i\mathbf{v}_i \longrightarrow ||\mathbf{A}^T\mathbf{u}_i||_2 = \sigma_i, \forall i \in [1, n]$$

again due to the orthonormality of the matrices $\mathbf{U}$ and $\mathbf{V}$. Because of the nonincreasing order of arrangement of the singular values, it is clear that the first singular value $\sigma_1 = ||\mathbf{A}||_2$ (see Appendix A.5.1 for definitions of matrix and vector norms). The singular values of $\mathbf{A}$ are the square roots of the nonzero positive eigenvalues of the associated symmetric, nonnegative normal matrices $\mathbf{A}\mathbf{A}^T$ or $\mathbf{A}^T\mathbf{A}$.

If for some index i, the corresponding singular value σ_i happens to be small, then from the first of the above equations, it is seen that the corresponding singular vector $\mathbf{v}_i$ acts as a sort of null-vector for the matrix $\mathbf{A}$. This means that $\mathbf{A}$ is close to being rank-deficient in which case its inverse will be unstable. In other words, the columns of $\mathbf{A}$ will display linear dependence.

Let us briefly introduce the *pseudo-inverse* also known as the *Moore-Penrose inverse* and *generalized inverse* (Groetsch, 1977; Bjorck, 1996). Let a matrix $\mathbf{A} \in R^{n \times n}$ be of full rank n. Then its inverse is given by

$$\mathbf{A}^{-1} = \mathbf{V}\Sigma^{-1}\mathbf{U}^T = \sum_{j=1}^{n}\mathbf{v}_j\sigma_j^{-1}\mathbf{u}_j^T,$$

as follows immediately from the SVD Eq. (5.2). The corresponding solution can, therefore, be written as

$$\mathbf{f} = \sum_{j=1}^{n}\mathbf{v}_j\sigma_j^{-1}(\mathbf{u}_j^T\mathbf{g}). \tag{5.5}$$

Now consider that $\mathbf{A}$ is rank-deficient. A straightforward inverse such as in Eq. (5.5) cannot, of course, exist in this case. However, from our discussions about the relations between the fundamental subspaces (also refer to Figure 5.1), we note that range[$\mathbf{A}$] and the orthogonal complement $N(\mathbf{A})^{\perp}$ of the null space $N(\mathbf{A})$ of $\mathbf{A}$ are isomorphic to each other. A mapping can thus be defined between these two spaces. It is this map from range[$\mathbf{A}$] (which is also $N(\mathbf{A}^T)^{\perp}$) to $N(\mathbf{A})^{\perp}$ that constitutes the pseudoinverse denoted by $\mathbf{A}^+$. The map is given by

$$\mathbf{A}^+ = \mathbf{V}\Sigma^+\mathbf{U}^T,$$

where the matrix Σ^+ is

$$\Sigma^+ = \begin{pmatrix} \frac{1}{\sigma_i} & \cdots & 0 & 0 & \cdots & 0 \\ & & & & & \cdot \\ \cdot & & & & & \cdot \\ \cdot & \cdots & \frac{1}{\sigma_r} & 0 & \cdot & 0 \\ 0 & \cdots & 0 & 0 & \cdots & 0 \\ \cdot & & & & & \cdot \\ \cdot & & & & & \cdot \\ 0 & \cdots & 0 & 0 & \cdots & 0 \end{pmatrix}.$$

Interestingly,

$$\mathbf{A}^+\mathbf{U} = \mathbf{V}\Sigma^+\mathbf{U}^T\mathbf{U} = \mathbf{V}\Sigma^+ = (\mathbf{v}_1,\ldots,\mathbf{v}_r,\mathbf{0},\ldots,\mathbf{v}_0)\Sigma^+.$$

We already know that $(\mathbf{u}_1,\ldots,\mathbf{u}_r)$ is the span of the range of $\mathbf{A}$ and $(\mathbf{v}_1,\ldots,\mathbf{v}_r)$ span the orthogonal complement $N(\mathbf{A})^{\perp}$. The above equation of A^+ then explicitly shows that the generalized inverse $\mathbf{A}^+$ maps the range of $\mathbf{A}$ to $N(\mathbf{A})^{\perp}$. Moreover, since $\Sigma^+\Sigma = \Sigma\Sigma^+ = \mathbf{I}$, it follows that $\mathbf{A}^+\mathbf{A} = \mathbf{V}\mathbf{V}^T$. $\mathbf{A}^+\mathbf{A}$ is the projector of the range of $\mathbf{A}$ onto the orthogonal complement of $N(\mathbf{A})$. Similarly, $\mathbf{A}\mathbf{A}^+$ is the projector of the orthogonal complement of $N(\mathbf{A})$ onto the range of $\mathbf{A}$.

For singular problems, the generalized inverse $\mathbf{A}^+$ is to be used. The discrete generalized solution is given by

$$\mathbf{f}^+ = \mathbf{A}^+\mathbf{g} = \sum_{j=1}^{r(\mathbf{A})} \mathbf{v}_j\sigma_j^{-1}(\mathbf{u}_j^T\mathbf{g}), \tag{5.6}$$

where the rank of $\mathbf{A}, r(\mathbf{A}) < n$. Equation (5.6) is the least-square (minimum 2-norm) solution obtained by minimizing the norm functional $\|\mathbf{A}\mathbf{f} - \mathbf{g}\|_2$. The instability is evident by the presence of the factor σ_j^{-1} in the above equation. Thus for pure mathematical rank-deficiency (that is, ill-posedness in the absence of errors and noise), all small singular values are simply ignored. Equation (5.6) is the analog of Picard's existence condition, namely, for the index $\forall j > n-r(\mathbf{A})$, the scalar product $| < \mathbf{u}_j^T\mathbf{g} > | < \sigma_j$.

Let us next consider Eq. (5.5). The two main factors that influence the inverse are: the decay of the singular values σ_j and that of the Fourier coefficients

$\mathbf{u}_j^T \mathbf{g}$. The former is associated with the matrix $\mathbf{A}$ and the latter with the data $\mathbf{g}$. Both $\mathbf{A}$ and $\mathbf{g}$ are corrupted by experimental and numerical errors, although errors in the data usually dominate. The behavior of the solution $\mathbf{f}$, therefore, depends on how the decay in σ_j compares with that of $\mathbf{u}_j^T \mathbf{g}$. Let the singular values decay to a level σ_A at $j = j_A$, while the Fourier coefficient $\mathbf{u}_j^T \mathbf{g}$ settles down to a level σ_g at $j = j_g$. Then the singular values which should be retained in extracting information regarding the exact object are those σ_j for which the index $j \leq \min(j_A, j_g)$. In practice, it is j_g which is the lesser of the two. Some filter must, therefore, be used in order to screen out the singular values with index j in the range $j_g < j \leq n$, n being the number of singular values. Using a filter is equivalent to multiplying the diagonal singular value matrix Σ^{-1} by some matrix $\mathbf{F}$. The regularized solution can then be written as $\mathbf{f}_{\text{reg}} = \mathbf{V}\mathbf{F}\Sigma^{-1}\mathbf{U}\mathbf{g}$.

Let us briefly comment on the determination of the singular system. The singular values of a symmetric $n \times n$ matrix $\mathbf{S}$ are given by the stationary values of the *Rayleigh quotient* $||\mathbf{S}\mathbf{x}||_2 \backslash ||\mathbf{x}||_2$, i.e., of the map $\mathbf{x} \to ||\mathbf{S}\mathbf{x}||_2 \backslash ||\mathbf{x}||_2$. The k-th eigenvalue $\lambda_k(\mathbf{S})$ is given by (Theorem 8.1.2, Golub and Van Loan, 1989; see also Stewart, 1973)

$$\lambda_k(\mathbf{S}) = \max_{\dim(W)=k} \min_{\mathbf{y} \neq 0 \in W} \frac{\mathbf{y}^T \mathbf{S} \mathbf{y}}{\mathbf{y}^T \mathbf{y}}$$

for $k \in [1, n]$. For $\mathbf{A} \in R^{m \times n}$, the singular values are obtained by applying the above theorem to the symmetric matrix $\mathbf{A}^T \mathbf{A}$ and the result is (Theorem 8.3.1, Golub and Van Loan, 1989)

$$\sigma_k(A) = \max_{\dim(W)} = k \min_{\mathbf{y} \neq 0 \in W} \frac{||\mathbf{A}\mathbf{y}||_2}{||\mathbf{y}||_2}.$$

In the above, $k \in [1, \min(m, n)]$.

In a continuous setting, the singular system can rarely be determined analytically except in a few simple cases. The usual method is to discretize the continuous problem into the linear system (cf. Eq. (4.8)), the Petrov–Galerkin discretization (Section 4.5) being most frequently used for the purpose. The singular system of the matrix $\mathbf{M}$ is obtained and the $n \to \infty$ limit of the discrete singular system is considered to be the approximation of the continuous singular system (see Hansen, 1998, for further details and convergence results). Let $\{\phi_i\}_{i=1}^n$ and $\{\psi_i\}_{i=1}^n$ be the orthonormal basis function sets and introduce the expansions

$$u_i(z) = \sum_{i=1}^n u_{ij}\phi_j(z),$$

and

$$v_i(z) = \sum_{i=1}^n v_{ij}\psi_j(z).$$

Then it is shown by Hansen (1998) that as the dimension of the system $n \to \infty$, the discrete singular values $\sigma_i \implies$ the continuous singular values μ_i uniformly and the singular vector $u_i, v_i \implies$ singular functions u, v in the mean-square sense. (See also Allen *et al.*, 1985).

Before leaving this section and proceeding to the discrete Tikhonov's regularization, let us introduce (albeit only briefly) the idea behind the frequently occurring *truncated SVD* or TSVD. TSVD is based on the definition of the so-called *numerical rank* of a matrix. Consider a mathematical matrix, that is, a matrix that would represent a problem in the absence of errors whether experimental and/or numerical. Let the mathematical matrix have a certain rank, say, r_m. This means that there are m-number of linearly independent columns of the matrix or m positive singular values. Now if there are errors (such as noise, rounding-off, discretization, approximation, and so on), then the elements of the matrix will be perturbed and some of the columns which were originally linearly independent may become linearly dependent or close to it. Alternatively, the original mathematical matrix $\mathbf{A}$ will be perturbed to a matrix $\tilde{\mathbf{A}} = \mathbf{A} + \Delta$. Δ is the perturbation and $||\Delta||_2 \le \epsilon$, where $\epsilon > 0$ is the level of the error. If there exist columns of $\mathbf{A}$ which are guaranteed to remain linearly independent for all Δ such that $||\Delta||_2 \le \epsilon$, then the number of these columns will be defined as the numerical rank of the mathematical matrix $\mathbf{A}$ and denoted by $r_\epsilon(\mathbf{A})$ in order to emphasize the dependence of the numerical rank upon the error level. The perturbed matrix $\tilde{\mathbf{A}}$ is replaced by a mathematical matrix $\mathbf{A_k}$ possessing the numerical rank $r_\epsilon = k$ and the numerics is performed with $\mathbf{A_k}$. SVD calculation based on this matrix $\mathbf{A_k}$ is known as TSVD. The filter function for TSVD is rectangular:

$$
\begin{aligned}
s(\lambda_i) &= 1 \text{ if } \lambda_i \le \lambda_k \\
&= 0 \text{ if } \lambda_i \le \lambda_k.
\end{aligned}
$$

This was pointed out earlier in Section 3.5. The corresponding minimum-norm least-square solution (the regularized solution) is called the *truncated SVD solution* and can be written as

$$
\mathbf{f}_{\text{TSVD}} = \sum_{i=1}^{k} \frac{\mathbf{u}_i^T \mathbf{g}}{\sigma_i^k} \mathbf{v}_i, \tag{5.7}
$$

where σ_i^k are the k singular values of the matrix $\mathbf{A_k}$. If the rank-deficiency, say, p, of the original mathematical matrix $\mathbf{A}$ were known, then the least-square solution could be written as

$$
f_{\text{LS}} = \sum_{i=1}^{p} \frac{\mathbf{u}_i^T \mathbf{g}}{\sigma_i} \mathbf{v}_i \tag{5.8}
$$

by simply ignoring the $n - p$ small singular values. It may be that the second norm $||f_{\text{TSVD}}||_2$ is smaller than the second norm $||f_{\text{LS}}||_2$. However, the residual will be higher (Hansen, 1998). *The TSVD calculation is, therefore, a minimum-norm least-square calculation performed by replacing the original matrix $\mathbf{A}$ with a new matrix $\mathbf{A}_k$ with numerical rank k.*

We now present a brief overview of Tikhonov's regularization for the discrete case.

5.3 The Discrete Tikhonov Regularization

Consider the Tikhonov functional (3.17). Its discrete form is

$$||\mathbf{A}\mathbf{f} - \mathbf{g}||_2 + \alpha||\mathbf{B}(\mathbf{f})||_2, \tag{5.9}$$

where the continuous stabilizer $\Omega(f)$ is replaced by the matrix-2 norm $||\mathbf{B}(\mathbf{f})||_2$. The matrix $\mathbf{B}$ is usually either the identity matrix, a diagonal weighting matrix or a matrix of a derivative operator. The minimization of the functional (5.9) is unique provided that the intersection of the null spaces of $\mathbf{A}$ and $\mathbf{B}$ is empty, that is, $N(\mathbf{A}) \cap N(\mathbf{B}) = \emptyset$ (see Section 5.5).

Now the discrete Tikhonov functional (5.9) contains two matrices $\mathbf{A}$ and $\mathbf{B}$. However, working with only one matrix is obviously desirable from the standpoint of numerical computations. Therefore, it stands to reason that the stabilizing matrix be absorbed into $\mathbf{A}$ by an appropriate transformation (Elden, 1982) thereby reducing the Tikhonov functional (5.9) to the discrete form of the functional (3.16) in which $\mathbf{B}$ is the identity matrix. Such a transformation is said to reduce Eq. (5.9) to the *standard form* (Hansen 1998) and for a square matrix, consists of: $\tilde{\mathbf{A}} = \mathbf{A}\mathbf{B}^{-1}, \tilde{\mathbf{f}} = \mathbf{B}^{-1}\mathbf{f}$, the data vector $\mathbf{g}$ remaining unchanged. The standard form of the functional (5.9) thus becomes

$$||\tilde{\mathbf{A}}\tilde{\mathbf{f}} - \mathbf{g}||_2 + \alpha||\tilde{\mathbf{f}}||_2. \tag{5.10}$$

If $\mathbf{B}$ is a $m \times n, m \geq n$, rectangular matrix, then the transformation uses the generalized inverse of $\mathbf{B}$ in place of $\mathbf{B}^{-1}$. If, on the other hand, $m < n$, then the transform may be somewhat involved. However, we do not go into the details here and refer the reader to the above cited paper of Elden (see also Hansen, 1998).

It was mentioned above that in order to obtain a regularized solution, it is necessary to filter out the singular values above a certain index threshold. For the standard form for the functional (5.9) with $\mathbf{B} = I_n, I_n$ a $n \times n$ unit matrix, a regularized solution f_{reg} can be written as

$$\mathbf{f}_{\text{reg}} = \sum_{i=1}^{n} f_i \frac{\mathbf{u}_i^T \mathbf{g}}{\sigma_i} \mathbf{v}_i, \tag{5.11}$$

where f_i is the i-th element of the $n \times n$ diagonal filter function matrix. In the general case, the filter can be obtained by considering the matrix Euler–Lagrange equation

$$(\mathbf{A}^T\mathbf{A} + \alpha\mathbf{B}^T\mathbf{B})\mathbf{f} = \mathbf{A}^T\mathbf{g} \tag{5.12}$$

from which the regularized solution is

$$\mathbf{f}_{\text{reg}} = (\mathbf{A}^T\mathbf{A} + \alpha\mathbf{B}^T\mathbf{B})^{-1}\mathbf{A}^T\mathbf{g}. \tag{5.13}$$

Let us introduce at this point the concept of the *generalized SVD* or GSVD in short. Briefly, a GSVD is as follows. Let $\mathbf{A} \in R^{m \times n}, m \geq n$ and $\mathbf{B} \in$

$R^{p \times n}$. Then there exists (Theorem 8.7.4, Golub and Van Loan, 1989) orthogonal matrices $\mathbf{U} \in R^{m \times m}$ and $\mathbf{V} \in R^{p \times p}$ such that

$$\mathbf{U}^T \mathbf{A} \mathbf{X} = \mathbf{C} = \text{diag}(c_1, c_2, \cdots, c_n), c_i \geq 0$$

and

$$\mathbf{V}^T \mathbf{B} \mathbf{X} = \mathbf{D} = \text{diag}(d_1, d_2, \cdots, d_q), d_i \geq 0,$$

where $\mathbf{X} \in R^{n \times n}$ an invertible matrix and $q = \min(p, n)$. Then the elements of the set

$$\sigma(\mathbf{A}, \mathbf{B}) \equiv (c_1/d_1, \cdots, c_n/d_q)$$

are called the *generalized singular values* of $(\mathbf{A}, \mathbf{B})$. Note that if $\mathbf{B} = \mathbf{I}$, then $\sigma(\mathbf{A}, \mathbf{B}) = \sigma(\mathbf{A})$. Going back to the matrix Euler–Lagrange equation (5.12), we have the following. For the Tikhonov regularization in standard form, the regularized solution f_i is given by

$$f_i = \frac{\lambda_i}{\lambda_i + \alpha}.$$

If not in the standard form, but $\Omega(f) = \mathbf{B} f$, f_i is

$$f_i = \frac{\gamma_i}{\gamma_i + \alpha},$$

where γ_i is the i-th *generalized singular value*.

5.4   An Example

In this section, we consider an example in order to demonstrate how discrete regularization can be carried out in practice. We choose inverse diffraction of Example 5, Chapter 2, for our example.[4] and consider only a planar geometry. Let the source plane be situated at a height z and the observation plane at $z', z' > z$, the intervening medium between the planes being homogeneous of impedance $\mathcal{Z}$. Suppose that the source plane has been set into a vibratory motion and let v_n be the normal component of its velocity. The pressure field (the hologram pressure) p on the observation plane at z' is related to the normal velocity v_n at z by

$$p = \mathcal{F}^{-1} G_{z \to z'} \mathcal{F} v_n. \tag{5.14}$$

In Eq. (5.14), $\mathcal{F}$ is the Fourier transform operator given by

$$\mathcal{F} = \int_{R^2} e^{-i \mathbf{k}_\perp \cdot \mathbf{x}_\perp} d\mathbf{x}_\perp, \tag{5.15}$$

---

[4]The author is indebted to Dr. E. G. Williams of the Naval Research Laboratory, Washington, D.C., for the material of this Section.

and the inverse Fourier transform operator $\mathcal{F}^{-1}$ is

$$\mathcal{F}^{-1} = \left(\frac{1}{2\pi}\right)^2 \int_{R^2} e^{i\mathbf{k}_\perp \cdot \mathbf{x}_\perp} d\mathbf{k}_\perp. \tag{5.16}$$

In Eqs. (5.15) and (5.16), $\mathbf{k}_\perp = (k_x, k_y)^T$ is a two-dimensional wave vector and $\mathbf{x}_\perp = (x, y)^T$ is the coordinate of a point on a plane. Green's function $G_{z \to z'}$ from the plane at z to that at $z', z' > z$, is given by

$$G_{z \to z'} = \frac{\mathcal{Z}k}{\sqrt{k^2 - |\mathbf{k}_\perp|^2}} e^{i\sqrt{k^2 - |\mathbf{k}_\perp|^2}(z - z')}. \tag{5.17}$$

$G_{z \to z'}$ can be considered to be Neumann's Green's function for the problem. If we discretize Eqs. (5.14) through (5.17), we obtain

$$\mathbf{p}_d = \mathcal{F}_d^{-1} \mathbf{G}_d \mathcal{F}_d \mathbf{v}_d, \tag{5.18}$$

the subscript d indicating discretization. In Eq. (5.18), $\mathbf{p}_d, \mathbf{v}_d \in \mathcal{C}^M$ are complex column vectors of length M and the $M \times M$ complex matrix $\mathbf{G}_d \in \mathcal{C}^{(M \times M)}$ is Green's function matrix. $\mathcal{F}_d$ and $\mathcal{F}_d^{-1}$ are the matrix DFT (digital Fourier transform) operators (see Press et al., 1988, for details of the DFT operators) corresponding to $\mathcal{F}$ and $\mathcal{F}^{-1}$, respectively.

Now, being the operators of direct and inverse Fourier transforms, the matrices $\mathcal{F}_d$ and $\mathcal{F}_d^{-1}$ are orthonormal. Therefore, $\mathcal{F}_d^{-1} = \mathcal{F}_d^T$ and $\mathcal{F}_d \mathcal{F}_d^{-1} = \mathcal{F}_d^{-1} \mathcal{F}_d = I$. Furthermore, the Green's function matrix $\mathbf{G}_d$ is diagonal. The latter reflects the fact that a wave of a particular frequency in the angular spectrum of v_n (emanating from the plane at z) carries information only about that frequency in the hologram pressure on the plane at z'. In the present context, this means that a particular Fourier component of the normal velocity determines exactly the same Fourier component of the pressure p. This matrix is, therefore, a diagonal matrix given by

$$\mathbf{G}_d = \text{diag}[\sigma_{11}, \sigma_{12}, \cdots, \sigma ij, \cdots],$$

with diagonal elements

$$\sigma_{ij} = \frac{\mathcal{Z}k}{\sqrt{k^2 - |(\mathbf{k}_\perp)_{ij}|^2}} e^{i\sqrt{k^2 - |(\mathbf{k}_\perp)_{ij}|^2}(z - z')}. \tag{5.19}$$

$(\mathbf{k}_\perp)_{ij} = (k_{xi}, k_{yi})^T, i = 1, 2, \cdots, M,$ and $|(\mathbf{k}_\perp)_j|^2 = k_{xi}^2 + k_{yi}^2.$
Equation (5.18) can be expressed in the form

$$\mathbf{p}_d = \mathbf{H}\mathbf{v}_d,$$

where the $M \times M$ matrix

$$\mathbf{H} = \mathcal{F}_d^{-1} \mathbf{G}_d \mathcal{F}_d \tag{5.20}$$

is the *transfer function* or *impulse response* matrix for the planar inverse diffraction problem. For the example under consideration, the ill-posedness was found

(Chapter 2) to arise from the inhomogeneous waves for which the magnitude $|\mathbf{k}_\perp|$ exceeded k. Now $|\mathbf{k}_\perp| = k$ defines the radiation circle. Therefore, outside this circle, $|\sigma_{ij}|$ decreases as $|(\mathbf{k}_\perp)_{ij}|$ increases. In that case, as can be seen from Eq. (5.19), the singular values are

$$\sigma_{ij} = \frac{\mathcal{Z}k}{i\sqrt{|(\mathbf{k}_\perp)_{ij}|^2 - k^2}} e^{-\sqrt{|(\mathbf{k}_\perp)_{ij}|^2 - k^2}}(z - z')$$

which fall off with the increase of $|(\mathbf{k}_\perp)_{ij}|$ monotonically. This indicates the ill-conditioning of the matrix $\mathbf{H}$ for large values of $|\mathbf{k}_\perp|$. It is interesting to point out the close resemblance of Eq. (5.20) to the SVD equation (5.2). The two are, however, not identical. The singular values in SVD are real, whereas these are complex in Eq. (5.20). It is more reasonable to consider Eq. (5.20) as an eigensystem decomposition of the transfer matrix $\mathbf{H}$.

Let us now consider the regularization of the holograph equation $\mathbf{p}_d = \mathbf{H}\mathbf{v}_d$. Toward that consider noisy data and let $\mathbf{p}_d^\epsilon$ be the observed pressure vector on the plane at z'. Therefore, $\mathbf{p}_d^\epsilon = \mathbf{p}_d + \epsilon$, and the expectation value of the second moment is given by $(\overline{\mathbf{p}_d^\epsilon - \mathbf{p}_d})^2 = \tau\sqrt{M}, \tau$ being the variance of the noisy data. We, therefore, consider the regularization of the approximate problem $\mathbf{p}_d^\epsilon = \mathbf{H}\mathbf{v}_d^\epsilon$. The discrete form of the Tikhonov functional Eq. (3.16) in this case is given by

$$T_\alpha(\mathbf{v}_d^\epsilon) = ||\mathbf{H}\mathbf{v}_d^\epsilon - \mathbf{p}_d^\epsilon||_2^2 + \alpha||\mathbf{v}_d^\epsilon||_2^2. \tag{5.21}$$

In Eq. (5.21), the identity matrix $\mathbf{I}$ replaced the matrix $\mathbf{B}$ of Eq. (5.9). The regularized solution is given by

$$\mathbf{v}_{d,\text{reg}} = \mathcal{R}_d\mathbf{p}_d^\epsilon. \tag{5.22}$$

The regularizing operator is the matrix

$$\mathcal{R}_d = [\mathbf{H}^T\mathbf{H} + \alpha I]^{-1}\mathbf{H}^T, \quad \alpha > 0. \tag{5.23}$$

(Compare with Eq. (5.12)).

Upon substituting $\mathbf{H}$ from Eq. (5.20) into Eq. (5.23), $\mathcal{R}_d$ is found to be

$$\mathcal{R}_d = \mathcal{F}_d^{-1}\tilde{\Sigma}_d\mathcal{F}_d, \tag{5.24}$$

where

$$\tilde{\Sigma}_d = \text{diag}[\{\sigma_1\}, \{\sigma_2\}, \cdots, \{\sigma_N\}], \tag{5.25}$$

where

$$\sigma_i = \frac{\sigma_i^*}{\alpha + |\sigma_i|^2},$$

* denoting complex conjugate. $\{\sigma_i\}$ represents the set $\{\sigma_{ij}, j = 1, 2, \cdots, M,$ of Eq. (5.19). In deriving Eqs. (5.24) and (5.25) we used the following.

$$\begin{aligned}
\alpha\mathbf{I} + \mathbf{H}^T\mathbf{H} &= \alpha\mathcal{F}_d^{-1}\mathbf{I}\mathcal{F}_d + \mathcal{F}_d^{-1}\mathbf{G}_d^T\mathcal{F}_d\mathcal{F}_d^{-1}\mathbf{G}_d\mathcal{F}_d \\
&= \mathcal{F}_d^{-1}[\alpha\mathbf{I} + \mathbf{G}_d^T\mathbf{G}_d]\mathcal{F}_d.
\end{aligned}$$

remembering that the Fourier transform matrices are orthonormal. Therefore,

$$[\alpha \mathbf{I} + \mathbf{H^T H}]^{-1} = \mathcal{F}_d^{-1}(\mathbf{G}_d^T(\mathbf{G}_d)^{-1})\mathcal{F}_d$$

If $\alpha = 0$, then Eq. (5.24) becomes

$$\mathcal{R}_{0d} = \mathcal{F}_d^{-1}\tilde{\Sigma}_{0d}\mathcal{F}_d.$$

The matrix $\tilde{\Sigma}_{0d}$ is

$$\tilde{\Sigma}_{0d} = \mathrm{diag}\left[\left\{\frac{1}{\sigma_1}\right\}, \left\{\frac{1}{\sigma_2}\right\}, \cdots \left\{\frac{1}{\sigma_N}\right\}\right].$$

(Just set α to zero in Eq. (5.25)). This is the unregularized operator. The regularized operator $\mathcal{R}_d$ of Eq. (5.24) can be written in terms of the unregularized $\tilde{\Sigma}_{0d}$ as

$$\mathcal{R}_d = \mathcal{F}_d^{-1}\mathbf{F}^\alpha \tilde{\Sigma}_{0d}\mathcal{F}_d, \tag{5.26}$$

where the filter is $\mathbf{F}^\alpha$ given by

$$\mathbf{F}^\alpha = \mathrm{diag}\left[\left\{\frac{\lambda_1}{\alpha + \lambda_1}\right\}, \left\{\frac{\lambda_2}{\alpha + \lambda_2}\right\}, \cdots, \left\{\frac{\lambda_N}{\alpha + \lambda_N}\right\}\right] \tag{5.27}$$

is a *filter*, $\lambda_i = |\sigma_i|^2$. Equation (5.27) clearly brings out the filtering property of Tikhonov regularization. When $\alpha = 0$, the filter factor is unity and as $\alpha \to \infty$ (or $\alpha \gg |\lambda_i| \, \forall i$), the filter goes to zero in accordance with the results of Chapter 3. From Eqs. (5.22), (5.26) and (5.27), the regularized solution is obtained as

$$\mathbf{v}_{d,\mathrm{reg}} = \mathcal{F}_d^{-1}\mathbf{F}^\alpha \tilde{\Sigma}_{0d}\mathcal{F}_d\mathbf{p}_d^\epsilon.$$

In the Tikhonov case, the quantity $|\mathbf{k}_\perp|$ is a measure of the regularization parameter α.

Let us next consider the iterative regularization of Landweber type. In this case, the regularized solution is obtained from the equation

$$\mathbf{v}_d = \left(I - \beta \mathbf{H^T H}\right)\mathbf{v}_d + \beta \mathbf{H^T}\mathbf{p}_d$$

and solving it in the iterative format

$$\mathbf{v}_d^{(n)} = \left(I - \beta \mathbf{H^T H}\right)\mathbf{v}_d^{(n-1)} + \beta \mathbf{H^T}\mathbf{p}_d, \, n = 1, 2, \cdots.$$

The solution is

$$\mathbf{v}_d^{(n)} = \mathcal{R}^{(n)}\mathbf{p}_d = \mathcal{F}_d^{-1}\mathrm{diag}\left(\{\hat{\sigma}_1\}, \{\hat{\sigma}_2\}, \cdots \{\hat{\sigma}_N\}\right)\mathcal{F}_d\mathbf{p}_d,$$

where

$$\hat{\sigma}_j = \frac{1 - (1 - \beta|\sigma_j|^2)^n}{\sigma_j}, \quad j = 1, 2, \cdots, M.$$

In analogy with the Tikhonov case (Eq. (5.27)), the iteration filter is given by

$$
\mathbf{F}^{(n)} = \operatorname{diag}\left(1 - \left(1 - \beta\,|\{\sigma_1\}|^2\right)^n, 1 - \left(1 - \beta\,|\{\sigma\}_2|^2\right)^n, \cdots, \right.
$$
$$
\left. 1 - \left(1 - \beta\,|\{\sigma_N\}|^2\right)^n\right).
$$

In order for convergence to occur, β must be chosen such that $\beta = |\sigma_{\max}|^{-2}$. If the discrepancy principle is to be applied for the determination of the regularization parameter n, then $||\mathbf{p}_d^{n,\epsilon} - \mathbf{p}_d^{\epsilon}|| = \epsilon$.

This concludes our brief introduction to Tikhonov's regularization in a discrete situation. For further details regarding the contents of the last two sections, the reader is referred to the previously mentioned monograph of Hansen (1998) which includes a wealth of information about the all-important algorithmic aspects. Instead we devote the rest of the chapter to the presentation of the discrete solution of the Tikhonov functional (3.17) from a slightly different perspective in which use is made of the Fredholm equation of the second kind and general type by considering a specific form of the stabilizer.

5.5   Discrete Solution of a Tikhonov Functional

In the particular case to be discussed in this section, the stabilizer in Eq. (3.17) is chosen to be the L_2-norm of the first derivative of the solution. Equation (3.17) is, therefore,

$$
T_\alpha(f) = ||Af - g||^2 + \alpha||f'||^2. \tag{5.28}
$$

Let us first derive the Euler–Lagrange equation. Write $f = f + \epsilon\eta$, f, $\eta \in X$. To the first order in ϵ, Eq. (5.28) becomes

$$
\begin{aligned}
T_\alpha(f + \epsilon n) = \ & < A(f + \epsilon\eta) - g, A(f + \epsilon\eta) - g > +\alpha < (f + \epsilon\eta)', (f + \epsilon\eta)' > \\
= \ & ||Af||^2 + ||g||^2 - 2 < Af, g > +2\epsilon(< A\eta, Af > - < Ag, \eta >) \\
& +\epsilon^2 < A\eta, A\eta > +\alpha||f'||^2 + 2\epsilon\alpha < f', \eta' > +\epsilon^2||\eta'||^2 \\
= \ & ||Af||^2 + ||g||^2 - 2 < Af, g > \\
& +2\epsilon(< \eta, A^*Af > - < \eta, A^*g >) + \\
& +\epsilon^2 < \eta, A^*A\eta > +\alpha||f'||^2 + 2\epsilon\alpha < f', \eta' > +\epsilon^2||\eta'||^2.
\end{aligned}
$$

Differentiating the above expression in ϵ and setting the result to zero yields

$$
< \eta, A^*Af > - < \eta, A^*g > +\alpha < f', \eta' > +\epsilon< \eta, A^*A\eta > +\epsilon||\eta'||^2 = 0.
$$

Reducing the inner product $< f', \eta' >$ *via* integration by parts, imposing the boundary condition $f'(0) = f'(1) = 0$ and, finally, setting ϵ to zero results in the desired Euler–Lagrange equation

$$
A^*Af - \alpha f'' = A^*g, \tag{5.29}
$$

where, as before,

$$(A^*Af)(x) = \int_0^1 \int_0^1 K(x', x)K(x', x'')f(x'')dx'dx'', \qquad (5.30)$$

$$A^*g(x) = \int_0^1 K(x', x)g(x')dx'. \qquad (5.31)$$

Equation (5.29) is of the form

$$[A^*A + \alpha D]f = A^*g, \qquad (5.32)$$

in which D is the operator of second derivative. Equations of the type (5.32) in which the usual identity operator I is replaced by some (possibly simple) differential operator, is called a Fredholm integral equation of the second kind of *general type*. A numerical minimization of Tikhonov's functional (5.28) can be obtained by solving the Euler–Lagrange equation (5.29) numerically.

Several issues arise at this point. First, how to solve a Fredholm integral equation of the second kind of general type numerically. Moreover, since discrete mathematics is involved, the solution will lie in some Euclidean space R^n which is not a subspace of the original infinite-dimensional function space. The second issue, therefore, involves the relation between the numerical (that is, in R^n) and the function space solution. This is to be contrasted with the previously obtained least-square, Galerkin or collocation solutions which were all function space solutions and in which infinitely precise mathematics was assumed when solving the linear system $\mathbf{Mf} = \mathbf{g}$. In the fully discrete case, errors due to the numerics are bound to occur and one is concerned about the relation between the numerical and the actual solution in the limit that the dimension n of R^n increases toward infinity. That this is not trivial can be seen by considering the identity operator on a subspace spanned by the polynomials x^{j-1}. In this case, the matrix $\mathbf{M}$ of the linear system $\mathbf{Mf} = \mathbf{g}$ is the *Hilbert matrix* $\mathbf{H}$ the elements of which are given by

$$H_{ij} = \int_0^1 x^{i+j-2}dx = [i + j - 1]^{-1},$$

assuming an interval of $(0, 1)$. Now, $\mathbf{f} = \mathbf{H}^{-1}\mathbf{g}$. For small values of n, the solution can be calculated reasonably accurately. However, as n increases, the precision deteriorates quickly. Indeed when $n \sim$ *number of decimal places carried by the computer*, the numerical result becomes meaningless (Kingcaid and Cheney, 1991).

In sum then, the original problem $Af = g$, $A : X \to X$, X an infinite-dimensional function space, is to be approximated in finite dimensions by $\mathbf{A}_n\mathbf{f}_n = \mathbf{g}_n$, $\mathbf{A}_n : X_n \to X_n$, X_n being a finite-dimensional space which is not necessarily a subspace of X. The problems involve the relationship between $\mathbf{f}_n$ and the actual solution f and the stability of the solution as $n \to \infty$. The details of a method for solving a Fredholm integral equation of the second kind of general type is outlined in Appendix 5.3 and is based on Linz (1978). The numerical treatment of the regularized solution is patterned after Lukas (1980). In addition, the discussion below uses concepts and results from difference calculus which are reviewed in Appendix 5.2.

Let I_Δ be a uniform net with spacing h. Furthermore, extend the net: x_{-1} to the left and to x_{n+1} to the right with $f_{-1} = f(x_1)$ and $f_{n+1} = f(x_{n-1})$. The discretized Euler–Lagrange equation corresponding to Eq. (5.29) can be written as

$$\sum_{j=0}^{n} \left[Q_{ij}f_j - \frac{\alpha}{h^2}\{f_{i+1} - 2f_i + f_{i-1}\}\right] = \sum_{j=0}^{n} q_{ij}g_j, \quad i = 0, 1, \cdots, n, \qquad (5.33)$$

where $Q_{ij} = (\mathbf{A^T W A W})_{ij}$, $P_{lm} = P(lh, mh)$, $q_{ij} = (\mathbf{A^T W})_{ij} = A_{ji}W_j$, and $\mathbf{W}$ is a diagonal matrix of the quadrature weights. Moreover, the second derivative f'' was approximated by its centered-difference formula

$$f'' = \frac{f_{i+1} - 2f_i + f_{i-1}}{h^2}.$$

In matrix form, the solution of Eq. (5.33) can be expressed by

$$\overline{f}_\alpha = \left[\mathbf{Q} - \frac{\alpha}{h^2}\mathbf{M}\right]^{-1}\mathbf{qg}. \qquad (5.34)$$

The matrix $\mathbf{M}$ in Eq. (5.34) is given by

$$M = \begin{bmatrix} 2 & -2 & & & & \\ -1 & 2 & -1 & & & \\ & & \cdots & \cdots & & \\ & & & -1 & 2 & -1 \\ & & & & -2 & 2 \end{bmatrix}.$$

As to the differential part of the equation, we have

$$f'' = \frac{f_{i+1} - 2f_i + f_{i-1}}{h^2} = B_n f, \qquad (5.35)$$

$$f(-1) = f(1), \ \ f(n+1) = f(n-1),$$

where $B_n f$ describes the action of some numerical operator on f. At this point, we refer to the example given in Appendix 5.3 where the stability and consistency of problems of the type (5.35) are demonstrated. Condition (a) of Theorem 5.3, Appendix 5.3, is, therefore, satisfied. Assume an appropriate quadrature rule in which the integrals of Eqs. (5.30) and (5.31) can be evaluated accurately. Especially, the weights are assumed to be bounded:

$$\sum_j |w_j| < \infty.$$

The integrals are then uniformly bounded and consistent (for details see Baker, 1977). The second and third conditions of Theorem 5.3 are also satisfied. It remains only to verify the collective compactness of the sequence of operators $\{P_\Delta K_\Delta\}$. This is done next.

Let $\{\psi_\Delta\}$ be a bounded sequence on I_Δ and represent the prolongation operator P_Δ by an operator of linear interpolation. Then

$$(P_\Delta K_\Delta \psi_\Delta)(x) = \sum_{i=0}^{N} \phi_{\Delta i}(x) \sum_{j=0}^{N} Q_\Delta(x_i, x_j)\psi_\Delta(x_j),$$

where $\{\phi_\Delta\}$ is a sequence of piecewise linear functions on the net. By virtue of the fact that $\{\psi_\Delta\}$ is bounded and the integrals are also uniformly bounded, it is clear that the sequence of functions $\{(P_\Delta K_\Delta \psi_\Delta)\}$ is uniformly bounded. The remaining task is to show that the sequence is *equicontinuous* so that the theorem of *Arzela–Ascoli* can be used to prove the compactness of $\{P_\Delta K_\Delta\}$. Toward that consider two points $x_1, x_2 \in [0,1]$. Then

$$|(P_\Delta K_\Delta \psi_\Delta)(x_1) \quad - \quad (P_\Delta K_\Delta \psi_\Delta)(x_2)|$$

$$\leq \left| \sum_{j=0}^{N} \psi_\Delta(x_j) \sum_{i=0}^{N} [\phi_{\Delta i}(x_1) \quad - \quad \phi_{\Delta i}(x_2)] Q_\Delta(x_i, x_j) \right.$$

Since $\{\phi_\Delta\}$ is piecewise linear, $|\phi_{\Delta i}(x_1) - \phi_{\Delta i}(x_2)| \leq |x_1 - x_2|$, $\forall i$. From the boundedness of the sequence $\{\psi_\Delta\}$ and Q_Δ, it immediately follows that $\{P_\Delta K_\Delta\}$ is equicontinuous and the compactness of $\{P_\Delta K_\Delta\}$ is established from the Arzela–Ascoli theorem. The numerical scheme (5.33) is thus consistent and stable and, the solution $\overline{f}_\alpha$ given by Eq. (5.34) converges to the actual solution as $\Delta \to 0$.

The error consists of two parts: one for the integrals and the other for the derivative. The local truncation error for the integrals is given by

$$\sup_{I_{\Delta n}} |(A^* A f)(x_i) - \mathbf{A}^* \mathbf{W} \mathbf{A} \mathbf{W} \overline{\mathbf{f}}(\mathbf{x_i})|.$$

The error for the derivative part is

$$\alpha \Big[\sup_{I_{\Delta n}} |f''(x_i) - \frac{1}{h^2} \mathbf{M} \overline{\mathbf{f}}(\mathbf{x_i})|$$

$$+ \max[|f'(0) - \frac{f(x_1) - f(x_{-1})}{2h}|, |f'(1) - \frac{f(x_{n+1}) - f(x_{n-1})}{2h}|].$$

For further discussions on the error analysis, we refer the reader to Lukas (1980).

The functional (5.28) can also be solved numerically directly. Let f_α denote this solution. Consider again a uniform net with a spacing h and let $\mathbf{W}$ be a diagonal matrix of quadrature weights as previously. Furthermore, assume a forward difference approximation for f'. The numerical minimization of the functional (5.28) then reduces to

$$\text{minimize}_{\underline{f} \in X_n} \sum_{i=0}^{n} \Big[\sum_{j=0}^{n} A_{ij} w_j \underline{f}_j \Big]^2 + \alpha \sum_{i=0}^{n-1} \frac{(f_{i+1} - f_i)^2}{h}.$$

$A_{ij} = A(ih, jh)$ and $\underline{f}_j = \underline{f}(jh)$. Formally, the solution can be written as

$$\underline{f}_\alpha = [\mathbf{A}^\mathbf{T}\mathbf{W}\mathbf{A}\mathbf{W} + \frac{\alpha}{h}\mathbf{W}^{-1}\mathbf{L}]^{-1}\mathbf{A}^\mathbf{T}\mathbf{W}\mathbf{f},$$

where the matrix $\mathbf{L}$ is the same as the matrix $\mathbf{M}$ in Eq. (5.34) with 2 in the first and the last row replaced by 1. Assume that the quadrature scheme is such that the numerical integrals in Eqs (5.30)-(5.31) converge to their exact values as the mesh is made finer and finer. The solution $\underline{f}_\alpha$ is then unique. However, the properties of consistency, stability and convergence of the numerical approximation are best analyzed in the framework of the Euler–Lagrange equation for the functional (5.28).

5.6   Appendix A.5.1

Vector-Matrix Norms and the Condition Number of a Matrix

Let $\mathbf{v} \in R^m$ be a vector in a Euclidean space of dimension m. The norm, $||\mathbf{v}||$, of $\mathbf{v}$ is a measure of the *size* of $\mathbf{v}$ which can be interpreted as its length. $||\mathbf{v}||$ is a functional $|| \cdot || : R^m \to R^1$ which satisfies the following conditions:

   i. $||\mathbf{v}|| \geq 0$, $||\mathbf{v}|| = 0$ iff $\mathbf{v} = 0$: positive-definiteness.

   ii. $||\alpha\mathbf{v}|| = ||\alpha||\,||\mathbf{v}||$, $\alpha \in R^1$: homogeneity.

   iii. $||\mathbf{v} + \mathbf{w}|| \leq ||\mathbf{v}|| + ||\mathbf{w}||$: the triangle inequality.

The *absolute-value norm* is simply $|\mathbf{v}|$. The finite-dimensional p-norm $||\mathbf{v}||_p, 1 \leq p < \infty$, is defined by

$$||\mathbf{v}||_p = \Big(\sum_{i=1}^{m} |v_i|^p \Big)^{1/p}.$$

The most frequently used $||\mathbf{v}||_p$ norms are the *1-norm* $||\mathbf{v}||_\mathbf{1}$ and the *2-norm* $||\mathbf{v}||_\mathbf{2}$:

$$||\mathbf{v}||_1 = \sum_{i=1}^{m} |v_i|,$$

and

$$||\mathbf{v}||_2 = \Big[\sum_{i=1}^{m} |v_i|^2 \Big]^{\frac{1}{2}},$$

respectively. If $p = \infty$, then the so-called *maximum* or ∞*-norm* is obtained, namely

$$||\mathbf{v}||_\infty = \max_{1 \leq i \leq n} |v_i|.$$

For a given vector $\mathbf{v}$, the ∞-norm is never greater than any finite p-norm and, therefore,

$$||\mathbf{v}||_\infty \leq ||\mathbf{v}||_p, \ p \ \text{finite}.$$

Moreover, the value of the p-norm is bounded in terms of the maximum norm $||\mathbf{v}||_\infty$ as

$$||\mathbf{v}||_p \leq m^{\frac{1}{p}} ||\mathbf{v}||_\infty.$$

The maximum norm $||\mathbf{v}||_\infty$ is, of course, computationally the simplest. The algorithm for computing a finite p-norm $||\mathbf{v}||_p$ is $\sim O(m)$ and hence is more expensive than that of the $||\mathbf{v}||_\infty$-norm. Note that although both the 1-norm $||\mathbf{v}||_1$ and the 2-norm $||\mathbf{v}||_2$ are each of the order of m, the former requires only m floating-point additions, whereas the latter involves m floating-point multiplications. The computation of the 2-norm is thus more expensive than that of the 1-norm.

The p-norms $||\mathbf{v}||_p$ are basis dependent except when $p = 2$. The 1-norm and the ∞-norm can be calculated in any basis. However, these depend upon the basis in which the vector $\mathbf{v}$ is expressed. One can verify these statements by considering the canonical basis $C = (\hat{e}_1, \hat{e}_2)$ and, say, a basis given by

$$\mathbf{g} = \left\{ \mathbf{g_1} = \frac{1}{\sqrt{2}} \begin{pmatrix} 1 \\ 1 \end{pmatrix}, \; \mathbf{g_2} = \frac{1}{\sqrt{2}} \begin{pmatrix} 1 \\ -1 \end{pmatrix} \right\}.$$

The norms of a $n \times n$ matrix $\mathbf{A}$ (over R or C) are also constructed in a similar way in accordance with conditions (i)–(iii) above. Again, as in the vector case, the ∞-norm $||\mathbf{A}||_\infty$ is the simplest to calculate. We, therefore, present this norm first.

$$
\begin{aligned}
||\mathbf{Ax}||_\infty &= \max_{i \in (1,n)} \left| \sum_{i=1}^{n} a_{ij} x_j \right| \\
&\leq \max_{i \in (1,n)} \sum_{i=1}^{n} |a_{ij}| |x_j| \\
&\leq \left[\max_{i \in (1,n)} \sum_{i=1}^{n} |a_{ij}| \right] \left[\max_{i \in (1,n)} |x_i| \right].
\end{aligned}
$$

From this, it follows that

$$||\mathbf{A}||_\infty = \max_{i \in (1,n)} \sum_{i=1}^{n} |a_{ij}|$$

since

$$\max_{i \in (1,n)} |x_i| = ||\mathbf{x}||_\infty.$$

The *operator* or p-norm $||\mathbf{A}||_p$ is defined as

$$||\mathbf{A}||_p = \max_{\mathbf{v} \neq 0} \frac{||\mathbf{Av}||_p}{||\mathbf{v}||_p}.$$

The 1-norm $||\mathbf{A}||_1$ is, therefore,

$$||\mathbf{A}||_1 = \max_{\mathbf{j}} \sum_{j=1}^{n} |a_{ij}|.$$

In accordance with the vector 2-norm, a matrix 2-norm is defined as (Atkinson, 1989)

$$||\mathbf{A}||_2 = (\max \beta : \mathbf{A}^T\mathbf{A}\mathbf{v} = \beta\mathbf{v})^{1/2}.$$

The 2-norm $||\mathbf{A}||_2$ is simply the maximum singular value $\sigma_{\max}$. That is,

$$||\mathbf{A}||_2 = \sigma_{\max}.$$

In general, if $\mathbf{A} \in R^{m \times n}$, then there exists a unit norm n-vector $\mathbf{u}$ such that

$$\mathbf{A}^T\mathbf{A}\mathbf{u} = \mu^2\mathbf{u},$$

where $\mu = ||\mathbf{A}||_2$. Moreover, a rule of thumb for the 2-norm is

$$||\mathbf{A}||_2 \le \sqrt{||\mathbf{A}||_1||\mathbf{A}||_\infty}.$$

The p-norm $||\mathbf{A}||_p$ satisfies the relation

$$||\mathbf{A}||_p \le ||\mathbf{A}||_p||\mathbf{v}||_p$$

for every $\mathbf{A} \in R^{m \times n}$ and $\mathbf{v} \in R^{n \times q}$. The p-norm depends on the size of the matrix. In other words, the norm functional $||\mathbf{A}||_p$ depends on the class $R^{m \times n}$ to which the matrix belongs. In general, the p-norm defines a *consistency relation* (Golub and Van Loan, 1989). In other words, if $\mathbf{A} \in R^{m \times n}, \mathbf{B} \in R^{n \times q}$, and, therefore, $\mathbf{AB} \in R^{m \times q}$, then

$$||\mathbf{AB}||_1 \le ||\mathbf{A}||_2||\mathbf{B}||_3.$$

There is no easy way to calculate the p-norm of a matrix. The best method is to determine its singular values which, for an $n \times n$ matrix, is of the order of n^2 floating-point multiplications. The same order holds for any matrix p-norm. However, again as in the vector case, the 1-norm and the ∞-norm require only floating-point additions instead of floating-point multiplications. For this reason, it is either the 1-norm or the ∞-norm that is used most frequently. These two norms are clearly basis-dependent.

Another important matrix norm is the so-called *Frobenius* norm, $||\mathbf{A}||_F$, defined for $\mathbf{A} \in R^{m \times n}$. $||\mathbf{A}||_F$ can be derived as follows.

$$
\begin{aligned}
||\mathbf{A}\mathbf{x}||_2 &= \left[\sum_{i=1}^{m} \left| \sum_{j=1}^{n} a_{ij}x_j \right|^2 \right]^{\frac{1}{2}} \\
&\le \left[\sum_{i=1}^{m} \left(\sum_{j=1}^{n} |a_{ij}|^2|x_j|^2 \right) \right]^{\frac{1}{2}} \\
&= \left[\sum_{j=1}^{m} \left(\sum_{i=1}^{n} |a_{ij}|^2|x_j|^2 \right) \right]^{\frac{1}{2}} \\
&= ||\mathbf{A}||_F||\mathbf{x}||_2.
\end{aligned}
$$

The Frobenius norm is then

$$||\mathbf{A}||_F = \left[\sum_{i=1}^{n} |a_{ij}|^2\right]^{\frac{1}{2}}.$$

Note that the Frobenius norm is the trace of the canonical inner product of $\mathbf{A}$ with itself. It also follows that

$$||\mathbf{A}||_2 \leq ||\mathbf{A}||_F.$$

However, the Frobenius norm may over estimate the 2-norm. For an $m \times n$ matrix, the evaluation of the Frobenius norm requires mn floating-point multiplications, $mn - 1$ floating-point additions and a square root. The calculation of $||\mathbf{A}||_F$ is, therefore, significantly more expensive compared to the 1 or ∞-norm.

Finally, we present the following summary of various norm relations (Golub and Van Loan, 1989) may prove to be useful. Let $\mathbf{A} \in R^{m \times n}$. Then

i. $||\mathbf{A}||_2 \leq ||\mathbf{A}||_F \leq \sqrt{n}\, ||\mathbf{A}||_2$.

ii. $\max_{ij} |\mathbf{A}_{ij}| \leq ||\mathbf{A}||_2 \leq \sqrt{mn}\, \max_{ij} |\mathbf{A}_{ij}|$.

iii. $\frac{1}{\sqrt{n}}||\mathbf{A}||_\infty \leq ||\mathbf{A}||_2 \leq \sqrt{mn}\, ||\mathbf{A}||_\infty$.

iv. $\frac{1}{\sqrt{m}}||\mathbf{A}||_1 \leq ||\mathbf{A}||_2 \leq \sqrt{n}\, ||\mathbf{A}||_1$.

In finite-dimensional vector spaces, the choice of the norm is not critical. If a problem converges in one norm, it converges in the other norms also. For example, if $\mathbf{u}, \mathbf{v} \in R^m$, then if $\mathbf{u} \to \mathbf{v}$, then $||\mathbf{u}-\mathbf{v}|| \to 0$ irrespective of the choice of the norm. This is to be contrasted to the case of infinite-dimensional vector spaces (which include function spaces) where the convergence may crucially depend on the choice of the norm. As an example, the inversion of the Fourier transform is well-posed in L_2-norm. That is, the transform preserves the norm of the source function. In contrast, the inversion of the Laplace transform is not well-posed in any norm.

The Condition Number of a Matrix

The condition number, $CN(\mathbf{A})$, of a matrix $\mathbf{A}$, is intimately connected to the bounds of the solution error due to perturbations in the data vector. Consider the linear system $\mathbf{A}\mathbf{f} = \mathbf{g}$. The unperturbed solution is $\mathbf{f} = \mathbf{A}^{-1}\mathbf{g}$. Let $\delta\mathbf{g}$ be the uncertainty in data $\mathbf{g}$ and $\delta\mathbf{f}$ the corresponding perturbation in the solution $\mathbf{f}$. Let the change $\delta\mathbf{g}$ be *small*. By this we mean that the ratio

$$\frac{||\delta\mathbf{g}||_2}{||\mathbf{g}||_2} = \epsilon \ll 1.$$

For example, if $\mathbf{g}$ is a computed vector, then ϵ is expected to be of the order of the relative error in rounding and we call this the machine precision and denote it by ϵ_{mach}. On the other hand, if the uncertainty is due to experimental errors, then ϵ is a measure of noise. In practice, it is the latter that dominates.

Our objective is to formulate a criterion for expressing the relative error $||\delta\mathbf{f}||/||\mathbf{f}||$ in the solution in terms of that in the data, namely, $||\delta\mathbf{g}||/||\mathbf{g}||$. We are actually looking for a relation such as

$$\sup_{\delta\mathbf{g}}\left\{\frac{||\delta\mathbf{f}||/||\mathbf{f}||}{||\delta\mathbf{g}||/||\mathbf{g}||}\right\}\leq C.$$

Given a vector $\mathbf{g}$, the norm and the level of uncertainty ϵ, the supremum is over the directions of $\delta\mathbf{g}$. The upper bound C depends on various factors such as the specification of the vector and matrix norms, the basis in which the norm is to be evaluated, the level of the data uncertainty ϵ and accuracies one is inclined to consider satisfactory regarding both numerical computations and the relative error.

Now

$$||\delta\mathbf{f}||\leq||\mathbf{A}^{-1}||||\delta\mathbf{g}||,$$

and

$$||\mathbf{g}||\leq||\mathbf{A}||||\mathbf{f}||.$$

It then follows that

$$\sup_{\delta\mathbf{g}}\left\{\frac{||\delta\mathbf{f}||\,/\,||\mathbf{f}||}{||\delta\mathbf{g}||\,/\,||\mathbf{g}||}\right\}=||\mathbf{A}||||\mathbf{A}^{-1}||=\ \mathrm{CN}(\mathbf{A})\leq\mathbf{C}.$$

The quantity $CN(\mathbf{A})=||\mathbf{A}||||\mathbf{A}^{-1}||$ is called the *condition number* of the matrix $\mathbf{A}$. Note that if the matrix $\mathbf{A}$ is singular, then the condition number is infinity and its inverse $\{CN(\mathbf{A})\}^{-1}$ is zero. Therefore, $\{CN(\mathbf{A})\}^{-1}$ is a quantitative measure of how close a matrix is to being singular.

Clearly, a relatively small error in the data can lead to a large error in the solution if the condition number $CN(\mathbf{A})$ is large. For example, a condition number of 10^4 can lead to a relative error of 100 in the solution for only 1% error in the data. Note that the condition number is the upper bound of the inequality and, consequently, gives the worst-case estimate of the relative error. The equality may be obtained for some vector $\mathbf{g}$ and some deviation $\delta\mathbf{g}$. It is worthwhile to point out that a matrix does not have to be large in order to be ill-conditioned. Even a matrix as small as merely 2×2 can exhibit severe ill-conditioning. An interesting example is given by Nievergelt (1991).

The condition number does depend on the choice of the norm used and the basis in which to evaluate the norm. However, the order of magnitude of the condition number is generally independent of the choice of the norm or basis. One can, therefore, calculate $CN(\mathbf{A})$ in whatever norm is most suitable for the task.

The condition number can be usefully connected to the singular values of the matrix. We know that the maximum singular value $\sigma_{\max}$ equals the 2-norm $||\mathbf{A}||_2$. Now the singular values of the inverse matrix $(\mathbf{A}\mathbf{A}^{\mathbf{T}})^{-1}$ are the reciprocals of the singular values of the matrix $(\mathbf{A}\mathbf{A}^{\mathbf{T}})$ and, therefore, $||\mathbf{A}^{-1}||$ is $1/\sigma_{\min}$. From this

$$CN(\mathbf{A})=||\mathbf{A}||_2||\mathbf{A}^{-1}||_2=\frac{\sigma_{\max}}{\sigma_{\min}}.$$

This shows that the conditioning of the problem is essentially determined by the spread in the spectrum of the matrix operator. For a problem which is originally ill-posed, we know that the singular values has a limit point around zero. Therefore, if an unreasonably high demand is made on the accuracy of the solution by a highly refined discretization, then clearly, the condition number is going to be very large causing instability in the numerics. In essence, therefore, indiscriminately finer discretization results in increasingly linear dependency of the adjacent rows and columns of the matrix. In addition, using a completely unpreprocessed matrix is usually inadvisable. Proper scaling of the matrix elements must be done in order to reduce the severity of the ill-conditioning which, in turn, depends upon how fast the spectrum decays. The discussion of this Appendix relied heavily upon Cantrell (2000).

5.7 Appendix A.5.2

Some Concepts from Difference Calculus and Linear Operator Theory

The objective of this Appendix is to briefly review the concepts from difference calculus and linear operator theory which were used in the main text. The materials are based on Isaacson and Keller (1966), Atkinson (1989), Richtmeyer and Morton (1967), and Gustafsson et al., (1995).

In order to fix ideas, let us consider the solution of the problem:

$$y' = f(x, y), \ y(a) = y_0, \ x \in [a, b]. \tag{1}$$

Let

$$I_\Delta : x_0 = a, x_j = x_0 + jh, h = \frac{b-a}{N}, j = 0, \cdots, N$$

be a uniform net over the interval [a,b] with spacing h The subscript Δ refers to the net. A forward-difference approximation for the derivative results in the Euler–Cauchy equation

$$u_{j+1} = u_j + hf(x_j, u_j), j = 0, \cdots, N. \tag{2}$$

The solution $\mathbf{u}$ of Eq. (2) is called the net function and exists if f is defined. It is well-known (Isaacson and Keller, 1966) that in the absence of errors, the Euler–Cauchy equation has a unique solution if f is bounded and continuous in x and Lipschitz continuous in y, that is, if $|f(x, y_1) - f(x, y_2)| \leq L|y_1 - y_2|$, $L > 0$.

Equation (2) assumes infinitely precise arithmetic. It should really be written as

$$U_{j+1} = [U_j + hf(x_j, U_j] + \rho_{j+1}, j = 0, \cdots, N,$$

where $\mathbf{U}$ is the net function in finite arithmetic, ρ_{j+1} being the rounding error in the evaluation of the quantity within the braces. Next suppose that u in Eq. (2) is replaced by the exact solution y. Then

$$y_{j+1} = y_j + hf(x_j, y_j) + h\tau_{j+1}, j = 0, \cdots, N. \tag{3}$$

The quantity τ_{j+1} in Eq. (3) is called the *local truncation* or *discretization* error. It is a measure of the amount by which the values of the exact solution y of Eq. (1) at the grid point of the net I_Δ fail to satisfy the difference equation (2) on the net. All this remains unchanged if a partial differential equation is considered instead of an ordinary one.

Now it is entirely possible that as the net is refined by letting $N \to \infty$, a difference scheme can end up in an equation which is different from the equation for which it was originally written. The finite-difference scheme is then said to be *inconsistent*. As an example, consider the parabolic equation

$$\dot{y} - y'' = 0, \ y(x,0) = f(x). \tag{4}$$

Define the net: $t_j = j\Delta t$, $x_k = k\Delta x$, $j = [0,T]$, $k = 0, \pm 1, \pm 2, \cdots \pm N$, and consider the following *weighted average* difference scheme

$$\frac{u_{i,j+1} - u_{i,j-1}}{2h} = \frac{u_{i+1,j} - 2[wu_{i,j+1} + (1-w)u_{i,j-1}] + u_{i-1,j}}{(\Delta x)^2}, \tag{5}$$

with $w, \frac{1}{2} \leq w \leq 1$, as the weight. Assume further that $\Delta t = r\Delta x$. If we expand the net function in Eq. (5) in a Taylor series around (i,j), replace u by y from Eq. (4) and then subtract the two, we obtain the following truncation error

$$\tau_{i,j} = \left[(\dot{y} - y'') + (2w - 1)\frac{2r}{\Delta x}\dot{y} + r^2\ddot{y} \right]_{i,j}. \tag{6}$$

If $w \neq \frac{1}{2}$ in Eq. (6), then $\tau_{i,j} \to \infty$ as $\Delta x \to 0$. On the other hand, if $w = \frac{1}{2}$, then the limit is a differential equation $\dot{y} - y'' + r^2\ddot{y} = 0$ which is different from the original equation (4). Therefore, when $\Delta t = r\Delta x$, the difference scheme (5) is always inconsistent with Eq. (4). However, if $\Delta t = r(\Delta x)^2$ and $w = \frac{1}{2}$, then the scheme is consistent. But if $w \neq \frac{1}{2}$, the scheme is again inconsistent. This shows that the truncation error and consistency are related, the former being a quantitative measure of the latter.

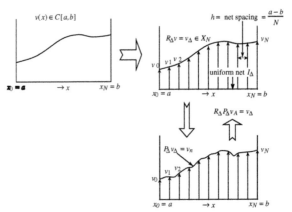

Figure A.5.2 – A schematic illustration of the restriction and propagation operators, R_Δ and P_Δ, respectively.

A more succinct analysis can be given by introducing the so-called *restriction* and *prolongation* operators mapping a Banach space V into itself. The restriction operator R_Δ transforms an element $v \in V$ to a net function $\mathbf{v}_\Delta \in X_n$, $\mathbf{v}_\Delta = R_\Delta v$, $\mathbf{v}_\Delta = \{v_1, \cdots, v_n\}^T$. Similarly, an inverse operation can be defined from $X_n \to V$ *via* a second linear operator P_Δ, the prolongation operator, such that $P_\Delta \mathbf{v}_\Delta = \tilde{v} \in X_n$ is a piecewise linear function which at any grid point assumes the value of the function v at that point. In other words, $(P_\Delta \mathbf{v}_\Delta)x_i = v_i$. An example is the familiar operator of linear interpolation. The operations of R_Δ and P_Δ are shown schematically in Figure A.5.2. Furthermore, the following conditions hold:

1. $\sup_\Delta \|R_\Delta\| < \infty$.

2. $\sup_\Delta \|P_\Delta\| < \infty$.

3. $(R_\Delta P_\Delta) = I$.

4. $P_\Delta R_\Delta v \to v$.

5. $\|R_\Delta v\| = \|\mathbf{v}\|$.

The first two conditions imply that R_Δ and P_Δ are bounded on the net. The last two tell us that as the grid is made finer and finer, the original function is recovered.

Thus the original problem $Af = g$ is approximated by $A_\Delta f_\Delta = g_\Delta$ in a finite-dimensional space X_n which bears no relation to the original space X. However, the connection between f and f_Δ is expressed through the introduction of the operators R_Δ, P_Δ. It is, therefore, reasonable to consider the triplet $\{A_\Delta, R_\Delta, P_\Delta\}$ as the approximation of A. However, for simplicity of notation, R_Δ, P_Δ will be omitted from the triple and we simply say that $\{A_\Delta\}$ is an approximation of A. In terms of the restriction and prolongation operators, the conditions of consistency, stability and convergence can be reformulated.

Definition 1. (Consistency). *An approximation $(A_\Delta, f_\Delta, g_\Delta)$ to the actual problem (A, f, g) is said to be consistent if for every $f \in X$*

$$\lim_{\Delta \to 0} \|R_\Delta Af - A_\Delta R_\Delta f\| = 0.$$

The difference $\tau_\Delta = R_\Delta Af - A_\Delta R_\Delta f$ is the consistency error.

Compare this with Eq. (3).

Definition 2. (Convergence). *A net function $\mathbf{f}_\Delta$ is said to converge discretely to f if*

$$\lim_{\Delta \to 0} \|\mathbf{f}_\Delta - R_\Delta f\| = 0.$$

The convergence is global if

$$\lim_{\Delta \to 0} \|\mathbf{f}_\Delta - P_\Delta f\| = 0.$$

In these definitions, the limit $\Delta \to 0$ means that the number N defining the spacing of the net tends to zero. Moreover, the discrete and the global convergence are equivalent. Indeed

$$
\begin{aligned}
\|f - P_\Delta f\| &\leq \|f - P_\Delta R_\Delta f\| + \|P_\Delta R_\Delta f - P_\Delta f_\Delta\| \\
&\leq \|f - P_\Delta R_\Delta f\| + \|P_\Delta\|\|R_\Delta f - f_\Delta\|.
\end{aligned}
$$

Therefore, if discrete convergence holds, then from property iv above, the equivalence follows. The same can be done by reversing the order, that is, by assuming local convergence first and then showing that the global convergence follows.

Definition 3. **(Stability)** *A sequence of operators A_Δ is considered to be stable if the sequence is uniformly bounded for a sufficiently fine grid. That is,*

$$
\sup_\Delta \|A_\Delta^{-1}\| < \infty.
$$

Here, $\sup_\Delta$ means that $N \geq N_0$, N_0 sufficiently large. The sequence $\{A_\Delta\}$ is then a stable approximation of the original operator A.

Essentially, $\{A_\Delta\}$ is a regularizing algorithm. An easy Corollary follows from this.

Corollary 1. *If $\{A_\Delta\}$ is a stable approximation of A, then for a sufficiently fine grid*

$$
\|R_\Delta f - f_\Delta\| \leq C(\|\tau_\Delta\| + \|R_\Delta Af - g(Af)\Delta\|),
$$

where, as earlier, τ_Δ is the consistency and the rounding error, respectively.

The corollary follows from the fact that

$$
\begin{aligned}
A_\Delta R_\Delta f - A_\Delta f_\Delta &= (A_\Delta R_\Delta f - R_\Delta Af) + (R_\Delta(Af) - (Af)_\Delta) \\
&= \tau_\Delta + (R_\Delta(Af) - (Af)_\Delta).
\end{aligned}
$$

From this

$$
R_\Delta f - f_\Delta = A_\Delta^{-1}\big\{(R_\Delta(Af) - g(Af)\Delta) + \tau_\Delta\big\}. \tag{7}
$$

More precisely, the result of the Corollary should include the round-off error term also which will be considered momentarily. Note that for a stable $\{A_\Delta\}$ and in view of property v of R_Δ, Eq. (7) is actually a convergence result. This immediately leads to the following important theorem.

Theorem 5.1 *If $\{A_\Delta\}$ is a stable and consistent approximation of A, then f_Δ converges to f.*

The theorem is a statement of the fact that consistency and stability imply convergence. This is a classical result originally credited to Lax and is known as *Lax's equivalence theorem* (Richtmeyer and Morton, 1967). Furthermore, from the above Corollary 4.1, the order of convergence can also be estimated.

Definition 4. (Order of Convergence) *Let $\{A_\Delta\}$ be an approximation of A. The approximation is said to be consistent with A to the order k for a given f if*

$$\|A_\Delta R_\Delta f - R_\Delta A f\| \leq cN^{-k},$$

where N is the mesh size and the constant c does not depend on N.

From Corollary 1, it follows immediately that the order of convergence is at least as fast as the order of consistency for a given f.

Let us examine some consequences of the boundedness of the inverse operator. Let $f_\Delta, \overline{f}_\Delta$ be two net functions with different round-off errors, say, ρ and $\overline{\rho}$, respectively. Then

$$R_\Delta A f - A_\Delta f_\Delta = \rho_\Delta$$

and

$$R_\Delta A f - A_\Delta \overline{f}_\Delta = \overline{\rho}_\Delta.$$

Subtracting, we have

$$A_\Delta(\overline{f}_\Delta - f_\Delta) = \overline{\rho}_\Delta - \rho_\Delta.$$

Taking norm on both sides of this equation yields

$$\|\overline{f}_\Delta - \overline{f}_\Delta\| \leq \|A_\Delta^{-1}\|\|\overline{\rho}_\Delta - \rho_\Delta\|.$$

Since A_Δ^{-1} is bounded, the above inequality implies that the grid function f_Δ is Lipschitz continuous in the data error and, by definition, this is stability. Hence, the boundedness of the inverse of A_Δ is a measure of the stability of the approximation.

Let us illustrate these concepts by considering an example the conclusions of which was used in Section 5.5..

Example

We consider a second-order *Sturm–Liouville* differential equation given by

$$-y'' + p(x)y' + q(x)y = f(x), \; y'(0) = y'(1) = 0. \tag{8}$$

Moreover, we assume that p and q are well-behaved functions and that $|p| \leq P$ and the function q is bounded above and below by $0 < q_{min} \leq |q| \leq q_{max}$. Let us approximate the derivatives by their centered-differences. The difference formula for Eq. (8) then becomes

$$a_j u_{j+1} + b_j u_j + c_j u_{j-1}, \tag{9}$$

where

$$a_j = -1 + p_j \frac{h}{2}$$

$$b_j = 2(1 + q_j \frac{h^2}{2})$$

$$c_j = -1 - p_j \frac{h}{2}.$$

It is easily seen that the scheme (9) is consistent. Indeed if we Taylor expand u_k around k and replace u_k by y_k, the grid values of the exact solution of (8), we obtain

$$-(y'')_j + p_j(y')_j + q_j y_j = f_j.$$

The truncation error is thus zero and the scheme is consistent.

Next is how to determine the stability of the solution. Let us extend the net by x_{-1} to the left and x_{n+1} to the right and assume that $u_{-1} = u_1$ and $u_{n-1} = u_n$. Then the linear system corresponding to the Euler–Cauchy equation (2) is given by $\mathbf{Mu} = \mathbf{f}$, where the matrix $\mathbf{M}$ is given by

$$\mathbf{M} = \begin{bmatrix} 1 & & \cdots & & & 0 \\ a_2 & b_2 & c_2 & & & \\ \vdots & a_3 & b_3 & c_3 & & \vdots \\ & \vdots & \vdots & \vdots & & \\ & & a_{n-1} & b_{n-1} & c_{n-1} & \\ 0 & & & & n \end{bmatrix}.$$

Assume that the maximum of u occurs in the net somewhere, say, at $j = J$. Then from Eq. (9)

$$b_J u_J = -(a_J u_{J+1} + c_J u_{J-1}) + f_J h^2.$$

Since $a_J + c_J = -2$, we have

$$(1 + q_J \frac{h^2}{2})|u_J| \le |u_J| |f_J| h^2$$

resulting in the bound

$$||\mathbf{u}||_\infty \le \frac{2}{q_{\min}} ||f_J||_\infty.$$

The inverse operator $\mathbf{M}^{-1}$ is, therefore, bounded and by Definition 3 the algorithm is stable.

5.8   Appendix A.5.3

Nyström's Method

Recall from the introduction of this chapter that one of our objectives is to solve a Fredholm integral equation of the second kind of general type. Toward that the following discussion of Nyström's method (Atkinson, 1997) is important. Moreover, the method deserves an exposition in its own right as it is frequently used in the numerical solution of the Helmholtz integral equation of scattering (see Colton and Kress, 1992, for an in-depth discussion). It is also necessary to introduce the concept of a sequence of operators being *collectively compact* (Anselone, 1971). To motivate the discussion, let us consider the integral equation

$$f(x) + \lambda \int_D K(x, x') f(x') dx' = g(x), \ x' \in D$$

and its quadrature reduction

$$f_n(x) + \lambda \sum_{j=1}^{n} w_j K(x, x_j) f_n(x_j) = g(x), \ x \in D. \qquad (10)$$

At a set of points $\{x_j\}$, Eq. (10) becomes

$$f_n(x_i) + \lambda \sum_{j=1}^{n} w_j A(x_i, x_j) f_n(x_j) = g(x_i), \ i = 1, 2, \cdots n. \qquad (11)$$

Equation (11) is a linear system of order n. Let us also assume that the quadrature converges to the true integral as $n \to \infty$. This requires (by virtue of the *principle of uniform boundedness*) that

$$\sup_{n \geq 1} \sum_{j=1}^{j_n} |w_j| < \infty.$$

The Nyström method is based on the following observation. Each solution of Eq. (10) furnishes a solution of Eq. (11): simply put $x = x_i$ in Eq. (10). We can also solve Eq. (10) by first solving Eq. (11) and then replacing the solution thus obtained in the quadrature part of Eq. (10) in order to obtain $f_n(x)$ at any $x \in D$. In essence, this is an interpolation result that has been found to be highly accurate. In the Nyström method, the actual calculations are performed using Eq. (11), but the error estimate is carried out within the framework of functional analysis using Eq. (10).

Let us write Eqs. (10) and (11) as

$$(I + \lambda A_n) f_n = g,$$

and

$$(I + \lambda \mathbf{A_n}) \mathbf{f}_n = \mathbf{g}.$$

$f_n \in X$ and $A_n : X \to X$. Contrarily, $\mathbf{f_n} \in X_n$ and $\mathbf{A_n} : X_n \to X_n$, where X_n is finite-dimensional.

The error analysis of the Nyström method is complicated by the fact that the norm convergence $\lim_{n \to \infty} \|A - A_n\| = 0$, cannot be guaranteed. As a simple example, consider the integral

$$Lf = \int_0^1 f(t) dt, \ f \in C[0, 1].$$

Let $f(t)$ be a piecewise linear function defined as

$$f(ih) = 0, i = 0, 1, \cdots, n$$
$$f((i + \frac{1}{2})h) = 1.$$

It is clear that $Lf = \frac{hn}{2} = \frac{1}{2}$. However, if we consider the Riemann sum defined by the operator L_n, that is,

$$L_n = h \sum_{j=0}^{n-1} f((i + \frac{1}{2})h)), \ \xi \in [0, h],$$

then the value of the integral $L_n f = 1 - \frac{1}{n}$. Considering L, L_n as operators on C[0,1] with supremum norm, we have

$$\|L - L_n\|_\infty = \sup_{f=1} \|(L - L_n)f\| \to \frac{1}{2}, \ n \to \infty.$$

Therefore, $\lim_{n \to \infty} \|L - L_n\| \neq 0$. However, it is easily seen that $\|(L - L_n)L\|$ or $\|(L - L_n)L_n\| \to 0$ as $n \to \infty$.

It is, therefore, an interesting question as to how to characterize operators, say, M (including $A_n, n \geq 1$) such that the norms $\|(A - A_n)M\| \to 0$ as $n \to \infty$. This is formulated *via* the concept of *collective compactness* of a sequence of operators.

Definition 5. (Collective Compactness) *Let X be a normed linear space and $\{K_n\}, n \geq 1$, be a set of linear operators on X into X. Then the set $\{K_n\}$ is called collectively compact if for each bounded set $U \subset X$, the image set $\{K_n v, v \in U, n \geq 1\}$ is relatively compact, that is, has a compact closure in X.*

The definition can be generalized to the case when the domains may not be the same for all members of the sequence $\{K_n\}$. Of course, each individual operator in a collectively compact set is certainly compact. This leads to the central theorem:

Theorem 5.2 *Let $\{K_n\} : X \to X$ be a set of uniformly bounded, pointwise convergent operators with the limiting operator $A : X \to X$. If $\{M_n\} : X \to X$ is a set of the collectively compact operators, then*

$$\lim_{n \to \infty} \|(A - A_n)M_n\| = 0.$$

The theorem holds also if M_n is replaced by A_n. In that case, we have the following important corollary.

Corollary 4.2. *Let $\{A_n\} : X \to X$ be a collectively compact set of operators with the limiting operator $A : X \to X$. Then*

$$\lim_{n \to \infty} \|(A - A_n)A\| = 0,$$

and

$$\lim_{n \to \infty} \|(A - A_n)A_n\| = 0.$$

We now return to the central problem of this section, namely, the solution of an integral equation of the second kind of general type. The following assumptions are made:

(a) $D : X \to X$ is a linear operator which is not necessarily bounded. We have a differential operator particularly in mind. Hence the symbol D.

(b) D^{-1} exists and is bounded in X.

(c) $A : X \to X$ is compact.

(d) $(D + \lambda A)$ has a bounded inverse on X.

Both D and A are approximated by the finite-dimensional operators D_n and A_n, respectively. We would like to formulate conditions so that $\{(D_n + \lambda A_n), r_n, p_n)\}$ is a consistent and stable approximation to $(D + \lambda A)$. Clearly, consistency is achieved if D_n and A_n are themselves consistent approximations of D and A, respectively. The condition for stability, however, must be worked out. We do not do this here, but instead refer the reader to Linz (1978). The conditions for stability are that A_n must be uniformly bounded and $P_n A_n$ must be collectively compact.

Theorem 5.3 (Linz) *(a)* $\{D_n\}$ *and* $\{A_n\}$ *are stable and consistent approximations to D and A, respectively. (c)* $\{A_n\}$ *is uniformly bounded. (d)* $P_n A_n$ *is relatively compact. Then* $\{(D_n + \lambda A_n)\}$ *is a stable and consistent approximation of* $(D + \lambda A)$.

Chapter 6

The Helmholtz Scattering

6.1 Introduction

The scattering of plane waves (acoustic, elastic or electromagnetic) from material objects is governed by the celebrated *Helmholtz equation* named after the German physicist Alexander von Helmholtz whom we already mentioned in Example 5 of Chapter 2. Of the material objects doing the scattering, there are essentially two types: obstacles and inhomogeneities. By obstacles is meant scatterers the material properties of which undergo jump discontinuities (relative to those of the ambient medium) across their boundaries. Contrarily, the change is gradual and smooth in the case of inhomogeneities. The scattering from inhomogeneities is discussed in the final section of this chapter. For obstacles, the fields are calculated through the well-known *Helmholtz representations* which are integral equations relating the scattered field to that on the surface of the obstacle. The calculation of obstacle scattering, therefore, involves the notions of surfaces and definitions of functions and their derivatives on the scatterer boundary. We begin with a brief synopsis of these.

Consider a domain Ω in R^n. By a *domain* is meant an open, bounded and connected set, the word *connected* implying that any two points in Ω can be joined by a curve without leaving the domain. The surface Γ bounding Ω is the set $\overline{\Omega}\backslash\Omega$. Moreover, Ω is assumed to be located on one side of Γ only. Usually, the manifold Γ is parameterized in terms of a mapping $\mathcal{M}$ of its local coordinates. The smoothness of the manifold is an important consideration and is determined by the smoothness of this mapping. If $\mathcal{M}$ is k-times continuously differentiable, then Γ is of class C^k. Mathematically, it means the following. Let x be a point on Γ and consider a neighborhood $\mathcal{O}(x) \in R^3$ of x. Then $\Gamma \in C^k$ means that there exists a k - times differentiable bijective mapping between the intersection $\mathcal{O}(x) \cap \Gamma$ and an open set $U \in R^2$ (see any text on differential geometry such as Do Carmo, 1976. See also Colton and Kress, 1983).

There are a few frequently used surfaces which serve as models for most physically relevant boundaries in scattering theory and which at the same time possess sufficient smoothness so as to yield meaningful analytical results. The most important are C^2, *Lyapunov* and *Lipschitz* boundaries. A Lyapunov surface is characterized by the condition

$$\cos^{-1}(\hat{n}(x), \hat{n}(y)) \leq L|x - y|^\alpha, \ 0 < \alpha \leq 1, \ L > 0,$$

where $\hat{n}(x), \hat{n}(y)$ are unit normals on Γ at x and y, respectively. A Lyapunov surface, therefore, possesses a tangent plane at every point on Γ, but not necessarily a curvature. An example is a circular cylinder with hemispherical end caps, in which case, the curvature is discontinuous at the intersection of the cylinder and the end caps. The mapping $\mathcal{M}$ for a Lyapunov surface has a Hölder type continuity, whereas for a Lipschitz domain, it is Lipschitz continuous in which case $\alpha = 1$. Note that a Lipschitz boundary is the smoothest class of Lyapunov surfaces. The smoothness characteristics of a surface is assumed to be valid everywhere on it except possibly at a finite number of points and curves of finite length. For further details on these points, see Jawson and Symm (1977), Kellogg (1953) and Mikhlin (1970).

In scattering theory, one defines functions on the closure of a domain. Let f be continuous in an open set Ω and let y be a point on its boundary Γ. Let $\{x_n\}$ be a sequence of points in Ω approaching y from the interior and $\{f_n\}$ a corresponding sequence of functions having a limit f at y. If the limit is independent of the sequence and the path along which y is approached, then f is defined on $\overline{\Omega}$. Thus if $f \in V(\overline{\Omega})$, where V is some function space, then

$$f(y \in \Gamma) = \text{limit}_{\Omega \ni x_n \to y \in \Gamma} f(x_n). \tag{6.1}$$

In other words, f is continuous in Ω and has a continuous extension to $\overline{\Omega}$. Similarly for the derivatives. If $f^{(k)}$ is the k-th derivative of f, then $f^{(k)} \in V(\overline{\Omega})$ means that there exists a function $\psi \in \Omega$ such that $f^{(k)} = \psi$ in Ω and ψ can be extended continuously to the boundary exactly in the manner described in Eq. (6.1). It is known from real analysis (Khinchin, 1960; Lass, 1957) that if f is continuous in a closed and bounded region Ω, then f is bounded in that region. Moreover, if $f \in C(\overline{\Omega})$, then f is uniformly continuous in $\overline{\Omega}$.

As an example of the differentiability of functions in a closed and bounded domain, let us consider (Hackbush, 1995) solving Laplace's equation $\Delta u = 0$ in a rectangle $\Omega = (0, \ell) \times (0, h)$ with boundary data: $\phi(x, y) = x^2$. If $u \in C^2(\overline{\Omega})$, then $\Delta u = 2$ if $(x, y) \in (0, \ell; 0)$ and $(0, \ell; h)$. Since $\Delta u = 0$ inside the rectangle, we see that u cannot be in $C^2(\overline{\Omega})$. However, it can be shown that u is in $C^1(\overline{\Omega})$.

In addition to the functions themselves, their normal derivatives on the boundaries are frequently required. Let us, therefore, explain what is meant by the normal derivative ∂_n. Consider a point y on $\Gamma \in C^k$ and let $\mathcal{O} \in \mathcal{R}^n$ be a neighborhood of y. Let $\phi \in C^k(\mathcal{O})$ be a real-valued function such that $\nabla\phi(y)$ is nonvanishing on $\Gamma \cap \mathcal{O}$. It is assumed that Γ has an *orientation* (positive or negative). This means that there exist continuously varying unit normals on Γ and this *normal* on Γ at $y \in \Gamma \cap \mathcal{O}$ is defined as

$$\hat{n}(y) = \pm \frac{\nabla\phi(y)}{|\nabla\phi(y)|}.$$

Note that $\hat{n}(y) \in C^{k-1}$ if $\Gamma \in C^k$. For $k \geq 2$, a neighborhood V of Γ itself can be defined such that the map $y \rightarrow y + t\hat{n}(y)$ is C^{k-1} *diffeomorphic* from $\Gamma \times (-\epsilon, \epsilon)$ onto $V, -\epsilon < t < \epsilon$. In other words, the map is *bijective*, i.e., one-to-one and onto, and $(k-1)$ times continuously differentiable. If u is a function which is differentiable in V, then

$$u_n^{\pm}(y) = (\partial_n u)^{\pm}(y) = \underset{t \to 0}{\text{limit}}\, \hat{n}(y) \cdot \nabla u(y \pm t\hat{n}(y)), \qquad (6.2)$$

the limit being uniform on Γ. Now consider a doughnut-shaped neighborhood V of Γ and let u be a function differentiable in V. Then the normal derivative $\partial_n u$ is defined by the directional derivative

$$\partial_n u = \hat{n} \cdot \nabla u.$$

Establishing Helmholtz's representations makes substantial use of *Green's identities* and *Gauss' integral theorem* or the *divergence theorem*[1]. The divergence theorem is essentially a generalization to higher dimensions of the *integration by parts* formula and the fundamental theorem of calculus well-known from the elementary calculus of a single variable. Green's identities, on the other hand, are relations involving a differential operator and its *adjoint*. These identities and the divergence theorem-the two important features of *potential theory*-play a fundamental role in scattering calculations. We, therefore, summarize them below (Kellogg, 1953; Günter 1968; Mikhlin 1970). Let us mention that surfaces (such as $C^k, k \geq 1$, or Lyapunov or Lipschitz) for which Gauss' theorem holds are given the name *normal* domains in the literature (Kellogg, 1953).

6.2   Gauss' or Divergence Theorem

Let ϕ_i be a function of $x_i, i = 1, 2, 3$, defined in a closed, bounded domain Ω. Gauss' theorem says that

$$\int_{\Omega} \sum_{i=1}^{3} \partial_i \phi_i(x) \mathrm{d}\Omega(x) = \int_{\Gamma} \sum_{i=1}^{3} \phi_i(x)\hat{n}_i(x)\mathrm{d}\Gamma, \quad x = \{x_1, x_2, x_3\},$$

where $\partial_i = \partial_{x_i}$ and $\hat{n}_i = \hat{n} \cdot x_i$ is the component of the outward unit normal $\hat{n}$ to Γ along the coordinate x_i. In order for Gauss' theorem to hold, it is sufficient that $\phi_i, i = 1, 2, 3$, be in $C^1(\overline{\Omega})$. With this assumption the integral over Ω can be replaced by integration over $\overline{\Omega}$. Moreover, if $\{\phi_i\}$ are the components of a vector field $\mathbf{V}$, i.e., $\mathbf{V} = \{\phi_1, \phi_2, \phi_3\}^{\mathbf{T}}$, then Gauss' theorem takes the form

$$\int_{\Omega} \nabla \cdot \mathbf{V} \mathrm{d}\Omega = \int_{\Gamma} \mathbf{V} \cdot \hat{n} \mathrm{d}\Gamma. \qquad (6.3)$$

---

[1]In Russian literature, the divergence theorem is referred to as the *Ostrogradskii or Ostrogradiskii–Gauss* theorem. (See, for example, Koshlyakov *et al.*, 1964)

In the form given in Eq. (6.3), Gauss' theorem is known as the *divergence theorem*.

The restriction $\phi_i \in C^1(\overline{\Omega})$ can be weakened (Kellog, 1953; Hellwig, 1960). Suppose that $\phi_i \in C^0(\overline{\Omega})$. In this case, the integral on the L.H.S. of Eq. (6.3) exists as an improper integral (see Appendix A.7.6) since $\nabla \cdot \mathbf{V}$ can be singular on Γ. Let Γ' be a boundary in Ω such that $|x - x'| > \delta > 0, \forall x' \in \Gamma', \ x \in \Gamma$. In other words, the new domain Ω' enclosed by Γ' consists of all points of Ω which are at a distance greater than δ from the boundary Γ. As is usual in dealing with improper integrals, it is required that the integral over Ω in Eq. (6.3) exists as the limit of the integral Ω' as $\delta \to 0$. In that case, the divergence theorem holds for $\phi \in C^0(\overline{\Omega}) \cap C^1(\Omega)$.

It may happen that there exist curves on Γ on which the normal is not defined. However, since they constitute sets of measure zero, the value of the integral will remain unchanged. One can, therefore, consider evaluating the integral only on the smooth portions of the surface and then add them up (see Koshlyakov *et al.*, 1964). We mention in passing that the divergence theorem also holds for any $\psi \in H^1(\Omega)$ and, therefore, for a Lipschitz domain (McLean, 2000).

Let us next apply the theorem to the integration by parts in multidimensions. Let $\phi, \psi \in C^1(\overline{\Omega})$ be two functions and consider the integral

$$\int_\Omega \phi \psi_j \mathrm{d}x, \ \psi_j = \frac{\partial \psi}{\partial x_j} = \partial_j \psi.$$

Applying Gauss' theorem to the integral, we obtain

$$
\begin{aligned}
\int_\Omega \phi \psi_j \mathrm{d}x &= \int_\Omega \left[(\phi\psi)_j - \psi \phi_j \mathrm{d}x \right] \\
&= -\int_\Omega \psi \phi_j \mathrm{d}x + \int_\Gamma \phi \psi \hat{n}_j \ \mathrm{d}\Gamma.
\end{aligned}
$$

It can be further generalized to more complex functions such as $\phi D^n \psi$, where

$$D^n = D^{k_1} D^{k_2} \cdots D^{k_n}, \ D^{k_i} = \frac{\partial^k}{\partial x_i{}^k}, \ n = \sum_1^n k_i$$

in the multi-index notation. The result is

$$\int_\Omega \phi D^n \psi \mathrm{d}x = (-1)^n \int_\Omega \psi D^n \phi \mathrm{d}x + \int_\Gamma \Phi(\phi, \psi) \mathrm{d}\Gamma. \qquad (6.4)$$

The quantity $\Phi(\phi, \psi)$ is a function of ϕ and ψ and their derivatives up to $(n - 1)$-th order. Just as in the one-dimensional case, Eq. (6.4) shows that the derivatives are shifted from one function to the other. If the function ϕ happens to be a test function (Stakgold, 1979) in Ω, one obtains the *generalized* derivative of ψ. Equation (6.4) will be required in obtaining Green's identities below.

6.3 Green's Identities

Let L be a partial differential operator given by

$$L = a_{ij}(x)\partial_{ij}u + b_j(x)\partial_j u + c(x)u, \tag{6.5}$$

where $\partial_{ij} = \partial_{x_i}\partial_{x_j}$. The functions a_{ij}, b_j are not the derivatives of a and b. They are simply the coefficients of the corresponding derivatives of u. The operator L is defined in a region of space Ω. It is assumed that $u \in C^2(\bar{\Omega})$ and a_{ij}, b_j and c are in $C^2(\bar{\Omega}), C^1(\bar{\Omega})$ and $C^0(\bar{\Omega})$, respectively. L can also be written in the form

$$L = \partial_i(a_{ij}\partial_j u) + b'_j\partial_j u + cu, \tag{6.6}$$

where $b'_j = b_j - \partial_i a_{ij}$.

The *formal adjoint*, L^*, of L in Eq. (6.5) is an operator defined by

$$L^*u = \partial_{ij}(a_{ij}u) - \partial_j(b_j u) + cu. \tag{6.7}$$

However, corresponding to L in Eq. (6.6), L^* given by Eq. (6.7) can be rewritten as

$$L^*u = \partial_i(a_{ij}\partial_j u) - \partial_j(b'_j u) + cu. \tag{6.8}$$

Comparing Eqs. (6.6) and (6.8) shows that it is only the middle term in these expressions that changes in going from L to L^*. This is in accordance with a general rule (Stakgold, 1979) for forming a formal adjoint, namely, if

$$L = \sum_{|n|\le m} \alpha_k(x)D^k u,$$

then its formal adjoint is given by

$$L^* = \sum_{|n|\le m} \alpha_k(x)D^k(\alpha_k u),$$

where we have used the multi-index notation again. Furthermore, L and L^* are reciprocal to each other meaning that each is the formal adjoint of the other. Indeed, express L^*u in Eq. (6.8) as

$$L^*u = \partial_i(a_{ij}\partial_j u_j) - b'_j\partial_j u + (c - \partial_j b'_j)u.$$

If M is the formal adjoint of L^*, then

$$\begin{aligned} Mu &= \partial_i(a_{ij}\partial_j u) + \partial_j(b'_j u) + (c - \partial_j b'_j)u \\ &= \partial_i(a_{ij}\partial_j u) + b'_j\partial_j u + cu, \end{aligned}$$

which is the same as Eq. (6.6). If it so happens that $b'_j \equiv 0$, then from Eqs. (6.6) and (6.8) $L^* = L$ and the operator L is called *formally self-adjoint*. In this case

$$L^*u = \partial_i(a_{ij}\partial_j u) + cu, \quad a_{ij} = a_{ji}.$$

Let $u, v \in C^2(\overline{\Omega})$ and integrate the product vLu over Ω. Using Eq. (6.6) for L and the integration by parts formula (6.4) results in *Green's first identity* for L:

$$\int_\Omega vLu \ dx = - \int_\Omega a_{ij}\partial_j u \partial_i v dx \ + \ \int_\Omega v(b'_j \partial_j u + cu) dx +$$

$$+ \ \int_\Gamma va_{ij}(\partial_j u)\hat{n}_i d\Gamma. \tag{6.9}$$

Repeating the calculations for uL^*v using (6.8) for L^* yields

$$\int_\Omega uL^*v \ dx = - \int_\Omega a_{ij}\partial_j u \partial_i v \ dx \ + \ \int_\Omega u\left\{ cv - \partial_j(b'_j v)\right\} dx$$

$$+ \ \int_\Gamma a_{ij}u(\partial_j v)\hat{n}_i d\Gamma. \tag{6.10}$$

Equation (6.10) is Green's first identity for the formal adjoint L^*. We next subtract Eq. (6.10) from (6.9) and note that

$$\int_\Omega ub'_j\partial_j v \ dx = \int_\Omega v\partial_j(ub'_j) dx - \int_\Gamma uvb'_j\hat{n}_j \ d\Gamma. \tag{6.11}$$

Using the identity (6.11) to simplify calculations and since $a_{ij} = a_{ji}$, we finally obtain

$$\int_\Omega (vLu - uL^*v) \ dx = \int_\Gamma [a_{ij}(v\partial_j u - u\partial_j v) + b'_j uv]\hat{n}_j \ d\Gamma. \tag{6.12}$$

Equation (6.12) is known as *Green's second identity*.

If L is self-adjoint, then $L = L^*$ and the identities simplify:

$$\int_\Omega vLu \ dx = - \int_\Omega (a_{ij}\partial_j u \partial_i v - cuv) dx + \int_\Gamma a_{ij}v(\partial_j u)\hat{n}_i \ d\Gamma$$

for the first and

$$\int_\Omega (vLu - uL^*v) d\Omega = \int_\Gamma a_{ij}[(v\partial_j u - u\partial_j v)\hat{n}_i]d\Gamma.$$

for the second.

A highly important special case for scattering theory is when L is the Laplacian Δ in which case

$$L = \sum_{i=1}^n \partial_{ii},$$

$a_{ij} = \delta_{ij}, b_j = c = 0, \forall j$. The Laplacian is, therefore, self-adjoint and $a_{ij} = a_{ji}$. In this case the identities reduce to

$$\int_\Omega v\Delta u = - \int_\Omega (\nabla u \cdot \nabla v) dx + \int_\Gamma v(\hat{n} \cdot \nabla u) d\Gamma. \tag{6.13}$$

for the first and

$$\int_\Omega (v\Delta u - u\Delta v)\mathrm{d}x = \int_\Gamma \left[v(\hat{n} \cdot \nabla u) - u(\hat{n} \cdot \nabla v) \right]\mathrm{d}\Gamma \qquad (6.14)$$

for the second.

Again the restrictions on u and v can be weakened. If it is assumed that the integrals of uL^*v and vLu exist in the sense explained earlier, then for the first identity, u can be in $C^1(\overline{\Omega})$ and v in $C^2(\Omega) \cap C^1(\overline{\Omega})$, and for the second, $u, v \in C^2(\Omega) \cap C^1(\overline{\Omega})$. Moreover, since the divergence theorem is valid for a Lipschitz domain for any $u \in H^1(\Omega)$, it follows that the first Green's identity also applies to a Lipschitz domain if $u \in H^2(\Omega)$ and $v \in H^1(\Omega)$ (See McLean, 2000).

6.4 The Helmholtz Equation

As already mentioned, the central equation that governs the scattering of waves from material objects is the Helmholtz equation. It only stands to reason that a derivation of this equation be given before anything else.

Consider a fluid medium with mass density $\rho(x, t)$ and bulk modulus $\lambda(x, t)$, where $x \in R^3$ and t is time. Let a time harmonic plane wave of small amplitude propagate through the medium. The presence of the wave perturbs ρ and λ to $\rho(x, t) + \delta\rho(x, t)$ and $\lambda(x, t) + \delta\lambda(x, t)$, respectively, where the perturbations $\delta\rho$ and $\delta\lambda$ are infinitesimally small: $\delta\rho << \rho, \delta\lambda << \lambda$. Similarly, any velocity of the unperturbed medium $\mathbf{v_0}$ is changed to $\mathbf{v} = \mathbf{v_o} + \delta\mathbf{v}, |\delta\mathbf{v}| << |\mathbf{v_o}|$. Assume that the fluid is inviscid and compressible. Its motion is governed by *Euler's* equation (Landau and Lifshitz, 1987):

$$\rho\frac{D\mathbf{v}}{Dt} = -\nabla p(x, t).$$

$D = \partial_t + (\mathbf{v} \cdot \nabla)\mathbf{v}$ is the *Eulerian* derivative (also variously known as the *total, substantial, Stokes'* or *material* derivative) and p is the pressure. Linearizing Euler's equation (the *convective* term $(\mathbf{v} \cdot \nabla)\mathbf{v}$ is nonlinear) for an initially quiescent fluid ($\mathbf{v_0} = 0$) and introducing the displacement vector $\mathbf{u}$ we have

$$\nabla p(x, t) = -\rho(x, t)\ddot{\mathbf{u}}(x, t), \qquad (6.15)$$

where $\ddot{\mathbf{u}}$ denotes the second derivative of $\mathbf{u}$ with respect to time. p in Eq. (6.15) is the pressure disturbance (over and above the hydrostatic pressure) due to the wave.

Now the stress tensor σ in a mechanical medium is given by (Landau and Lifshitz, 1987)

$$\sigma = \lambda(\nabla \cdot \mathbf{u})\mathbf{I} + \mu(\nabla\mathbf{u} + \mathbf{u} \cdot \nabla).$$

$\mathbf{I}$ is the identity tensor, λ the bulk modulus and μ is the modulus of shear. For a fluid, $\mu = 0$, and the stress tensor reduces to just $\lambda(\nabla \cdot \mathbf{u})\mathbf{I}$. The average stress which is one-third of the trace of this tensor is related to pressure by the formula $\bar{\sigma} = -p$. Hence the second relation follows, namely

$$p(x, t) = -\lambda(x, t)\nabla \cdot \mathbf{u}(x, t). \tag{6.16}$$

From Eqs. (6.15) and (6.16), the wave equation for the pressure is obtained:

$$(\Delta p(x, t) - c^{-2}\ddot{p}(x, t)) = \frac{1}{\rho(x)}\nabla\rho(x) \cdot \nabla p(x, t), \tag{6.17}$$

where $c = \sqrt{\lambda\rho^{-1}}$ is the wave speed. The material parameters ρ and λ were assumed to be time independent. Considering a time harmonic wave with time dependence as $e^{-i\omega t}$, namely

$$p(x, t) = p(x)e^{-i\omega t},$$

Eq. (6.17) becomes

$$(\Delta + k^2(x))p(x) = \frac{1}{\rho(x)}\nabla\rho(x) \cdot \nabla p(x, t). \tag{6.18}$$

In Eq. (6.18)

$$k(x) = \omega c(x)^{-1} = \frac{2\pi}{\lambda}$$

is the wavenumber, λ being the wavelength.[2]

   If the medium is homogeneous except for a bounded domain $\Omega \in R^3$, then λ_0 and ρ_0 are independent of positions, the subscript zero indicating the ambient medium. Consequently, Eq. (6.18) reduces to two equations:

$$(\Delta + k_0^2)p(x) = 0, \ x \in \Omega_e \tag{6.19}$$

and

$$(\Delta + k_0^2)p(x) = \{k_0^2 - k^2(x)\}p(x) + \frac{1}{\rho(x)}\nabla\rho(x) \cdot \nabla p(x, t), \ x \in \Omega. \tag{6.20}$$

In Eq. (6.19), $\Omega_e = R^3 \setminus \overline{\Omega}$ is the domain exterior to Ω (the scatterer) and, k_0, k are the wavenumbers in Ω_e and Ω, respectively.

   Let us define a velocity potential $U(x, t)$ by

$$\mathbf{v} = \frac{1}{\rho}\nabla U,$$

leading to $p = -\dot{U}$ in accordance with Eq. (6.15). Again assuming a time harmonic U: $U(x, t) = u(x)e^{-i\omega t}$, the wave equations for $u(x)$ are also described by Eqs. (6.19) and (6.20). Therefore,

$$(\Delta + k_0^2)u(x) = 0, \ x \in \Omega_e, \tag{6.21}$$

---

[2]We have used λ to denote both bulk modulus and wavelength. However, since the bulk modulus will not appear further in the sequel, no confusion will arise from now on.

and

$$(\Delta + k_0^2)u(x) = \{k_0^2 - k^2(x)\}u(x) + \frac{1}{\rho(x)}\nabla\rho(x) \cdot \nabla u(x), \quad x \in \Omega. \qquad (6.22)$$

From now on we will work with u instead of the pressure p. If the mass density is uniformly distributed throughout the region Ω, then Eqs. (6.21) and (6.22) reduce to

$$(\Delta + k_0^2)u(x) = 0, \quad x \in \Omega_e, \qquad (6.23)$$

and

$$(\Delta + k^2(x))u(x) = 0, \quad x \in \Omega. \qquad (6.24)$$

Equation (6.23) or (6.24) is referred to as the *Helmholtz equation* which plays a key role in acoustics, optics, heat flow, diffusion, wave processes, and so on.[3]

In a general way, the Helmholtz equation is a special case of the so-called *telegrapher's equation* (Koshlyakov *et al.*, 1964) which is

$$\Delta\phi = \alpha_0\ddot{\phi} + 2\alpha_1\dot{\phi} + \alpha_2\phi. \qquad (6.25)$$

Separating the variables, i.e., writing the solution as $\phi(x,t) = \psi(x)\zeta(t)$ and then substituting back in Eq. (6.25) leads to the Helmholtz equation

$$(\Delta + k^2)\psi(x) = 0,$$

and

$$\alpha_0\ddot{\zeta} + 2\alpha_1\dot{\zeta} + (\alpha_2 - k^2)\zeta = 0.$$

The constant k is the usual constant that arises when the equations are separated in their individual variables. Various equations are obtained by manipulating the coefficients α_0, α_1 and α_2. For example, the wave equation is recovered

---

[3]In many texts, especially the ones that are more mathematically inclined, the Helmholtz operator is written as $-\Delta - k^2$ instead of $\Delta + k^2$. There are certain theoretical reasons for writing $-\Delta$ instead of Δ. It automatically ensures the positivity of many quantities. For example, for a positive source function f in Poisson's equation written as $-\Delta u = f$, $u = 0$ on Γ, the solution is also positive. The corresponding Green's function is positive as also the eigenvalues and the *Fourier symbol*. Also an elliptic equation often describes the steady-state of an otherwise transient system in the asymptotic limit of large t. The principle of limiting amplitude formulation of the radiation condition (see Section 6.6) serves as an example. The time-dependent equation of which the elliptic equation is the steady-state limit, is often of the form $\dot{u} - \Delta u = f$, resulting in Poisson's equation with a negative Laplacian.

One further point is worth mentioning. If the Helmholtz operator is written as $\Delta - k^2$ instead of $\Delta + k^2$, then the corresponding Helmholtz equation $(\Delta - k^2)u = 0$ is called the *modified* Helmholtz equation. Its characteristics are entirely different from the regular Helmholtz equation (6.23) or (6.24). Its solutions grow or decay at infinity exponentially instead of algebraically as in the case of the ordinary Helmholtz equation. Thus in two dimensions, Green's function for the modified case is given by the expression $-(1/2\pi)K_0(kr)$ instead of $(i\pi/2)H_0^{(1)}(kr)$, whereas the three-dimensional Green's function for the modified equation is $-\exp(kr)/4\pi r$ instead of $-\exp(ikr)/4\pi r$ for the regular Helmholtz equation (Ockendon *et al.*, 1999).

by setting both α_1 and α_2 to zero so that only α_0 is nonzero. Similarly, if $\alpha_0 = \alpha_2 = 0$, but $\alpha_1 > 0$, then Eq. (6.25) reduces to the equations of heat conduction and diffusion.

Before proceeding further, let us introduce the *Helmholtz representation* of the solution of the Helmholtz equation in the interior of a region.

6.5 The Helmholtz Representation in the Interior

Let $\Omega \subset R^3$ be a domain of class C^2 and u a function in $C^2(\overline{\Omega})$. Consider Green's second formula Eq. (6.14) with v as the free-space Green's function $G^0(x, x'; k)$. Then using the identity given by Eq. (6.4), we have

$$\int_\Omega [u(x')\Delta G^0(x, x'; k) - G^0(x, x'; k)(\Delta u)(x')]dx'$$

$$= \int_\Gamma [u(y)\nabla G^0(x, y; k) - G^0(x, y; k)(\nabla u)(y)] \cdot \hat{n}(y)dy. \qquad (6.26)$$

$\hat{n}(y)$ is the outward unit normal to the surface Γ bounding Ω. At $x = x'$, $G^0(x, x'; k)$ has a singularity. Let us exclude this point by surrounding it with a ball Ω_ϵ of radius ϵ having a surface Γ_ϵ. Let Ω^* be the region $\Omega \setminus \Omega_\epsilon$. In Ω^*, Green's formula (6.26) becomes

$$\int_{\Omega^*} [u(x')\Delta G^0(x, x'; k) \quad - \quad G^0(x, x'; k)(\Delta u)(x')]dx'$$

$$= \int_{\Gamma^*} [u(y)\nabla G^0(x, y; k)$$
$$-G^0(x, y; k)(\nabla u)(y)] \cdot \hat{n}(y)dy, \qquad (6.27)$$

where $\Gamma^* = \Gamma \cup \Gamma_\epsilon$. $\hat{n}$ is an outwardly pointing normal vector on Ω^*. On Γ_ϵ, $\hat{n}$ is, therefore, pointing away from Ω^* into Ω_ϵ. Since the Green's function does not have any singularity in Ω^*, we have

$$\Delta G^0(x, x'; k) + k^2 G^0(x, x'; k) = 0, \ x \in \Omega^*.$$

Using the above Helmholtz equation for G^0 in Eq. (6.27) and after minor algebra, we obtain

$$\int_{\Gamma^*} [u(y)\nabla G^0(x, y; k) \quad - \quad G^0(x, y; k)(\nabla u)(y)] \cdot \hat{n}(y)dy$$

$$= -\int_{\Omega^*} G^0(x, x'; k)(\Delta u + k^2 u)(x')dx'. \qquad (6.28)$$

For $y \in \Gamma_\epsilon$, we obtain

$$\nabla G^0(x, y; k) = -(ik - \epsilon^{-1}) \frac{e^{ik\epsilon}}{4\pi\epsilon} \, \hat{n}(y). \tag{6.29}$$

Next we break up the integral over Γ^* in Eq. (6.28) into its component parts over Γ and Γ_ϵ and use Eq. (6.29) for the derivative of Green's function in the Γ_ϵ integral which reduces to

$$\int_{\Gamma_\epsilon} \left[\frac{\partial u}{\partial y} + (ik - \epsilon^{-1})u \right] \frac{e^{ik\epsilon}}{4\pi\epsilon} \, dy. \tag{6.30}$$

Since u is $C^2(\Omega)$ by assumption, then letting $\epsilon \to 0$ in Eq. (6.30) the integral over Γ_ϵ vanishes.

In the same way, we break up the volume integral in (6.28) into its component parts over Ω and Ω_ϵ. Again letting $\epsilon \to 0$, we see that the Ω_ϵ part results in $-u(x)$ since

$$\Delta G^0(x, x' : k) + k^2 G^0(x, x'; k) = -\delta(x - x').$$

The Helmholtz representation in the interior of Ω is then obtained

$$u(x) = \int_\Gamma [G^0(x, y; k)(\nabla u)(y) - u(y)\nabla G^0(x, y; k)] \cdot \hat{n}(y) dy$$

$$- \int_\Omega G^0(x, x'; k)(\Delta u + k^2 u)(x') dx'. \tag{6.31}$$

If, in addition, u satisfies Helmholtz's equation, then the representation, Eq. (6.31), reduces to

$$u(x) = \int_\Gamma [G^0(x, y; k)(\nabla u)(y) - u(y)\nabla G^0(x, y; k)] \cdot \hat{n}(y) dy.$$

Note that if $x \in R^3 \setminus \bar{\Omega}$, i.e., x is outside the obstacle, then the integral is identically zero. This is because Green's function in this case has no singularity and the integral over Ω_ϵ is zero. The above conclusions are summarized in the theorem below.

Theorem 6.1 (The Helmholtz Representation in the Interior) *Let* Ω *be a bounded domain of class* C^2. *Let* u *be a solution of the Helmholtz equation* $(\Delta + k^2)u = 0$ *in* Ω *such that* u *is continuously differentiable in* $\bar{\Omega}$ *and has bounded second derivatives in* Ω. *Then*

$$\int_\Gamma [G^0(x, y; k)(\nabla u)(y) \quad - \quad u(y)\nabla G^0(x, y; k)] \cdot \hat{n}(y) dy =$$

$$= \quad u(x), \ x \in \Omega$$

$$= \quad 0, \ x \in R^3 \setminus \bar{\Omega},$$

where $\hat{n}$ *is an outwardly pointing unit normal.*

Prior to the derivation of an analogous representation in the exterior region, we need the *radiation condition* and various estimates involving the exterior solution. To these we now turn.

6.6   The Radiation Condition

Consider the Helmholtz equation in a domain Ω_0 which is the exterior of a sphere of radius r_0 with center at the origin. Assuming spherical symmetry, the equation reduces to

$$(ru)'' + k_0^2(ru) = 0. \tag{6.32}$$

$r = |x|$ is the radial coordinate and the prime represents differentiation in r. Also assume that k_0^2 is real. The general solution of Eq. (6.32) can be written as

$$u(r) = A\frac{\cos k_0 r}{r} + B\frac{\sin k_0 r}{r},$$

where A and B are constants. The solution is regular if $r \geq r_0$ and vanishes as r goes to ∞. Introducing time harmonicity, the solution becomes[4]

$$A\frac{e^{i(k_0 r - \omega t)}}{r} + B\frac{e^{-(ik_0 r + \omega t)}}{r}. \tag{6.33}$$

Since ωt increases with time, the surfaces of constant phase are of increasing radii (as time progresses) for the first term and *vice versa* for the second in Eq. (6.33). The first term thus describes waves propagating away from the sphere, whereas the second exponential represents spherical phase fronts converging towards the sphere from infinity. If the surface of the sphere of radius r_0 contains sources as in scattering problems, the second term must be rejected on physical grounds. The solution is, therefore, given by the first exponential alone. In addition, due to the presence of the factor r^{-1}, the amplitude of the waves must vanish at infinity as pointed out above.

In a general, i.e., non-spherically symmetric situation, there will exist an arbitrary system of spherical waves propagating along the radii with an angular distribution of amplitudes. However, each system of waves must obey the same general rule of vanishing amplitude at infinity. In the vicinity of the scatterer, the resultant field may be highly complex. But at a great distance from the sources, the system will disentangle and become spherical again.[5] Hence for an arbitrary bounded region containing the sources of the waves, the physical solution must also be given by the outgoing wavefunction alone.

The physics involved in the scattering of plane waves in an unbounded exterior is illustrated schematically in Figure 6.1. Ω_w with bounding surface Γ_w and Ω_s bounded by Γ_s represent regions containing the wave and the scattering sources, respectively.

---

[4]If the time harmonicity is taken to be $e^{i\omega t}$ instead of $e^{-i\omega t}$, the complex conjugate of the Helmholtz equation is obtained and the solution is also complex conjugate. However, only the real component of the field (pressure or displacement) being of interest, the two time variations are equivalent and one may use either one. In addition, k_0^2 may in general be complex.

[5]Remember that our problem is linear and no nonlinear wave-wave or wave-medium interactions can occur.

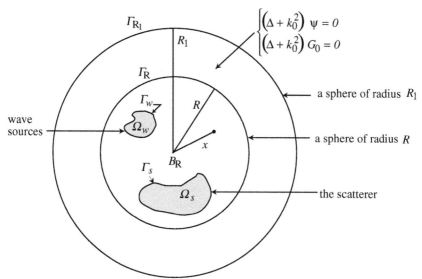

Figure 6.1 – A schematic illustrating the nonphysical nature of incoming wave contributions to the field at x. $x \in B_R$, a ball of radius R enclosing all source regions.

Let us show that the contribution from Γ_R (Figure 6.1) to the field at any interior point cannot vanish even if the radius R recedes to infinity. Let $\Gamma_{R_1}, R_1 > R$, be a second sphere surrounding Γ_R and let $u^R(x), u^{R_1}(x)$ be the fields at x contributed by Γ_R and Γ_{R_1}, respectively. Now outside Γ_R, the field and the free-space Green's function satisfy homogeneous Helmholtz equations. Calculating $u^R(x)$ and $u^{R_1}(x)$ by the interior formula (Theorem 6.11), subtracting one from the other, and applying the divergence theorem to the vector $(u\nabla G^0 - G^0\nabla u)$ in the annular region between Γ_R and Γ_{R_1}, it is found that $u^R(x) = u^{R_1}(x)$, i.e., $u(x)$ is invariant to the radius of these nonphysical spheres. Therefore, a nonvanishing contribution from any of these spheres remains nonvanishing even in the limit $R \to \infty$ implying thereby the existence of sources of energy at infinity which must be rejected on physical grounds. In other words, the scattered field must consist of outgoing waves only.

It is necessary to formulate an analytical criterion which will ensure the vanishing of $u^{\text{sc}}(R)$ as $R \to \infty$, that is, those solutions of the Helmholtz equation that behave as outgoing waves. Such a criterion is the much-celebrated *radiation condition* which is originally due to Sommerfeld (1949). Let us make some comments at this point. There are distinct differences between an interior and an exterior Helmholtz problem. For an interior problem in a bounded region $\Omega \in R^3$, the operator $-\Delta$ has a compact resolvent in $L_2(\Omega)$ (Nédélec, 2001). This means that the interior problem has a unique solution. It will be seen later (Chapter 7) that this is indeed so provided that a certain condition is met, namely, k must not belong to the interior Dirichlet spectrum. Moreover, the interior solution is a standing wave solution, which can be considered to be real without any loss of generality. The exterior problem, on the other hand, is entirely different. The operator $-\Delta$ does not have a compact resolvent in $L_2(\Omega)$ in this case and conditions must be imposed in order to enforce uniqueness. These are usually imposed at infinity. The solutions in the exterior are complex, although k^2 may be real. We also note that plane waves are solutions of the homogeneous problem and these do not decay at infinity. Therefore, it is necessary to require that the solutions decay at infinity at least as r^{-1} in order to eliminate plane waves. However, even this is not sufficient to force uniqueness as can be seen by considering the functions $u = \sin kr/r$ which is a nontrivial solution of the Helmholtz equation in free space. The extra condition that indeed leads to uniqueness is furnished by the well-known *Sommerfeld's radiation condition* to which we now turn.

Consider first the case where $k > 0$ is real. If we differentiate the exponentials in Eq. (6.33) in r, the following relations result:

$$r(u' - ik_0 u) = -u, \qquad r(v' + ik_0 v) = -v,$$

where

$$u = \frac{e^{ik_0 r}}{r}, \quad v = \frac{e^{-ik_0 r}}{r}.$$

Since u, $v \to 0$ as $r \to \infty$, the above relations can be reformulated as

$$\lim_{r\to\infty}(u' - ik_0 u) \sim o(r^{-1}); \ \lim_{r\to\infty} ru \sim O(1). \tag{6.34}$$

$$\lim_{\to\infty}(v' + ik_0 v) \sim o(r^{-1}); \ \lim_{r\to\infty} rv \sim O(1). \tag{6.35}$$

'O' and 'o' are the so-called *Landau symbols* (Nayfeh, 1973) defined as:

$$f(x) \sim O(g(x)) \implies \frac{f(x)}{g(x)} \to C,$$

$$f(x) \sim o(g(x)) \implies \frac{f(x)}{g(x)} \to 0,$$

C being a constant. The order symbol $o(\zeta)$ implies a small quantity of order higher than ζ as $\zeta \to 0$. Equations (6.34) and (6.35) constitute the analytic criteria for distinguishing between the incoming and outgoing waves. The first condition of Eq. (6.34) is called Sommerfeld's *radiation condition*, whereas the second condition of the same equation is known as the *finiteness condition*.

For propagations in real media, the waves are damped as they travel away from the sources, the damping being small. The propagation constant is thus complex and, therefore, $k = k' + ik''$. The fundamental solutions in this case are

$$u = r^{-1}e^{-k''r}e^{ik'r}; \ v = r^{-1}e^{k''r}e^{-ik'r}$$

and the energy in the outgoing wave is $|u|^2 = A^2 r^{-2} e^{-2k''r}$. In order to prevent an unphysical catastrophic growth of energy as the wave propagates to infinity, the sign of the complex absorption k'' must be positive. It is also easy to see that the sign of k'' is the same as that of the imaginary part of k^2, since k' is real and positive. This is typical of real media.

The wave damping can be utilized in discriminating against the incoming wave v and separating out the outgoing wave u as the solution of the Helmholtz equation. Toward this, note that for the outgoing wave to vanish at infinity, it is sufficient that

$$e^{-k''}\frac{e^{ik'r}}{r} \to 0, \; r \to \infty$$

since the sign of k'' is positive. But the same cannot be said of the incoming wave v. It is singular as $r \to \infty$. The radiation conditions, therefore, take the form

$$\operatorname*{limit}_{r \to \infty} e^{|k''|r}(u' - ikr) \sim o(1),$$

$$\operatorname*{limit}_{r \to \infty} e^{|k''|r}u \sim O(1).$$

Note that if $k'' = 0$, Sommerfeld's conditions are recovered.

In summary, if we solve Helmholtz's equation for a complex k with $k'' > 0$ and then take the limit of this solution as $k'' \to 0$, then the physically correct wavefunction possessing an outgoing property will be obtained. This is called the *principle of limiting absorption*. In Agmon (1975), the principle is formulated essentially as (see Nachman, 1989)

$$||G^0(k) * f||_{L_2(-\delta)} \le \frac{c(\delta, a)}{|k|}||f||_{L_2(\delta)}, \; \delta > \frac{1}{2}, \; \forall k \in R^n, \; |k| \ge a,$$

where the Hilbert space $L_2(\delta)$ in R^n is defined as

$$L_2(\delta)(R^n) = \left\{ v: \; ||v||_\delta = \left[\int (1 + x^2)^\delta |v(x)|^2 dx \right]^{1/2} \right\}.$$

A unique solution of the Helmholtz equation with complex wavenumber exists in $H^2(R^3)$ and the passage to the limit $k'' \to +0$ is possible in the sense of $H^1_{loc}(R^3)$, that is, in every compact subset of R^3 in H^1.

There is yet another, third formulation of the radiation condition, the so-called *principle of limiting amplitude*. It arises out of the fact that the solution of the Helmholtz equation is the steady-state amplitude of an oscillating system subject to a harmonic forcing function. Let us, therefore, consider the Cauchy problem of the hyperbolic wave equation

$$\Delta w - c^{-2}\ddot{w} = H(t)S(x)e^{-i\omega t},$$

where $w = u(x)e^{-i\omega t}$, $\omega^2 = c^2k^2$ and $H(t)$ is the *Heaviside function* which is zero if $t < 0$, and is unity if $t > 0$. S is a source function. Originally, the medium is quiescent, i.e., $w = \dot{w} = 0$ for $t < 0$. Upon applying the forcing function at time $t = 0$, there will take place an initial, transient build-up of oscillations during which the solution will not be periodic. However, as $t \to \infty$, the response will tend to a time-periodic limit. A steady vibration will be established with

$u(x) = \mathrm{limit}_{t \to \infty} w(t, x) e^{i\omega t}$. This limiting solution $u(x)$, if the limit exists, will behave as an outgoing wave and constitute the physical solution of the Helmholtz equation. The opposite result is obtained by changing the sign of the frequency ω. The principle of limiting amplitude then says that the amplitude of the stationary vibrations can be obtained by passing to the limit $t \to \infty$ (if the limit exists) in the nonstationary oscillation and this steady-state amplitude is the solution of the Helmholtz equation. Again it can be shown that the limit exists in the sense of $H^1_{\mathrm{loc}}(R^3)$. For further details about the principle of limiting amplitude, we refer the reader to Tikhonov and Samarskii (1963) and Smirnov (1964).

It is worth pointing out that the principle of limiting absorption can be obtained from the Cauchy problem for the limiting amplitude if a damping term $\beta \dot{w}$ is added to the hyperbolic wave equation. This term, which is proportional to velocity, accounts for the resistance of the motion and guarantees the absorption of kinetic energy. It gives rise to the complex wavenumber $k^2 = k'^2 + i\beta\omega$ and the passage to the limit as $\beta \to +0$ yields the solution to the Helmholtz equation. For an excellent discussion of these questions, the reader is referred to Vainberg (1996).

The formulation of the radiation conditions via the principle of vanishing absorption or limiting amplitude involves solutions of auxiliary problems. In contrast, Sommerfeld's conditions are analytic in the sense that they are expressed directly in terms of the solution of the wave equation being solved. In general, Sommerfeld's radiation conditions are applicable to unbounded domains with infinity as an interior point. For the nonapplicability of Sommerfeld's conditions in bounded and semibounded regions such as in a strip $0 \le z \le 1$, see Smirnov (1964). It should be pointed out that the limits in the radiation conditions are uniform in $\hat{x} = x|x|^{-1}$, and, therefore, the solution decreases at the same rate along any radius.

Note that Eq. (6.34) contains not one but two conditions, namely, the radiation and finiteness condition. It turns out that the latter is redundant and, the radiation condition automatically implies the condition of finiteness. Atkinson (1949) did combine the two conditions into a single one

$$r \exp(ikr)[(ik - r^{-1}) - u_r] \to 0, \ \text{as} \ r \to \infty.$$

Later, Wilcox (1956) dispensed with the second condition altogether. Let us show how this comes about. We first consider the case where k is real and $k > 0$ and start with the spherical wave expansion of the solution and write

$$u(x) = \sum_{\ell=0}^{\infty} \sum_{m=-\ell}^{\ell} u_{\ell m}(r) Y_{\ell m}(\hat{\theta}), \tag{6.36}$$

where $\hat{\theta} = (\theta, \phi)$ are the angular coordinates in a spherical coordinate system, and $u_{\ell m}(r) = <u, Y_{\ell m}>$. Replacing Eq. (6.36) in the Helmholtz equation $(\Delta + k^2)u = 0$, the radial equation for $u_{\ell m}$ is obtained

$$u''_{\ell m} + \frac{2}{r} u'_{\ell m} + \left(k^2 - \frac{\ell(\ell+1)}{r^2} \right) u_{\ell m} = 0,$$

the solution being

$$u_{\ell m}(r) = \alpha_{\ell m} h_\ell^{(1)}(kr) + \beta_{\ell m} h_\ell^{(2)}(kr), \tag{6.37}$$

where $\alpha_{\ell m}$ and $\beta_{\ell m}$ are constants and $h_\ell^{(1)}, h_\ell^{(2)}$ are the spherical Hankel functions of the first and second kind (Gradshteyn and Rhyzhik, 1980), respectively.

Refer to Figure 6.1 and consider the norm $||u_R - iku(R)||^2_{L_2(\Gamma_R)}$, where u_R is the radial derivative of u on Γ_R. Thus

$$||u_R - iku||^2_{L_2(\Gamma_R)} = ||\sum_{\ell=0}^{\infty} \sum_{m=-\ell}^{\ell} \{(u_{\ell m})_R - iku_{\ell m}(R)\}Y_{\ell m}(\hat{\theta})||^2_{L_2(\Gamma_R)}.$$

From (6.34) it follows that as $R \to \infty$,

$$0 = \sum_{\ell=0}^{\infty} \sum_{m=-\ell}^{\ell} |(u_{\ell m})_R - iku_{\ell m}(R)|^2 R^2,$$

implying that $R\{(u_{\ell m})_R - iku_{\ell m}(R)\} \to 0$ as $R \to \infty$. Now the spherical Hankel functions $h_\ell^{(1)}$ and $h_\ell^{(2)}$ have the following asymptotic properties (Gradshteyn and Rhyzhik, 1980):

$$h_\ell^{(1,2)}(z) \sim z^{-1} e^{\pm i(z - \frac{\ell\pi}{2} - \frac{\pi}{2})}\{1 + O(z^{-1})\}, z \to \infty \tag{6.38}$$

$$(h_\ell^{(1,2)})'(z) \sim z^{-1} e^{\pm i(z - \frac{\ell\pi}{2})}\{1 + O(z^{-1})\}, z \to \infty. \tag{6.39}$$

It is readily seen from Eqs. (6.38) and (6.39) that for $n \geq 1$, $h_\ell^{(1)'}(z) - ikh_\ell^{(1)}(z) \sim O(z^{-2})$, whereas $h_\ell^{(2)'}(z) - ikh_\ell^{(2)} \sim O(z^{-1})$. The coefficient $\beta_{\ell m}$, therefore, must be set to zero. The solution u of the Helmholtz equation is then

$$u(x) = \sum_{\ell=0}^{\infty} \sum_{m=-\ell}^{\ell} \alpha_{\ell m} h_\ell^{(1)}(kr) Y_{\ell m}(\hat{\theta}). \tag{6.40}$$

From the asymptotic behavior of the Hankel functions given above, it is clear that $u(x)$ in (6.40) consists of outgoing waves. Thus the radiation condition $u_r - iku \sim o(r^{-1})$ alone separates out the outgoing wavefunction. Furthermore, from the orthonormality of the spherical harmonics, it follows that

$$||u||^2_{L_2(\Gamma_R)} = R^2 \sum_{\ell=0}^{\infty} \sum_{m=-\ell}^{\ell} |\alpha_{\ell m}|^2 |h_\ell^{(1)}(kR)|^2. \tag{6.41}$$

From Eqs. (6.38) and (6.41) it follows that $||u||^2_{L_2(\Gamma_R)}$ remains bounded as $R \to \infty$ and

$$\lim_{R \to \infty} ||u||^2_{L_2(\Gamma_R)} = k^{-2} \sum_{\ell=0}^{\infty} \sum_{m=-\ell}^{\ell} |\alpha_{\ell m}|^2. \tag{6.42}$$

This is Parseval's theorem. Since the surface area varies as R^2, Eq. (6.42) implies that $u(x) \sim O(|x|^{-1})$, $|x| \to \infty$. The radiation condition alone, therefore, implies the condition of finiteness.

Next consider k to be complex. From the radiation condition

$$
\begin{aligned}
0 &= \lim_{R \to \infty} \|u_R - iku\|_{L_2^2(\Gamma_R)} \\
&= \lim_{R \to \infty} [\|u_R\|_{L_2(\Gamma_R)}^2 + |k|^2\|u\|_{L_2(\Gamma_R)}^2 - 2\mathrm{Im} \int_{\Gamma_R} \overline{k}\overline{u}u_R \, d\Gamma_R. \quad (6.43)
\end{aligned}
$$

Strictly speaking, the left-hand side of Eq. (6.43) should be written as $\epsilon(R)$, $R \to \infty$, instead of zero. Let us apply Green's first identity (Eq. (6.13)) to the interior domain $\Omega_R = B_R \setminus \Omega_s$ of Figure 6.1 assuming plane-wave incidence. The unit normal on $\Gamma_R \cup \Gamma_s$ is assumed to point outward and obtain:

$$
\begin{aligned}
\int_{\Gamma_R} \overline{u}u_R d\Gamma_R &= \int_{\Omega_R} \left(\nabla\overline{u} \cdot \nabla u + \overline{u}\Delta u \right) d\Omega + \int_{\Gamma_s} \overline{u}u_r d\Gamma_s \\
&= \int_{\Gamma_s} \overline{u}u_r d\Gamma_s + [\|\nabla u\|^2 - k^2\|u\|^2]_{L_2(\Omega_R)}. \quad (6.44)
\end{aligned}
$$

Therefore,

$$
-\mathrm{Im} \int_{\Gamma_R} \overline{k}\overline{u}u_R d\Gamma_R = \mathrm{Im} \int_{\Omega_R} [k^2\overline{k}|u|^2 - \overline{k}|\nabla u|^2]d\Omega_R - \mathrm{Im} \int_{\Gamma_s} \overline{k}\overline{u}u_s d\Gamma_s, \quad (6.45)
$$

u_s denoting the normal derivative of u on Γ_s. Since $\mathrm{Im}k \geq 0$, $\mathrm{Im}(k^2\overline{k}) = |k|^2 k'' \geq 0$. The integral over Ω_R in the above Eq. (6.45) reduces to

$$
\mathrm{Im} \int_{\Omega_R} [k^2\overline{k}|u|^2 - \overline{k}|\nabla u|^2]d\Omega_R = \int_{\Omega_R} k'' \left[|k|^2|u|^2 - |\nabla u|^2\right]d\Omega_R \quad (6.46)
$$

since $\mathrm{Im}\overline{k} \leq 0$ (because $\mathrm{Im}k \geq 0$). The integral in Eq. (6.46) is, therefore, nonnegative. From Eqs. (6.43) through (6.46) we have

$$
\begin{aligned}
0 &= \lim_{R \to \infty} \left[\|u_R\|^2 + |k|^2\|u\|^2\right]_{L_2(\Gamma_R)} - 2\mathrm{Im} \int_{\Gamma_s} \overline{k}\overline{u}u_s d\Gamma_s \\
&\quad + 2\int_{\Omega_R} k'' \left[|k|^2|u|^2 + |\nabla u|^2\right]d\Omega_R \quad (6.47)
\end{aligned}
$$

resulting in

$$
\begin{aligned}
2\mathrm{Im} \int_{\Gamma_s} \overline{k}\overline{u}u_s d\Gamma_s &= \lim_{R \to \infty} \left[\|u_R\|^2 + |k|^2\|u\|^2\right]_{L_2(\Gamma_R)} \\
&\quad + 2\int_{\Omega_R} k'' \left[|k|^2|u|^2 + |\nabla u|^2\right]d\Omega_R. \quad (6.48)
\end{aligned}
$$

Each term on the right-hand sides of Eq. (6.48) is nonnegative, whereas the left-hand side is a finite quantity. Therefore,

$$
\int_{\Omega_R} [|u_R|^2 + |k|^2|u|^2]d\Gamma_R \sim O(1)
$$

as $R \to \infty$ and so is the volume integral. It then follows that

$$\lim_{R \to \infty} ||u||^2_{L_2(\Gamma_R)} \sim O(1). \tag{6.49}$$

Again because the surface area of Γ_R is proportional to R^2, Eq. (6.49) implies that the field must fall-off at least as fast as R^{-1}. This in turn implies that the condition of finiteness is redundant and is a consequence of the radiation condition itself. In modern usage, therefore, only the radiation condition is used.

Before leaving this section, let us briefly mention the so-called *approximate* radiation conditions. Strictly speaking, Sommerfeld's condition applies at infinity. Needless to say the condition must be tailored so as to suit the needs of a practical numerical solution of the Helmholtz equation. The approximate radiation conditions are derived precisely to fulfill this demand. There are several formulations of this problem (Bayliss and Turkel, 1980; Engquist and Majda, 1977, 1979; Kang, 1983) along with the review articles by Turkel (1983) and Givoli (1991). For a recent application of the approximate conditions to finite-element calculations of Helmholtz scattering problems, see Shirron (1995). Here we only present a brief outline of the Bayliss and Turkel scheme.

It will be shown in Section 6.8 that the scattered field can be expanded in a series

$$u(x) = \frac{e^{ik|x|}}{|x|} \sum_{n=0}^{\infty} \frac{U_n(\hat{x})}{|x|^n}, \quad x = (|x|, \hat{x}), \tag{6.50}$$

where the unit vector $\hat{x}$ denotes the direction of scattering. The series (6.50) is uniformly convergent beyond the smallest sphere of radius R_0 circumscribing the scatterer. Now, Sommerfeld's condition can be interpreted as giving a null operator $(\partial_r - ik)$ on u as $r = |x| \to \infty$. The idea behind the Bayliss and Turkel method is to construct operators which will perform the same function, but for a finite $|x| > R_0$. By direct differentiation of the series (6.50) (it can be differentiated term by term because of uniform convergence there) it follows that

$$\left(\partial_r - ik + \frac{1}{r}\right) u(x)\Big|_{|x|=R} = -\frac{e^{ikR}}{R^2} \sum_{n=0}^{\infty} n \frac{U_n(\hat{x})}{R^n}. \tag{6.51}$$

The application of the operator $(\partial_r - ik + \frac{1}{r})$, therefore, makes the $n = 0$ term of Eq. (6.50) vanish. Similarly, differentiating Eq. (6.51) once more yields

$$\left\{\partial_r^2 - \left(2ik - \frac{4}{r}\right)\partial_r - \left(k^2 + \frac{4ik}{r} - \frac{2}{r^2}\right)\right\} u(x)\Big|_{|x|=R} = \frac{e^{ikR}}{R^3} \sum_{n=0}^{\infty} n(n-1)\frac{u_n}{R^n}.$$

Rearranging, we have

$$\left(\partial_r - ik + \frac{3}{r}\right)\left(\partial_r - ik + \frac{1}{r}\right) u\Big|_{x=R} = \frac{e^{ikr}}{R^3} \sum_{n=0}^{\infty} n(n-1)\frac{u_n}{R^n}.$$

This operator product annihilates the $n = 1$ term. A hierarchy of operators defined by

$$L_m = \prod_{j=1}^{m} \left(\partial_r - ik + \frac{2j - 1}{r} \right)$$

can be used in order to continue the process of annihilation. This hierarchy of operators then approximates the radiation condition on a sphere of finite diameter R.

It is interesting to note that Sommerfeld's condition, if written as $\partial_r u = iku$, can be looked upon as the *Dirichlet-to-Neumann* operator (which is just ik in this case). By definition, the Dirichlet-to-Neumann operator converts the field on a boundary surface to its normal derivative on the same surface. To the first-order operator given by Eq. (6.51) there corresponds the Dirichlet-to-Neumann operator $\{ik - (R)^{-1}\}$. In the same way, the second-order Dirichlet-to-Neumann operator can also be defined, namely

$$D_N = -\frac{1}{2} \left(ik - \frac{1}{R} \right)^{-1} \left(2k^2 + \frac{4ik}{R} - \frac{2 + n(n + 1)}{4R^2} \right).$$

In order to obtain the second order D_N, it was assumed that the solution of the Helmholtz equation on the sphere of radius R could be written as $u(R, \theta) = u(R)P_n(\cos \theta)$. The second derivative $\partial_r^2 u$ was eliminated by using the Helmholtz equation $\Delta u + k^2 u = 0$. This led to the second-order boundary condition:

$$-2 \left(ik - \frac{1}{R} \right) \partial_r u = \left\{ 2k^2 + \frac{4ik}{R} - \frac{2 + n(n + 1)}{R^2} \right\} u.$$

It is pointed out by Shirron (1995) that the higher-order conditions are not compatible with the standard finite-element formulation. Another analysis of radiation boundary conditions of a finite distance is given by Hagstrom (1997). Hagstrom also considers the location of the surface fixed and increases the order of the approximation.

This concludes our presentation of the radiation conditions and we are now ready to discuss Helmholtz's representation in the exterior. The derivation is presented below.

6.7   The Helmholtz Representation in the Exterior

Let us first obtain some order of magnitude estimates of the free-space Green's function

$$G^0(x, x'; k) = \frac{e^{ik|x - x'|}}{4\pi|x - x'|}$$

and its derivative. For the distance function $|x - x'|, |x'| << |x|$, we have

$$\begin{aligned} |x - x'| &= [|x|^2 - 2x \cdot x' + |x'|^2]^{\frac{1}{2}} \\ &= |x| - \hat{x} \cdot x' + O(|x|^{-1}), \end{aligned} \qquad (6.52)$$

where $\hat{x} = x|x|^{-1}$. Similarly,

$$|x - x'|^{-1} = \frac{1}{|x|}[1 + O(|x|^{-1})]. \tag{6.53}$$

From Eqs. (6.52) and (6.53), we have

$$G^0(x, x'; k) = \frac{e^{ik|x|}}{4\pi|x|}[e^{-ik\hat{x}\cdot x'} + O(|x|^{-1})]. \tag{6.54}$$

For the gradient of Green's function, we have

$$\nabla G^0(x, x'; k) = \frac{ike^{ik|x|}}{4\pi|x|}e^{-ik\hat{x}\cdot x'}[1 + O(|x|^{-1})]\frac{x - x'}{|x - x'|}. \tag{6.55}$$

From Eq. (6.55) it also follows that for $|x| \gg |x'|$,

$$\nabla G^0(x, x'; k) \cdot \hat{x} - ikG^0(x, x'; k) \sim O(|x|^{-2}). \tag{6.56}$$

Consider again the physical situation described in Figure 6.1. In that figure, let $\Gamma_s \in C^2$ and assume that $u \in C^2(\Omega_e) \cap C^1(\bar{\Omega}_e)$ is a solution of Helmholtz's equation in the exterior $\Omega_e = R^3 \setminus \bar{\Omega}_s$ and apply the Helmholtz representation (Theorem 6.1) in $\Omega_R = B_R \setminus \bar{\Omega}_s$. Then

$$\begin{aligned} u(x) &= -\int_{\Gamma_R} [u(y)\frac{\partial G^0}{\partial y} - G^0(x, y; k)\frac{\partial u}{\partial y}]dy \\ &+ \int_{\Gamma_s} [u(y)\frac{\partial G^0}{\partial y} - G^0(x, y; k)\frac{\partial u}{\partial y}]dy. \end{aligned} \tag{6.57}$$

In the Γ_s-integral of Eq. (6.57), the normal is pointing into the exterior domain. Using the estimates (6.54)–(6.56) for Green's function and its derivative, Eq. (6.49) for $|u|^2$, and from the radiation condition, it follows that

$$\int_{\Gamma_R} u(y)\left\{\nabla G^0(x, y; k) \cdot \hat{y} - ikG^0(x, y; k)\right\}dy \to 0, R \to \infty,$$

and

$$\int_{\Gamma_R} G^0(x, y; k)\left\{\nabla u(y) \cdot \hat{y} - iku(y)\right\}dy \to 0, R \to \infty.$$

The integral over Γ_R, therefore, vanishes as $R \to \infty$, and we obtain Helmholtz's representation in the exterior

$$u(x) = \int_{\Gamma_s} \left[u(y)\frac{\partial G^0}{\partial y} - G^0(x, y; k)\frac{\partial u}{\partial y}\right]dy, \tag{6.58}$$

with an outwardly pointing normal. Again if $x \in \Omega_s$, the integral in Eq. (6.58) vanishes by the interior representation. Combining all this, we obtain

Theorem 6.2 (The Helmholtz Representation in the Exterior) *Let*
Ω_e *be an unbounded domain exterior to a regular closed surface* Γ *enclosing*
a domain $\Omega \subset R^3$. *Let u be a solution of the Helmholtz equation* $(\Delta + k_0^2)u = 0$
in Ω_e *which is continuously differentiable in* $\bar{\Omega}_e$ *and has bounded continuous*
second derivatives in Ω_e. *Then*

$$\int_\Gamma \left[u(y) \frac{\partial G^0}{\partial y} - G^0(x, y; k) \frac{\partial u}{\partial y} \right] dy \; = \; u(x), \; x \in \Omega_e$$
$$= \; 0, \; x \in \Omega.$$

$\hat{n}$ *is outward on* Γ *and pointing into* Ω_e.

The Helmholtz representation in the exterior is, therefore, established. The
derivation hinged upon the fact that $\|u\|^2_{L_2(\Gamma_R)} \to 0$ as $R \to \infty$, and this in
turn followed from the single radiation condition $u_r - iku \sim o(r^{-1})$. Note
that neither in the derivation of the Helmholtz representation in the exterior
nor in establishing the redundancy of the finiteness condition, have we actually
used Sommerfeld's condition $u_r - iku \sim o(r^{-1})$ directly, but only its norm
$\|u_R - iku\|_{L_2(\Gamma_R)}$. Hence this norm itself may serve as the radiation condition.
After all, the entire purpose of this condition is to separate out the outgoing
part of the solution of the Helmholtz equation in the exterior which, as we have
just seen, the norm does quite well. The norm radiation condition is naturally
weaker than the actual function $(u_r - iku)$.

6.8   Some Properties of the Scattering Solutions

Classically, the solution of the Helmholtz equation must be at least twice continuously differentiable because of the presence of the Laplacian. The Laplacian
part then yields a continuous function (not generally valid except in one space
dimension), whereas k^2u is twice continuously differentiable. This situation cannot, of course, occur if u is infinitely differentiable, i.e., $u \in C^\infty$. Indeed the
solution of the Helmholtz equation turns out to be not only C^∞, but also in
C^ω, and is analytic in all its variables in every compact set of the domain of its
definition.[6] There exist a number of results in the theory of elliptic partial differential equations that establish analyticity of the solution from the sufficiently
often differentiability of the coefficients and solutions. One such theorem is
given below.

Theorem 6.3 (Hellwig, 1960) *If* $u(x) \in C^2(\Omega) \cap C(\overline{\Omega})$ *solves the Helmholtz*
equation $(\Delta + k^2)u = 0$, *then u is analytic in* $\Omega, k > 0$.

The theorem can be considered to be a generalization of a similar result for
harmonic functions and is a consequence of the analyticity of Green's function

---

[6]Recall that a function is analytic in a variable in a domain if it has a power series expansion
in that variable at each point of the domain.

everywhere except at the point of singularity. Another interesting consequence of u being analytic is this. It is known (Morse and Feshbach, 1953) that if two functions have a common region of analyticity and if they coincide with each other on an arbitrary arc (no matter how small as long as it is not a mere point) of a continuous curve in the region, then the functions are identical in their common region of analyticity. Now suppose that u and the function zero fulfill these conditions. Then u must be identically zero. Hence the generalization

Theorem 6.4 *If a solution of the Helmholtz equation vanishes in an open subset in its domain of definition, then it must vanish identically everywhere.*

This reflects the fact that an analytic function is "rigid" (see for example, Needham, 1997). There exist a number of classical results concerning the null solution of the Helmholtz equation. These are indispensable in establishing the uniqueness of the solutions of various scattering problems. The main point of departure here is Atkinson's (1949) and Wilcox's (1956) expansions of the scattering amplitude stated in the theorem below.

Theorem 6.5 . *Let $u \in C^2(\Omega) \cap C^1(\bar{\Omega})$ be a solution of the Helmholtz equation $(\Delta + k^2)u = 0$ in $R^3 \setminus \bar{\Omega}$. Let $r(u_r - iku) \to 0$ as $r \to \infty$ uniformly with respect to the angular variables $\hat{\theta}$. Then u can be expressed as*

$$u(x) = \frac{e^{ikr}}{r} \sum_0^\infty u_n(\hat{\theta}) r^{-n}, \quad x = (r, \hat{\theta}). \tag{6.59}$$

Moreover, $\exists$ a sphere of radius R such that the series is absolutely convergent for all $r \geq R$, the convergence being uniform with respect to $\hat{\theta}$.

The expansion in the theorem is not to be confused with that in terms of the spherical basis functions as, for example, in Eq. (6.40). The individual terms $e^{ikr} u_n(\hat{\theta}) r^{-n}$ do not satisfy the Helmholtz equation. Moreover, the expansion is valid for complex k. Also the coefficients u_n satisfy a recurrence relation (Colton and Kress, 1983)

$$2iknu_n = n(n-1)u_{n-1} + Bu_{n-1}, \tag{6.60}$$

where

$$B = \frac{1}{\sin\theta} \partial_\theta (\sin\theta \partial_\theta) + \frac{1}{\sin^2\theta} \partial_\phi^2$$

is the *Laplace–Beltrami operator*. The solution $u(x)$ can be written as

$$u(x) = \frac{e^{ikr}}{r} u_0(\hat{\theta}) + O\left(\frac{1}{r^2}\right). \tag{6.61}$$

In the asymptotic limit of large r, the coefficient $u_0(\hat{\theta}) : \hat{\Omega} \to C$ is the *scattering amplitude* or the *far-field pattern* of u (will also be written as u_∞). Now if u_0 is identically zero, then from the expansion (6.59) and the recurrence relation (6.60), $u = 0$ outside a sphere enclosing the scatterer. Now by the analyticity property, u can be continued analytically inside the sphere excluding the singularity on the boundary. Hence it follows that

Theorem 6.6 . *Let $u \in C^2(R^3 \setminus \overline{\Omega})$ be the scattering solution of the Helmholtz equation. If the corresponding scattering amplitude vanishes identically, then $u = 0$ in the exterior $R^3 \setminus \overline{\Omega}$.*

A closely related theorem is:

Theorem 6.7 (Copson, 1975) *Let u be a solution of the Helmholtz equation having continuous bounded second derivatives everywhere and satisfying Sommerfeld's radiation condition at infinity. Then u is identically zero.*

Theorem 6.7 can be considered to be a generalization of *Liouville's theorem* in complex analysis (Churchill, 1960) according to which if a function is entire and bounded in the complex plane, then it is constant. An interesting conclusion follows from this theorem. Note that the solution u is the velocity potential of a wave motion. Sommerfeld's radiation condition at infinity implies that this wave motion is harmonic (being a spherically propagating wave) and expanding. Then Theorem 6.7 states that the wave motion must be due to sources at which the differentiability conditions are violated (Copson, 1975).

From Eq. (6.49)

$$\lim_{R \to \infty} \|u\|^2_{L_2(\Gamma_R)} \sim O(1).$$

Now suppose that the order 'O' is replaced by 'o' giving

$$\lim_{R \to \infty} \|u\|^2_{L_2(\Gamma_R)} \sim o(1).$$

Then from the expansion (6.59) and the recurrence relation (6.60) it follows that

$$\|u_0\|^2_{L_2(\hat{\Omega}_3)} \sim o(1).$$

Consequently, from Theorem 6.6, $u \sim o(1)$, i.e., u vanishes in $R^3 \setminus \overline{\Omega}$. Hence the following well-known result.

Lemma 6.1 (Rellich's Lemma). *Let $u \in C^2(R^3 \setminus \overline{\Omega})$ be a solution of the Helmholtz equation satisfying the radiation condition. If*

$$\lim_{R \to \infty} \|u\|^2_{L_2(\Gamma_R)} \sim o(1),$$

then u vanishes in $R^3 \setminus \overline{\Omega}$.

That u must vanish in the exterior to a sphere containing the obstacle if Rellich's Lemma applies, follows at once from Eq. (6.42). If $k \neq 0$, then Eq. (6.42) shows that the coefficients $\alpha_{\ell m}$ must vanish for all ℓ, m. Then from Eq. (6.41) u must vanish. Note that Theorem 6.8 is also a consequence of Rellich's Lemma. This follows immediately from Eq. (6.59) and the lemma itself. Note that for Rellich's lemma to hold, k can be neither imaginary nor zero.

The conditions under which $\sim O(1)$ can be replaced by $\sim o(1)$ can also be formulated. It is a straightforward consequence of Eq. (6.48) that for u to be zero in $R^3 \setminus \overline{\Omega}$,

$$Im\overline{k} \int_{\Gamma_R} \overline{u} u_r d\Gamma_R \leq 0.$$

Hence the following lemma.

Lemma 6.2. *Let $(\Delta + k^2)u = 0$ in Ω_e, $k = k' + ik''$, $k', k'' > 0$, and u satisfies the radiation condition. If*

$$\lim_{R \to \infty} \mathrm{Im} \bar{k} \int_{\Gamma_R} \bar{u} u_r \mathrm{d}\Gamma_R \le 0,$$

then u is zero in $R^3 \setminus \bar{\Omega}$.

An excellent discussion of Rellich's lemma appears in Hellwig (1960).

Finally, we note an interesting property of the scattering solution of the Helmholtz equation. Because of the bounds (6.39), it follows from Eq. (6.41) that the quantity $r^2|h_\ell^1(kr)|^2$ cannot increase with r. Therefore,

$$r^2|h_\ell^1(kr)|^2 \ge R^2|h_\ell^1(kR)|^2$$

if $r \le R$. This shows that the norm $\|u\|_{L_2(\Gamma_R)}^2$ must remain bounded by $\|u\|_{L_2(\Gamma_{\bar{R}})}^2$, where $\bar{R}$ is the radius of the smallest sphere circumscribing the scatterer.

6.9 The Helmholtz Scattering from Inhomogeneities

The analysis so far was for obstacles. The Helmholtz representations were obtained with reference to a physically bounding surface Γ of the scatterer which divided the space *distinctly* into an inside and an outside. By *distinctly* is meant that the two domains had their common points only on the boundary surface. In this sense, the boundary Γ (which is a normal domain, see Section 6.1) can be thought of as a *Jordan surface* in three space dimensions. Also note that the interior of the scatterer did not explicitly appear in the Helmholtz representations. This was, however, implicit in that the fields in these representations were those on the boundary and their determinations require that conditions be imposed on the boundary which may involve the scatterer interior depending upon the nature of the obstacle. These are the hallmarks of scattering by obstacles in which the material parameters undergo jump discontinuities at the boundary.

We now alter all this and consider a situation where the gradient of the velocity potential is continuous across the scatterer. We then obtain an *inhomogeneity*. One can visualize this by considering that the surface Γ is blended smoothly to the ambiance by being mollified with a smooth function. As a matter of fact, if the width of the mollifier is allowed to vanish, then the obstacle is recovered and an integral representation of the obstacle scattering can indeed be obtained in this limit (see Ström, 1991). There is no boundary condition in this case, but the radiation condition at infinity still applies. In addition, Rellich's lemma remains valid.

There are two fundamental equations for scattering from an inhomogeneity. These are the *classical* and the *plasma* wave equations (Friedlander, 1958; Balanis, 1997; see also Rose *et al.*, 1985; Cheney and Rose, 1988). The basic equation is (6.17) which is reproduced below for the sake of convenience.

$$[\Delta - c^{-2}(x)\partial_t^2]p(x,t)) = \frac{1}{\rho(x)}\nabla\rho(x) \cdot \nabla p(x,t).$$

Assume that the density be constant, but the bulk modulus is inhomogeneous in the scattering region. Equation (6.17) then reduces to

$$[\Delta - c^{-2}(x)\partial_t^2]p(x,t)) = 0. \tag{6.62}$$

This is the classical wave equation also called the *variable velocity* wave equation. The classical wave equation is, therefore, realized if the density is constant throughout the medium, but the bulk modulus or the compressibility remains inhomogeneous. Considering a time harmonic pressure wave as before, Eq. (6.63) reduces to

$$[\Delta + \omega^2 c^{-2}(x)]p(x,t)) = 0.$$

Noting that

$$\omega^2 c^{-2}(x) = (\omega^2 c_0^{-2})(c_0^2 c^{-2}(x)) = k_0^2 \epsilon(x), \epsilon(x) = n^2(x),$$

we obtain

$$[\Delta + k_0^2 \epsilon(x)]p(x) = 0 \tag{6.63}$$

in the exterior. In Eq. (6.63), ϵ is the *dielectric constant* and $n = c_0 c^{-1}(x)$ is the *refractive index* of the inhomogeneity. Since for any real medium, the refractive index is greater than unity, it follows that the variable velocity $c(x)$ must be less than the velocity c_0 in the host medium.

Next consider the transformation (Coen *et al.*, 1986) $p(t,x) = \sqrt{\rho(x)}\psi(t,x)$. In the new variable, Eq. (6.17) reduces to

$$[\Delta - v(x) - c^{-2}(x)\partial_t^2]\psi(t,x) = 0, \tag{6.64}$$

where

$$v(x) = \frac{3}{4}\left[\frac{|\nabla p|^2}{\rho^2} - \frac{2}{3}\frac{\nabla p}{\rho}\right].$$

Let the velocity be constant this time throughout the medium while the density varies with x. Equation (6.64) then assumes the form

$$[\Delta - v(x) - c_0^{-2}\partial_t^2]\psi(t,x) = 0. \tag{6.65}$$

Equation (6.66) is the plasma wave equation. Again assuming time harmonicity and upon Fourier transforming in time, the plasma wave equation becomes Schrödinger's equation

$$[\Delta + \lambda - v(x)]\psi(t,x) = 0.$$

We will say more on these equations later, in Chapter 8, where the uniqueness questions are discussed. Instead we turn our attention in the rest of this section to certain aspects of the classical wave equation (6.63).

For an inhomogeneity which is compactly supported in a regular domain $\Omega \subset R^3$, Eq. (6.63) can be written as

$$(\Delta + k_0^2 \epsilon(x))u(x) = 0, \quad x \in \Omega, \tag{6.66}$$

$$\lim_{r \to \infty} r[u'_{\text{sc}} - iku_{\text{sc}}] = 0, \quad r = |x|,$$

where the total field $u = u^{\text{inc}} + u_{\text{sc}}$. We can rewrite Eq. (6.66) in the form

$$(\Delta + k_0^2)u(x) = k_0^2(1 - \epsilon(x))u(x), \quad x \in \Omega. \tag{6.67}$$

In terms of the free-space Green's function, the integral equation of scattering corresponding to Eq. (6.67) is given by

$$u_{\text{sc}}(x) = k_0^2 \int_\Omega (1 - \epsilon(x))G^0(x, y; k_0)u(y)dy, \tag{6.68}$$

$\epsilon(x)$ being equal to unity beyond the support Ω. Equation (6.68) is of the form

$$
\begin{aligned}
u_{\text{sc}}(x) &= \int_\Omega \phi(x)G^0(x, y; k_0)u(y)dy, \\
&= \int_{R^3} \phi(x)G^0(x, y; k_0)u(y)dy, \quad x \in R^3,
\end{aligned}
\tag{6.69}
$$

where $\phi(x) = k_0^2(1 - \epsilon(x))$. The last line in Eq. (6.69) follows because the support of $\epsilon(x)$ is compact. Equation (6.69) is the same as that for a single-layer potential (discussed in Chapter 7) except for the fact that the integral is now a volume instead of a boundary integral. Accordingly, the solution defined by Eq. (6.69) is a *volume potential*.

Now Green's function $G^0(x, y; k_0)$ has $|x - y|^{-1}$ singularity. The potential, therefore, exists as an improper integral. Clearly, if the density $\phi(x)$ is Hölder continuous with Hölder exponent $\alpha \in (0, 1]$, then Eq. (6.69) can be converted into a regular integral. As a matter of fact, the regularity is given by a well-known result (Gilbarg and Trudinger, 1977, Theorems 4.11–4.13) in the theory of elliptic partial differential equations according to which if $\phi(x) \in C^{(0,\alpha)}(\Omega)$ and $u \in C^2(\Omega)$, then $u \in C^{(2,\alpha)}(\Omega)$ assuming that the refractive index $\epsilon \in C^2(\Omega)$. Then $u \in C^2(R^3)$ is a solution of the original equation (6.67) as can be verified by direct differentiation which is possible because of the regularity of the integral. The smoothness of the solution u, of course, increases with that of the density ϕ. Thus if $\phi \in C^{(1,\alpha)}(\Omega)$, then $u \in C^{(3,\alpha)}(\Omega)$. Contrarily, if $\phi \in C^0(\Omega)$, then $u \in C^{(1,\alpha)}(\Omega)$. Equation (6.68) which can be considered to be the Helmholtz representation in the exterior in the case of an inhomogeneity is the celebrated *Lippmann–Schwinger* equation originally derived in potential scattering in quantum physics (see, for example, Roman, 1965).

It is interesting to determine the norm of the integral operator of the Lippmann-Schwinger equation (6.69). If T_{LS} denotes the operator, then

$$\|T_{LS}\| \leq k_0^2 \|1 - \epsilon\|_\infty \max_{x \in B(0,R)} \int_{B(0,R)} |G^0(|x-y|; k_0)| dy, \quad x \in B(0,R)$$

$$\leq 4\pi k_0^2 \|1 - \epsilon\|_\infty \max_{x \in B(0,R)} \int_{B(0,R)} \||x-y|^{-1}| dy$$

$$= \frac{(k_0 R)^2}{2} \|1 - \epsilon\|_\infty, \tag{6.70}$$

where $B(0,R)$ is a ball of radius R centered at the origin, the support Ω of the inhomogeneity being enclosed by the ball. The calculation of integrals such as the above will be done in detail in Chapter 7. From Eq. (6.70), $\|T_{LS}\| < 1$ if

$$\frac{(k_0 R)^2}{2} \|1 - \epsilon\|_\infty < 1. \tag{6.71}$$

The condition can also be written in terms of the L_2-norm of $(1 - \epsilon)$ and we have

$$\overline{1 - \epsilon} \leq \frac{1}{(k_0 R)^2},$$

where $\overline{1 - \epsilon}$ represents the average of $(1 - n)$, that is

$$\overline{1 - \epsilon} = \sqrt{\frac{\|1 - \epsilon\|_{L_2}}{V}}, \tag{6.72}$$

where V is the volume of the region Ω. In this case, the solution can be written in terms of a Neumann series in the operator T_{LS} (see Chapter 4, Theorem 4.1). In other words, one can write the solution as

$$u_{sc}(x) = \sum_{j=0}^{\infty} T_{LS}^j u^{inc}.$$

Keeping only the zero-th and the first-order term in the above expansion and in the limit $|x| \to \infty$, the following interesting expression for the scattering amplitude results, namely

$$u_{B\infty}(\hat{x}; x^{\hat{inc}}) = \frac{k_0^2}{4\pi} \int_{R^3} (1 - \epsilon(y)) e^{ik_0(x^{\hat{inc}} - \hat{x}) \cdot y} dy, \tag{6.73}$$

where $\hat{x}, x^{\hat{inc}}$ denote unit vectors along the scattering and the incident field direction, respectively. The approximation made above by keeping terms only to the first-order is the well-known *Born approximation*. Equation (6.73) is the scattering amplitude in the Born approximation and this is why the subscript B was added to u_{sc} in the above equation.

Thus the Born approximation holds when scattering inside the inhomogeneity is negligible compared to the incident wave. The measure of the smallness is characterized by either of the equations (6.71) or (6.72). Note that it is only the product of $\overline{1 - \epsilon}$ and $(k_0 R)^2$ that is required to be small, not the quantities themselves. Therefore, arbitrary large values of $\overline{1 - \epsilon}$ are allowed provided that either k_0 and/or R is sufficiently small so as to satisfy the conditions in Eq. (6.71) or (6.72) and *vice versa*. In short, all that is required is for the total phase change $k_0 R$ to be negligible as the wave traverses the inhomogeneity. For a scatterer for which the deviation $(1 - \epsilon)$ is not negligible compared to unity, the Born condition can still be satisfied provided that the size of the scatterer is sufficiently small as to produce a negligible phase-shift. However, in most practical applications, the wavelength is many times larger than the object size. Therefore, from Eqs. (6.71) or (6.72) which characterize the approximation, it follows that the quantity $||1 - \epsilon||_\infty$, the magnitude of the deviation of the dielectric constant from its free-space value, must be negligible. This is often expressed by the statement that the scatterer must be *weak* and that the Born approximation is a *weak scattering approximation*. It also expresses the fact that only single scattering events are allowed and no multiple scattering can be considered.

It is interesting to note from Eq. (6.73) that if the Born approximation applies, then the scattering amplitude is simply the Fourier transform of the inhomogeneity in the conjugate variable $k_0(x^{\text{inc}} - \hat{x})$. This can be very useful in inverse solutions provided, of course, that the approximation is valid. However, for a given incident wave, the Ewald sphere (see example 5, Chapter 2) is of radius k_0 centered at the origin. So only frequencies up to $2k_0$ are available. Note that the center of the Ewald sphere which is at the origin cannot be shifted in the Born approximation. Any shift in the origin is tantamount to multiple scattering inside the inhomogeneity. It is also interesting to note that the condition of validity of the Born approximation for the Helmholtz equation is contrary to that of the Schrödinger equation. The former holds when the frequency is low whereas the opposite is the case for the Schrödinger equation.

As Eqs. (6.71) and (6.72) indicate, the Born solution is a volume-dependent solution. Within the framework that the scattered field is small compared to the incident field, there exists the well-known *Rytov approximation* that is not dependent upon the volume of the inhomogeneity. However, Rytov's is a high-frequency approximation and, therefore, is not of direct interest to the analysis in the resonance region which is what we are concerned with here. We, therefore, do not go into the details of the Rytov case. Analogously for the obstacle problem, there is the so-called *physical optics far-field approximation* for high frequencies and is also not discussed for the same reason. The approximations such as those of Born or Rytov are called *linear* because of the fact that replacing the total field in the scattering integral (6.68) by the incident field is equivalent to neglecting the nonlinearity of the problem which arises from the dependence of the field (inside the integral) on the inhomogeneity. The linear approximations have given rise to a number of important inversion techniques of which the most well-known is the *diffraction tomography* and are applied to numerous problems in nondestructive evaluation and medical imaging. Finally, we

would like to mention that there exist methods that take into account stronger
scattering situations. These go by the generic name of the *distorted wave Born
approximation*. It is assumed that the scatterer consists of a known (not necessarily weak) background on which is superimposed a weakly varying inhomogeneity. The free-space Green's function in the Lippmann-Schwinger Eq. (6.68)
is replaced by the one corresponding to the background medium. The unknown
weak inhomogeneity is, however, considered in the Born approximation. The
literature on the linearized approximations, their theories, applications and related topics is vast. We only mention the review articles by Bates *et al.* (1991)
and Langenberg (1987) and the text by Carrion (1987) which contain extensive
bibliographies on the subject.

Remarks

Thus far, we have obtained two representations of a scattering solution
of the Helmholtz equation. In three space dimensions, these are Eq. (6.40),
which is also known as the *Rayleigh series*, and the Atkinson–Wilcox expansion,
Eq. (6.59), along with the recursion relation, Eq. (6.60). In their respective domains of definition, the two are equivalent. In three dimensions, we consider
this domain to be the exterior of the smallest sphere circumscribing the scatterer
(with respect to a center located in its interior). The sphere, of course, becomes
a circle if the scattering is considered in two space dimensions. However, when
we enter the interior of the circumscribing sphere or circle, the validity of the
representations becomes questionable. The main issues that arise in the latter
case are discussed in Section 9.4 and in Appendix A.9.3.

Now the spherical harmonics $Y_{\ell m}(\hat{\theta})$ constitute a complete orthonormal set
in $L_2(\hat{\Omega}_3)$ (see Appendix A.8.1). The scattering function $f(\hat{\theta})$ in the Atkinson–
Wilcox expansion can, therefore, be expressed as

$$f(\hat{\theta}) = \sum_{\ell=0}^{\infty} \sum_{m=-\ell}^{\ell} f_{\ell m} Y_{\ell m}(\hat{\theta}).$$

The coefficients $f_{\ell m}$ above and those given by $\alpha_{\ell m}$ of Eq. (6.40) are the same.
In order to see this, it is only necessary to substitute the expansion of the
spherical Hankel function $h_\ell^{(1)}(z)$ (Abramowitz and Stegun, 1964; Gradshteyn
and Rhyzik, 1980), namely

$$h_\ell^{(1)}(z) = \sum_{m=0}^{\ell} \frac{(\ell+m)!}{m!(\ell-m)!} \frac{1}{2^m} \frac{e^{\left(i\left[z - \frac{(\ell-m-1)\pi}{2}\right]\right)}}{z^{m+1}}$$

in Eq. (6.40), simplify and then compare the result with Eq. (6.59). Moreover,
to the order of $(k|x|)^{-2}$, the far-field ($k|x| \gg 1$) is completely determined by
the scattering amplitude $f_\infty(\hat{\theta})$.

Note that in three space dimensions, the terms in the series given by Eq. (6.40)
can be arranged in powers of $(k|x|)^{-1}$. In two dimensions, however, this turns
out not to be the case. Consider the 2-D expression corresponding to the 3-D

Eq. (6.40) which takes the form

$$u(x) = \sum_{\ell=-\infty}^{\infty} \alpha_\ell H_\ell^{(1)}(k|x|)e^{i\ell\theta}.$$

Let us next consider the expansion for the cylindrical Hankel function $H_\ell^{(1)}(z)$. It is given by (Abramowitz and Stegun, 1964; Gradshteyn and Rhyzik, 1980)

$$H_\ell^{(1)}(z) = (-i)^\ell \left[\frac{2}{\pi z}\right]^{\frac{1}{2}} e^{i(z-\frac{\pi}{4})}$$
$$\times \sum_{m=0}^{\infty} \frac{(-1)^m \Gamma\left(\ell+m+\frac{1}{2}\right)}{(2iz)^m m! \Gamma\left(\ell-m+\frac{1}{2}\right)}.$$

Now the ratio of the Gamma functions appearing in the above expression can be written as (Gradshteyn and Rhyzik, 1980)

$$\frac{\Gamma\left(\ell+m+\frac{1}{2}\right)}{\Gamma\left(\ell-m+\frac{1}{2}\right)} = 2^{2(1-m)} \prod_{p=1}^{m} \left\{\ell^2 - \left(p-\frac{1}{2}\right)^2\right\}.$$

Next define the operator Kyurchan et al., (1996)

$$D_\theta = \prod_{p=1}^{\infty} \frac{(-1)^m}{2^{2(m-1)}} \frac{1}{(2i)^m} \frac{1}{m!} \left[\partial_\theta^2 - \left(p-\frac{1}{2}\right)^2\right].$$

In terms of D_θ, we recast the scattering solution $u(x)$ as

$$u(x) = \left[\frac{2}{\pi z}\right]^{\frac{1}{2}} e^{i(z-\frac{\pi}{4})} \sum_{\ell=-\infty}^{\infty} (-i)^\ell \frac{1}{z^\ell} D_\theta e^{i\ell\theta}.$$

This clearly shows that the two-dimensional solution cannot be arranged entirely as a series in inverse powers of $k|x|$ as opposed to the three-dimensional case.

As pointed out by Kyurkchan et al. (1996), there is also a third representation of a scattering solution of the Helmholtz equation. This is obtained using the *Weil representation* (also called the *Sommerfeld–Weil representation*) which we have already encountered in Example 5 in Chapter 2 on the inverse source problem. In 3-D, the Sommerfeld-Weil representation gives

$$u(x) = \frac{1}{2i\pi} \int_0^{2\pi} \int_0^{\frac{\pi}{2}+i\infty} f(\theta_1,\theta_2)$$
$$\times e^{\{ik|x|[\sin\theta\cos\theta_1\cos(\phi-\theta_2)+\cos\theta\cos\theta_1]\}} \sin\theta_1 d\theta_1 d\theta_2,$$

where $x = (|x|, \theta, \phi)$. The corresponding 2-D version is

$$u(|x|, \theta) = \frac{1}{\pi} \int_{-i\infty}^{\pi+i\infty} f(\theta)e^{[-ik|x|\cos(\phi-\theta)]}d\phi.$$

The integrations are over the paths in the complex $\hat{\theta}$- and complex θ-plane. For further details, see Kyurkchan *et al.* (1996).

Finally, we would like to point out that in this text, we will consider three boundary conditions for obstacle scattering. These are:
The Dirichlet Boundary Condition:

$$(\Delta + k^2)u^{\mathrm{sc}}(x) = 0 \text{ in } x \in \Omega,$$

$$u^{\mathrm{sc}}(x) = f(x), \ x \in \Gamma.$$

The Neumann Boundary Condition:

$$(\Delta + k^2)u^{\mathrm{sc}}(x) = 0 \text{ in } x \in \Omega,$$

$$\partial_n u^{\mathrm{sc}}(x) = g(x), \ x \in \Gamma$$

and the Transmission Boundary Condition:

$$(\Delta + k^2)u^{\mathrm{sc}}_+(x) = 0 \text{ in } x \in \Omega_e,$$

$$(\Delta + k^2)u^{\mathrm{sc}}_-(x) = 0 \text{ in } x \in \Omega_i,$$

$$u^{\mathrm{sc}}_+(x) - u^{\mathrm{sc}}_-(x) = -u^{\mathrm{inc}}, \ x \in \Gamma$$

$$\partial_n[u^{\mathrm{sc}}_+(x) - u^{\mathrm{sc}}_-](x) = -\mathcal{F}u^{\mathrm{inc}}, \ x \in \Gamma.$$

In Dirichlet and Neumann problems, Ω is either the interior (Ω_i) or exterior (Ω_e) domain giving rise to the interior and exterior problem, respectively. The quantity $\mathcal{F}$ in the transmission problem depends on the material parameters of the interior and exterior medium. For an acoustic medium, this parameter is the mass density, whereas in an electrodynamical problem, this would be the dielectric constant. We also mention in passing that for scalar problems, the analysis presented in this book can be applied to both acoustical and electrodynamical scattering.

We conclude our remarks by pointing out the following two things. On many occasions, the scattering solution was assumed to be in $C^2(\overline{\Omega})$ or $C^2(\Omega) \cap C(\overline{\Omega})$, and so forth. Now, the Helmholtz equation, if written as $\Delta u = -k^2 u$, must require u to be infinitely differentiable since the Laplacian Δ reduces the order of smoothness by two, whereas that of $-k^2 u$ on the R.H.S. remains unchanged. This seeming contradiction, however, is not real. In other words, solutions of the Helmholtz equation which are in the restricted function spaces as in the above are also in C^∞. This is by virtue of the so-called *regularity theorems* in elliptic partial differential equations. Essentially, this can be looked upon as a generalization to the Helmholtz equation of one of the fundamental properties of a harmonic function in a bounded domain. The reader may recall that if a function is harmonic (i.e., a solution of Laplace's equation $\Delta u = 0$) in a bounded domain, it is also C^∞ there. The analysis of the question of regularity of solutions is perhaps the most technical in the mathematics of the PDE's, and the interested reader is referred to the classical texts such as Gilbarg and

Trudinger (1977) for details. A lucid and highly readable account of the subject appears in Evans (1991).

We already mentioned that the scattering solutions of the Helmholtz equation are analytic in all their arguments. Moreover, they belong to the class of the so-called *real analytic functions*. Let us recall the definition of a real analytic function.

Definition (Real analytic function). *A function* $f : R^n \to R^1$ *is called real analytic at a point* $x_0 \in R^n$ *if* $\exists$ *a quantity* $\rho > 0$ *and constants* $\{f_\alpha\}$ *such that*

$$f(x) = \sum_\alpha f_\alpha(x - x_0)^\alpha, \quad |x - x_0| < \rho.$$

The sum is taken over all multiindices α *and the constants* $\{f_\alpha\}$ *are given by*

$$f_\alpha = \left(\frac{1}{\alpha} D^\alpha f\right)\Big|_{x=x_0},$$

where $\alpha! = \Pi_1^n \alpha_i$.

In other words, the function must have a Taylor series, as is expected, if it must be analytic. The analyticity may be over the entire real axis or an interval of it. However, the radius of convergence of the series may vary with the location of the coordinate x_0 on the real axis. Functions such as $\cos x$ and e^x have infinite radius of convergence. Put alternatively, these functions can be obtained at any x_0 from their values at the origin. This may not hold in general where the radius of convergence may depend upon x_0. This is an important point and will be discussed further in Section 9.5 in connection with the recent work of Bruno and Reitich (1992).

Chapter 7

The Direct Scattering Problems

7.1 Introduction

As already mentioned, solving an inverse problem requires knowledge of the associated direct problem. In the present case, it consists of solving the Helmholtz equation. The primary concerns are about the existence and uniqueness of the solutions. That is, does a solution exist to the Helmholtz equation with a given boundary condition, and if so, is it unique? These are addressed in this chapter. If the existence is assumed, then establishing uniqueness is relatively straightforward. Indeed much of what is needed in this regard already appeared in Chapter 6. Proving existence, however, turns out to be surprisingly involved. The analysis of the existence of a solution is traditionally carried out within the framework of integral equations *via* the so-called *layer potentials*. It may be recalled that these potentials are traditionally used in the mathematical treatment of Laplace's equation and form what is known as the *potential theory* (Günter, 1968; Kellogg, 1953; Landkof, 1972). Rigorous modern treatments of the subject are Colton and Kress, 1983, and McLean, 2000). As a matter of fact, much of the potential theory for Helmholtz's equation parallels that of Laplace's equation for harmonic functions. This is not surprising since Green's functions for the two equations differ only in an exponential factor $\exp(\mathrm{i}kr)$. Otherwise, the fundamental solutions for both equations have the same r^{-1} singularity. The exponential factor, however, does not alter things in a fundamental way. Indeed it can be dispensed with in much of the calculations for the Helmholtz equation as will be shown in the sequel.

The present chapter has two objectives: (a) the analysis of the potentials and (b) their applications to the solution of the Helmholtz equation. Our objective here is not to embark on an all-encompassing introduction to the subject which has innumerable subtleties and nuances and is charged with deep mathematics. Instead we aim at presenting only those details as are necessary for answering the questions addressed here. The discussions below are based primarily on Colton and Kress (1983), Folland (1995) and DiBenedetto (1995).

7.2   The Layer Potentials

In order to motivate the discussion, let us consider the interior Dirichlet problem for the Laplace equation in a sufficiently regular region $\Omega \in R^3$: $\Delta w(x) = 0$, $x \in \Omega$, $w = f$ on Γ. It is known (see, for example, Folland, 1995) that if the solution $w \in C^1(\Omega)$ is harmonic in Ω, then it can be written as

$$w(x) = \int_{\Gamma} [w(y)\partial_y F(x,y) - F(x,y)\partial_y w(x)]d\Gamma(y), \ x \in \Omega.$$

$$F(x,y;k) = (4\pi|x-y|)^{-1}, \ x,y \in R^3, \ x \neq y$$

is the fundamental solution of Laplace's equation.

$$\partial_y = \hat{n}(y) \cdot \nabla_y F = \underset{t \to 0}{\text{limit}}(\hat{n}(y) \cdot \nabla F(x, y - t\hat{n}(y))).$$

$t > 0$ is real. $\hat{n}(y)$ is the outward normal to Γ at y. It is also known that if a solution of the Dirichlet problem exists, then it is determined uniquely by the boundary data f alone. Let us, therefore, write the solution as

$$w(x) = -\int_{\Gamma} f(y)\partial_y F(x,y)d\Gamma(y), \ x \in \Omega. \tag{7.1}$$

Now w is harmonic because $\partial_y F(x,y;k)$ is harmonic. However, Eq. (7.1) will not reduce to the boundary data f in general because $\partial_y F(x,y)$ is not identical to a delta distribution. However, we can try to replace f in Eq. (7.1) by a function ψ which is defined continuously on Γ, consider instead the equation

$$v(x) = \int_{\Gamma} \psi(y)\partial_y F(x,y)d\Gamma(y), \tag{7.2}$$

and determine ψ from a boundary integral equation satisfying the boundary data f. Indeed it turns out that this equation is given by $v|_{\Gamma} = -\frac{1}{2}\psi + T\psi$. T is a compact operator. The solution to the above interior Dirichlet problem for the Laplace equation can, therefore, be obtained by requiring that ψ satisfy the boundary integral equation $-\frac{1}{2}\psi + T\psi = f$. Since T is compact, there exist techniques for handling such operator equations. The solution $v(x)$ given by Eq. (7.2) is called the *double-layer potential* and $\psi(y)$ its *moment* or *density*.

The physical meaning of $v(x)$ is as follows. Let γ be a surface parallel to Γ, that is, $\gamma = \{y|y = Y + t\hat{n}(Y), \forall Y \in \Gamma\}$, $t > 0$ real, and define

$$v_t(x) = t^{-1}\left[\int_{\gamma} F(x,y)\psi(y)d\gamma(y) - \int_{\Gamma} F(x,Y)\psi(Y)d\Gamma(y)\right].$$

Suppose that $-t^{-1}\psi(Y)$, $t^{-1}\psi(y)$ represent the densities of charge distributions on Γ and γ, respectively. Then $\text{limit}_{t \to 0} v_t(x)$ is the potential at x due to a distribution of dipoles of moment $\psi(Y)$ on Γ, the dipoles being oriented normally.

As an aside, let $\psi = 1$ in Eq. (7.2) and denote the resulting double-layer potential by DL1(x). Then viewing from the "inner" side of the layer, we have

$$
\begin{aligned}
\text{DL1}(x) &= \int_\Gamma \partial_y F(x,y) \mathrm{d}\Gamma(y) = \frac{1}{4\pi} \int_\Gamma \hat{n}(Y) \cdot \nabla_y |x-y|^{-1} \mathrm{d}\Gamma(Y) \\
&= \frac{1}{4\pi} \int_\Gamma \frac{(x-y) \cdot \hat{n}_y}{|x-y|^3} \mathrm{d}\Gamma(y)) = -\frac{1}{4\pi} \int_\Gamma \frac{\cos \theta}{|x-y|^2} \mathrm{d}\Gamma(Y) \\
&= \frac{1}{4\pi} \omega(x),
\end{aligned}
$$

where $\omega(x)$ is the solid angle subtended at x by Γ. For $x \in \Gamma$, this is just 2π and, consequently, DL1(x) = $\frac{1}{2}$.

In the context of the Helmholtz equation, the fundamental solution F is to be replaced by the free-space Green's function $G^0(x,y;k)$ and the double-layer potential of Eq. (7.2) becomes

$$
v(x) = \int_\Gamma \psi(y) \partial_y G^0(x,y;k) \mathrm{d}\Gamma(y). \tag{7.3}
$$

Analogously, a *single-layer potential* $u(x)$ is defined as

$$
u(x) = \int_\Gamma G^0(x,y;k) \phi(y) \mathrm{d}\Gamma(y). \tag{7.4}
$$

$u(x)$ is the potential at any point x in space induced by a charge distribution on Γ with density $\phi(y)$. Reversing the sign of ϕ and differentiating yields the field.

Physically, the fields are either emissions from a vibrating structure or diffraction of waves by a body. In either case, the physical sources creating the fields are idealized by a distribution of monopoles and dipoles placed on the objects. The structure is considered to be broken down into infinitesimal elemental surface areas on each of which is concentrated a monopole or a dipole with strength proportional to the elemental surface area. In the continuum limit, these are summed up into the integrals of Eqs. (7.3) and (7.4). For more about the physics of the potentials, the interested reader is referred to the well-known text on electrodynamics by Jackson (1975).

Later it will be necessary to calculate the normal derivative of the single-layer potential u. Hence we need to deal with the surface integrals on Γ the kernels of which are the derivatives of the Green's function with respect to x. It is thus necessary to consider integrals of the type:

$$
I = \int_\Gamma \partial_z G^0(x,y;k) \mathrm{d}\Gamma(y),
$$

where z is either x or y, $x,y \in \Gamma$. Since $G^0(x,y;k) = G^0(|x-y|;k)$, it follows that $\partial_y G^0(x,y;k) = \partial_x G^0(x,y;k)$ evaluated at (y,x) instead of (x,y). The functions $\partial_z G^0$ are so important that we designate them by special symbols and write $\partial_y G^0(x, y; k) = K(x, y)$ and $\partial_x G^0(x, y; k) = K^*(y, x) = K(y, x)$ to be used as kernels of the corresponding integral operators

$$(T_K f)(x) = \int_\Gamma K(x, y) f(y) d\Gamma(y) \tag{7.5}$$

and

$$(T_{K^*} f)(x) = \int_\Gamma K^*(x, y) f(y) d\Gamma(y), \tag{7.6}$$

respectively.   Let us also introduce an operator S in order to represent the single-layer potential $u(x)$. That is,

$$u(x) = (S\phi)(x) = \int_\Gamma G^0(x, y; k)\phi(y) d\Gamma(y). \tag{7.7}$$

Furthermore, it must be pointed out that the potentials u and v are solutions to Helmholtz's equation (interior and exterior) and satisfy the radiation condition at infinity.

Before proceeding, let us make some remarks regarding the replacement of the free-space Green's function

$$G^0(x, y; k) = \frac{1}{4\pi} \frac{e^{ik|x-y|}}{|x - y|}$$

by the potential theoretical fundamental solution

$$g^0(x, y) = \frac{1}{4\pi} \frac{1}{|x - y|} \tag{7.8}$$

(which is simply $G^0(x, y; 0)$) in much of the calculations involving the layer potentials.

7.3   Replacing $G^0(x, y; k)$ by $g^0(x, y)$.

An important objective in the analysis of the potentials is to determine their regularity or smoothness. One would like to know how smooth these functions are given the smoothness of the boundary and the moment functions. The smoothness, of course, depends upon how $|w(x_1) - w(x_2)|$ (w is either u or v) behaves with $|x_1 - x_2|$. In the final analysis, this reduces to examining the behavior of $|A(x_1, y) - A(x_2, y)|$ with x_1 and x_2, A being the kernel of w. Let us show that the calculations below can be carried out with $g^0(x, y)$ instead of the free-space Green's function $G^0(x, y; k)$.

Consider first the single-layer operator. Now

$$
\begin{aligned}
4\pi |G^0(x_1,y;k) - G^0(x_2,y;k)| \;\le\; & \left| |x_1-y|^{-1}e^{ik|x_1-y|} - |x_2-y|^{-1}e^{ik|x_2-y|} \right| \\
= \;& \left| \left(|x_1-y|^{-1} - |x_2-y|^{-1} \right) e^{ik|x_2-y|} \right. \\
& \left. - |x_1-y|^{-1} \left(e^{ik|x_2-y|} - e^{ik|x_1-y|} \right) \right| \\
\le \;& \left| |x_1-y|^{-1} - |x_2-y|^{-1} \right| \\
& + |x_1-y|^{-1} \left| e^{ik|x_2-y|} - e^{ik|x_1-y|} \right|.
\end{aligned}
\tag{7.9}
$$

From the mean value theorem of differential calculus[1] (Hille, 1964; Taylor and Mann, 1983) and the *triangle inequality*[2]

$$
\left| e^{ik|x_1-y|} - e^{ik|x_2-y|} \right| \le k|x_1-x_2|.
\tag{7.10}
$$

Substituting Eq. (7.10) in Eq. (7.9) yields

$$
\begin{aligned}
4\pi |G^0(x_1,y;k) \;-\; & G^0(x_2,y;k)| \le \left| |x_1-y|^{-1} - |x_2-y|^{-1} \right| \\
& + k|x_1-x_2||x_1-y|^{-1}.
\end{aligned}
\tag{7.11}
$$

Now

$$
\begin{aligned}
\left| |x_1-y|^{-1} - |x_2-y|^{-1} \right| \;=\; & \frac{\big| |x_1-y| - |x_2-y| \big|}{|x_1-y||x_2-y|} \\
\le \;& \frac{|x_1-x_2|}{|x_1-y||x_2-y|}
\end{aligned}
\tag{7.12}
$$

using the triangle inequality to $\big| |x_1-y| - |x_2-y| \big|$.

Let $x_1 \in \Gamma$ and consider a surface patch $\Gamma_\epsilon(x_1) = \{x \in \Gamma; |x_1-x| \le \epsilon\}$ of Γ.[3] Let x_2 be a point in the neighborhood V of Γ and let $y \in \Gamma \setminus \Gamma_\epsilon(x_1)$. In addition, restrict x_2 within the ball $B_{\frac{1}{2}\epsilon}(x_1)$. Then from simple geometric arguments, we have

$$
|x_2-y| \ge \frac{1}{2}|x_1-y|.
$$

---

[1]The mean value theorem states that if f is a real or complex-valued function of a variable z, is continuous in an interval $[a,b]$ and has a unique finite derivative at each point in (a,b), then $\exists$ at least one point z^* on the line joining a and b such that

$$
|f(z_1) - f(z_2)| \le |f'(z^*)||z_1-z_2|, \; z_1, z_2 \in (a,b),
$$

where $f'(z^*)$ is the derivative of f at $z^* \in (z_1,z_2)$. Alternately, one can write $z^* = \lambda z_1 + (1-\lambda)z_2$ and the mean value is simply the differential (with respect to λ) of $f(\lambda z_1 + (1-\lambda)z_2)$.

[2]By triangle inequality: $|a-b| \ge ||a|-|b||$. (Gradshteyn and Rhyzhik, 1980)

[3]A geometrical visualization of the patch can be described thus. Consider a ball $B_\epsilon(x_1)$ of radius ϵ centered around x_1. Then the patch $\Gamma_\epsilon(x_1)$ is the intersection of this sphere with Γ. In other words, $\Gamma_\epsilon(x_1) = B_\epsilon(x_1) \cup \Gamma$. Alternatively, one can consider a circle $C_\epsilon(x_1)$ of radius ϵ on the tangent plane T_{x_1} at x_1 with its center at x_1. Then the patch $\Gamma_\epsilon(x_1)$ is that vicinity of Γ around x_1 the projection of which is on the tangent plane T_{x_1} is the circle $C_\epsilon(x_1)$.

Using the above result in Eq. (7.12), it follows that

$$\left||x_1 - y|^{-1} - |x_2 - y|^{-1}\right| \leq \frac{2|x_1 - x_2|}{|x_1 - y|^2}. \tag{7.13}$$

Finally, upon combining Eqs. (7.11), (7.13) and the inequality just above Eq. (7.13), yields

$$4\pi|G^0(x_1, y; k) - G^0(x_2, y; k)| \leq \frac{2|x_1 - x_2|}{|x_1 - y|^2} + k|x_1 - x_2||x_1 - y|^{-1}. \tag{7.14}$$

A geometrical illustration of the above inequality (see also Ström, 1991) is shown in Figure 7.1.

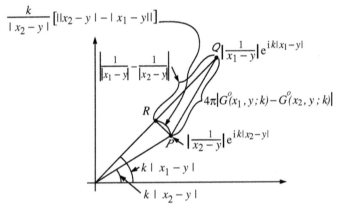

Figure 7.1 A geometric illustration of the inequality (7.14): $PQ \leq PR + PQ$.

In the computations afterwards, it will be necessary to integrate the quantity $|G^0(x_1, y; k) - G^0(x_2, y; k)|$ over surface patches $\Gamma_\epsilon(x_1)$ and $\Gamma_{\epsilon c}(x_1) = \Gamma \backslash \Gamma_\epsilon(x_1)$, the subscript c denoting complement. The integral involving $|x - y|^{-1}$ over Γ_ϵ goes to zero as $\epsilon \to 0$ as demonstrated below. Let Γ be smooth enough (e.g., $\Gamma \in C^2$) so that a tangent plane can be defined at each point x on it. There exists a one-to-one and onto map between $\Gamma_\epsilon(x)$ and its projection on the tangent plane at x. In the vicinity of x, the surface can be parameterized by $z = z(\rho, \theta)$, where (ρ, θ) define a polar coordinate on the tangent plane at x. Moreover, we have the following result from Appendix A.7.3, namely

$$|\hat{n}(x) \cdot \hat{n}(y)| \geq \frac{1}{2},$$

where $\hat{n}(y)$ is the unit normal at a point y on Γ. Let $d\Gamma$ be the oriented surface element at y and dS be the magnitude of its projection on the tangent plane. Then $dS(z) = \hat{n}(x) \cdot d\Gamma = |\hat{n}(x) \cdot \hat{n}(y)| d\Gamma \geq \frac{1}{2}d\Gamma$. Therefore, $d\Gamma \leq 2dS =$

$2\rho d\rho d\theta$. This results in

$$\int_{\Gamma_\epsilon(x)} |x - y|^{-1} d\Gamma(y) \;\leq\; 2 \int_0^{2\pi} \int_0^\epsilon \rho |x - y|^{-1} d\theta d\rho$$

$$\leq\; 2 \int_0^{2\pi} d\theta \int_0^\epsilon d\rho = 4\pi\epsilon. \qquad (7.15)$$

The integral, therefore, vanishes as $\epsilon \to 0$. Note that Eq. (7.15) describes an improper integral which exists uniformly in ϵ (see Appendix 7.6). Over $\Gamma_{\epsilon c}(x)$, where $x \neq y$, the integral is regular. Thus in so far as establishing the regularity of $u(x)$ is concerned, it is sufficient to consider only the first term on the R.H.S. of Eq. (7.14). However,

$$4\pi |g^0(x_1, y) - g^0(x_2, y)| \leq \frac{2|x_1 - x_2|}{|x_1 - y|^2}$$

obtained simply by setting $k = 0$ in Eq. (7.14) Consequently, it suffices to consider only the potential theoretical fundamental solution $g^0(x, y)$ of Eq. (7.8).

An identical conclusion holds for the double-layer potential $v(x)$ also following essentially the same line of reasoning as above. Toward that, define

$$A_i = (x_i - y) \cdot \hat{n}(y) |x_i - y|^{-3}$$
$$B_i = (x_i - y) \cdot \hat{n}(y) |x_i - y|^{-2}$$
$$D_{ij} = e^{ik|x_i - y|} - e^{ik|x_j - y|}.$$

In the case of the double-layer potential, we replace G^0 by K (see section 7.2), and obtain

$$4\pi[K(x_1, y) - K(x_2, y)] \;=\; \left[(A_1 - A_2)e^{ik|x_2 - y|}\right] + A_1 D_{12}$$

$$-ik\left[(B_1 - B_2)e^{ik|x_2 - y|}\right] - ik B_1 D_{12}. \;(7.16)$$

Using Eq. (7.10), the fact (Appendix A.7.2) that $|A_i| \leq C|x_i - y|^{-1}$ and $|B_i| \leq C$, Eq. (7.16) reduces to

$$4\pi |K(x_1, y) - K(x_2, y)| \leq |A_1 - A_2| + Ck|x_1 - x_2| \left[|x_1 - y|^{-1} + k\right]. \quad (7.17)$$

By the same reasoning as for the single-layer potential above, it is sufficient to consider only the term containing $|A_1 - A_2|$ on the R.H.S of Eq. (7.17). Again since

$$4\pi |g^0(x_1, y) - g^0(x_2, y)| \leq |A_1 - A_2|,$$

the conclusion follows. The same applies to the estimation of the important integral identity

$$\int_\Gamma |\hat{n}(y) \cdot \nabla_y G^0| d\Gamma \leq C$$

as demonstrated explicitly in Appendix A.7.3.

Before concluding this section, let us demonstrate the compactness and adjoint character of the operators T_K, T_K^* and S, which were defined in Section 7.2. The compactness of K can be shown as follows. By definition

$$(T_K f)(x) = \int_\Gamma K(x, y) f(y) d\Gamma(y).$$

Using Appendix A.7.2, it follows that $|K(x, y)| \leq C|x - y|^{-1}$, where C is a generic constant. We now use Eq. (7.15) and find that $||T_K f||_\infty \leq C\epsilon ||f||_\infty$. Now consider the surface patch $\Gamma_\epsilon(x)$ and define a function $K_\epsilon(x, y)$ such that $K_\epsilon(x, y) = K(x, y)$, $y \in \Gamma_\epsilon(x)$, and is zero otherwise. Consider now the equation

$$
\begin{aligned}
(T_K - T_{K_\epsilon})f &= \int_\Gamma (K - K_\epsilon)(x, y) f(y) d\Gamma(y) \\
&= \left(\int_{\Gamma_\epsilon(x)} + \int_{\Gamma_{\epsilon c}(x)} \right)(K - K_\epsilon)(x, y) f(y) d\Gamma(y).
\end{aligned}
$$

Then

$$||(T_K - T_{K_\epsilon})f||_\infty \leq C\epsilon ||f||_\infty.$$

Thus $||T_K - T_{K_\epsilon}|| \to 0$ as $\epsilon \to 0$. This proves the compactness of T_K in $C(\Gamma)$.

Consider $K : C(\Gamma) \to C(\Gamma)$, and define the bilinear form on $C(\Gamma) \times C(\Gamma)$ as

$$< \phi, \psi > = \int_\Gamma \phi(y) \psi(y) d\Gamma(y).$$

Then

$$
\begin{aligned}
< K\phi, \psi > &= \int_\Gamma (K\phi)(y) \psi(y) d\Gamma(y) \\
&= \int_\Gamma \int_\Gamma K(x, y) \phi(x) \psi(y) d\Gamma(x) d\Gamma(y).
\end{aligned}
$$

Since $\mathrm{limit}_{n \to \infty} K_n z \to Kz$ uniformly (just demonstrated above), we can write

$$
\begin{aligned}
< K\phi, \psi > &= \int_\Gamma \left[\int_\Gamma \left\{ \lim_{n \to \infty} K_n(x, y) \right\} \phi(x) d\Gamma(x) \right] \psi(y) d\Gamma(y) \\
&= \int_\Gamma \left[\lim_{n \to \infty} \int_\Gamma K_n(x, y) \phi(x) d\Gamma(x) \right] \psi(y) d\Gamma(y) \\
&= \int_\Gamma \phi(x) \left[\lim_{n \to \infty} \int_\Gamma K_n(x, y) \psi(y) d\Gamma(y) \right] d\Gamma(x) \\
&= \int_\Gamma \phi(x) \left[\int_\Gamma K(x, y) \psi(y) d\Gamma(y) \right] d\Gamma(x) = < \phi, K\psi > .
\end{aligned}
$$

The interchange of the order of integration is mathematically justified because of the uniform (i.e. operator norm) convergence of K_n to K. Thus K^* is adjoint of K, and since K is compact, so is K^*. Moreover, by definition of the bilinear form (see above), S is self-adjoint.

We now discuss the potentials beginning with the double-layer potential v.

7.4   The Double-layer Potential

The double-layer potential $v(x)$ is defined as:

$$v(x) = \int_\Gamma \hat{n}(y) \cdot \nabla_y \, G^0(x,y;k)\psi(y)\mathrm{d}\Gamma(y), \ x \in R^3 \setminus \Gamma,$$

and

$$v(x) = \int_\Gamma K(x,y)\psi(y)\mathrm{d}\Gamma(y) = (T_K\psi)(x), \ x \in \Gamma.$$

where $\psi(y) \in C(\Gamma)$. The potential is well-defined. We show that $v(x)$ can be extended continuously to Γ from anywhere in its toroidal neighborhood V (see Section 6.1). That is, $\exists \ \delta, \ \eta$ such that

$$|v(x) - v(x_0)| < \eta \text{ whenever } |x - x_0| < \delta, \ x \in V, \ x_0 \in \Gamma,$$

although the extension may not be continuous across Γ where it may have a jump. This is certainly true for a constant density as is evident from the lemma of Appendix A.7.1. In this case, $v = 0$ in $\Omega_e, -1$ in Ω and $-1/2$ on Γ.

In order to extend the result for a general moment function ψ, consider first a point $x \in \Omega$. Thus $x = x(t) = x_0 - t\hat{n}(x_0), t > 0, x_0 \in \Gamma, \hat{n}(x)$ being the outward normal. From the definition in Eq. (7.3) and replacing G^0 by g^0, we have

$$
\begin{aligned}
v(x(t)) &= \int_\Gamma \hat{n}(y) \cdot \nabla_y g^0(x(t),y)\psi(y)\mathrm{d}\Gamma(y) \\
&= \int_\Gamma \hat{n}(y) \cdot \nabla_y g^0(x(t),y)[\psi(y) - \psi(x_0)]\mathrm{d}\Gamma(y) \\
&\quad +\psi(x_0)\int_\Gamma \hat{n}(y) \cdot \nabla_y g^0(x(t),y)\mathrm{d}\Gamma(y) \\
&= -\psi(x_0) + \int_\Gamma \hat{n}(y) \cdot \nabla_y g^0(x(t),y)[\psi(y) - \psi(x_0)]\mathrm{d}\Gamma(y). \quad (7.18)
\end{aligned}
$$

In deriving Eq. (7.18) use was made of the result of Appendix A.7.1, namely

$$\int_\Gamma \hat{n}(y) \cdot \nabla_y g^0(x,y)\mathrm{d}\Gamma(y) = -1 \text{ if } x \in \Omega.$$

We next show that the boundary integral in Eq. (7.18) is continuous as $t \to 0$, i.e., as $x \to x_0$. Toward this, we again write Γ as the union of two surface patches: $\Gamma_\epsilon(x_0) = \{y | |x_0 - y| < \epsilon, y \in \Gamma\}$ and $\Gamma_{\epsilon c} = \Gamma \setminus \Gamma_\epsilon(x_0)$. Therefore, $\Gamma = \Gamma_\epsilon(x_0) \cup \Gamma_{\epsilon c}(x_0)$. Since ψ is continuous on Γ, $\psi(y) - \psi(x_0) \to 0$ as $y \to x_0$. Let $\tilde{\psi} \in C(\Gamma)$ be a moment such that $\tilde{\psi} = 0$ at $x_0 \in \Gamma$ and consider the difference

$$
\begin{aligned}
|\tilde{v}(x) - \tilde{v}(x_0)| &= \Big| \int_\Gamma \hat{n}(y) \cdot \nabla_y \left[g^0(x,y) - g^0(x_0,y)\right] \tilde{\psi}(y)\mathrm{d}\Gamma(y)\Big| \\
&\leq \int_{\Gamma_\epsilon} \left(|\hat{n}(y) \cdot \nabla_y g^0(x,y)| + |\hat{n}(y) \cdot \nabla_y g^0(x_0,y)|\right)|\tilde{\psi}(y)|\mathrm{d}\Gamma(y) \\
&\quad + \int_{\Gamma_{\epsilon c}} |\hat{n}(y) \cdot \nabla_y \left[g^0(x,y) - g^0(x_0,y)\right]||\tilde{\psi}(y)|\mathrm{d}\Gamma(y), \quad (7.19)
\end{aligned}
$$

where $\tilde{v}$ denotes the potential corresponding to the moment $\tilde{\psi}$. From Appendix A.7.3,

$$\int_{\Gamma_\epsilon} |\hat{n}(y) \cdot \nabla_y g^0(x, y)| \mathrm{d}\Gamma(y) \le C$$

and

$$\int_{\Gamma_\epsilon} |\hat{n}(y) \cdot \nabla_y g^0(x_0, y)| \mathrm{d}\Gamma(y) \le C' < \infty$$

because of the integrability of the kernel. Since $\tilde{\psi}(x_0) = 0$ and $\psi \in C(\Gamma)$, we can choose an ϵ such that $|\tilde{\psi}(y)| \le \frac{\eta}{3(C+C')}, y \in \Gamma_\epsilon, \eta > 0$. The first integral over Γ_ϵ on the R.H.S. of Eq. (7.19) is then bounded by

$$\int_{\Gamma_\epsilon} \left(|\hat{n}(y) \cdot \nabla_y g^0(x, y)| + |\hat{n}(y) \cdot \nabla_y g^0(x_0, y)| \right) |\tilde{\psi}(y)| \mathrm{d}\Gamma(y) \le \frac{2\eta}{3}$$

for a sufficiently small ϵ since each of the integrals is less than $\eta \backslash 3$. The second integral over $\Gamma_{\epsilon c}(x_0)$ contains the difference $\hat{n}(y) \cdot \nabla_y \{g^0(x, y) - g^0(x_0, y)\}$ and vanishes uniformly as $x \to x_0$. We can, therefore, choose a δ such that $|x - x_0| < \delta$ and the second integral is smaller than $\frac{\eta}{3}$. Hence $|\tilde{v}(x) - \tilde{v}(x_0)| < \eta$ if $|x - x_0| < \delta$ showing that $v(x)$ is continuous at x_0. Note that δ depends only on η and the norm of $\tilde{\psi}$. The convergence is, therefore, uniform. Returning to Eq. (7.18), this implies that the integral in this equation remains continuous as $t \to 0$. Thus $v(x(t))$ can be extended continuously from within Ω to Γ and the limit $t \to 0$ can be taken inside the integral. Combining this with the result of Appendix A.7.1 and the definition (7.3) of the double-layer potential, it follows that as $t \to 0$ from within Ω, that is $t \to 0^-$,

$$
\begin{aligned}
\lim_{t \to 0} v(x(t)) \;=\; & v_-(x_0) \\
=\; & -\psi(x_0) + \int_\Gamma \hat{n}(y) \cdot \nabla_y g^0(x_0, y) \psi(y) \mathrm{d}\Gamma(y) \\
& -\psi(x_0) \int_\Gamma \hat{n}(y) \cdot \nabla_y g^0(x_0, y) \mathrm{d}\Gamma(y) \\
=\; & -\psi(x_0) + \int_\Gamma K(x_0, y) \psi(y) \mathrm{d}\Gamma(y) + \frac{1}{2}\psi(x_0) \\
=\; & -\frac{1}{2}\psi(x_0) + \int_\Gamma K(x_0, y) \psi(y) \mathrm{d}\Gamma(y).
\end{aligned}
$$

In the same manner, if $t > 0$ and $t \to 0^+$, then

$$\lim_{t \to 0} v(x(t)) = v_+(x_0) = \frac{1}{2}\psi(x_0) + \int_\Gamma K(x_0, y) \psi(y) \mathrm{d}\Gamma(y).$$

In summary, the double-layer potential $v(x)$ with moment $\psi \in C(\Gamma)$ can be continuously extended from anywhere in the toroidal neighborhood V to Γ. The traces[4] $v_+(x)$ and $v_-(x)$ are, however, not identical and v has a jump

---

[4]In the literature, the trace on the boundary is frequently expressed *via* a trace operator γ. In this notation $v_\pm = \gamma_\pm v$.

discontinuity across the boundary:

$$[[v]]_\Gamma = v_+(x) - v_-(x) = \psi(x).$$

We use the symbol $[[\cdot]]_\Gamma$ to indicate the jump of any quantity across Γ.

Let us next determine how the normal derivative of the double-layer potential behaves across the boundary. That is, we would like to find out what happens to the quantity

$$(\partial_x v)(x) = (\hat{n} \cdot \nabla v)(x), \quad x \in \Gamma,$$

as Γ is crossed. Let us introduce the following notations. We will use the symbol $[[a]]$ without the subscript Γ to indicate simply the difference $a^e - a^i$ of some quantity a, $a^e \in \Omega^e$, and $a^i \in \Omega$. The symbol $[[a]]_\Gamma$ will be reserved to emphasize that the difference $[[a]]$ is being calculated at the boundary Γ. It is assumed that $\Gamma \in C^2$ and the moment $\psi \in C(\Gamma)$. Furthermore, consider a sequence of points $x^i = x - t\hat{n}(x)$ and $x^e = x + t\hat{n}(x)$, $x \in \Gamma$ and $t > 0$. $\{x^i\}$ is thus in Ω and $\{x^e\} \in R^3 \setminus \bar{\Omega}$ are points in the exterior. At this point let us recall that if $\hat{n}(x)$ is the outward normal to Γ at x and z a point in the toroidal neighborhood V of Γ, then the normal derivative $\partial_x a$ of a at z along $\hat{n}(x)$ is described by the following uniform limit

$$\partial_x a = (\partial_n a)(z) = \lim_{t \to 0^+} \hat{n}(x) \cdot (\nabla a)(z).$$

We are interested in the quantity

$$[[\partial_x v]]_\Gamma = \lim_{t \to 0^+} \partial_x v(x^e) - \lim_{t \to 0^+} \partial_x v(x^i),$$

the jump in the x derivative of the double-layer potential across Γ.

In terms of the above notations, we have

$$[[\partial_x v]] = \int_\Gamma \psi(y) \partial_y [[\partial_x g^0(\cdot, y)]] d\Gamma(y). \tag{7.20}$$

Rewrite Eq. (7.20) as

$$
\begin{aligned}
[[\partial_x v]] \;=\; & \int_\Gamma \psi(x) \partial_y [[\partial_x g^0(\cdot, y)]] d\Gamma(y) \\
& + \int_\Gamma \{\psi(y) - \psi(x)\} \partial_y [[\partial_x g^0(\cdot, y)]] d\Gamma(y).
\end{aligned}
\tag{7.21}
$$

The first integral in Eq. (7.21) is

$$\psi(x) \int_\Gamma \partial_x \left[\partial_y g^0(x^e, y) - \partial_y g^0(x^i, y) \right] d\Gamma(y).$$

By Appendix A.7.1, the above integral is identically zero. This is due to the fact that the first integral involving x^e is identically zero and the other involving x^i is the x-derivative of a constant. The first integral of Eq. (7.21), therefore, vanishes.

Next consider a ball $B_\epsilon(x)$ of radius ϵ centered at x and define by $S_\epsilon(x)$ the intersection of Γ and $B_\epsilon(x)$, i.e., $S_\epsilon(x) = \Gamma \cap B_\epsilon(x)$. In terms of $S_\epsilon(x)$, Eq. (7.21) becomes

$$
\begin{aligned}
[[\partial_x v]] &= \int_{S_\epsilon(x)} \{\psi(y) - \psi(x)\} \partial_y [[\partial_x g^0(\cdot, y)]] d\Gamma(y) \\
&+ \int_{S_{\epsilon c}(x)} \{\psi(y) - \psi(x)\} \partial_y [[\partial_x g^0(\cdot, y)]] d\Gamma(y), \quad (7.22)
\end{aligned}
$$

where, as usual, $S_{\epsilon c}(x) = \Gamma \setminus S_\epsilon(x)$. Again as earlier, $t \to 0$, that is, as $x^i, x^e \to x \in \Gamma$, the integral over $S_{\epsilon c}(x)$ in Eq. (7.22) vanishes since g^0 does not have a singularity there. We are thus left only with the integral over $S_\epsilon(x)$ and obtain

$$
|[[\partial_x v]]| \le \sup_{|x-y| \le \epsilon} |\psi(y) - \psi(x)| \int_{S_\epsilon(x)} |\partial_y [[\partial_x g^0(\cdot, y)]]| d\Gamma(y). \quad (7.23)
$$

Let us calculate $\partial_x \partial_y g^0$ directly and obtain

$$
\partial_x \partial_y g^0(z(x), y) = \sum_{\ell=1}^{3} \sum_{m=1}^{3} \hat{n}_\ell(x) \hat{n}_m(y) \partial_{\ell m} g^0(z(x), y), \quad (7.24)
$$

where $z(x)$ is either x^e or x^i, and $\partial_{\ell m} = \partial_{x_\ell} \partial_{y_m}$. Let us introduce further notations, namely

$$
\Phi^\xi = \frac{x^\xi - y}{4\pi}, \ \xi = e, i; \ \Phi = \frac{x - y}{4\pi},
$$

and

$$
\Psi = \frac{[(z - x) \cdot \hat{n}(x)][(z - y) \cdot \hat{n}(y)]|\zeta|^{-5}}{4\pi}.
$$

In the above notation, Eq. (7.24) takes the form

$$
\partial_x \partial_y g^0(\zeta, y) = |\Phi^\zeta|^{-\zeta} \delta_{ij} \hat{n}_i(x) \hat{n}_j(y) + 3\Psi(\zeta)
$$

and from this we obtain

$$
[[\partial_x \partial_y g^0]] = [[|\Phi|^{-3}]] \delta_{ij} \hat{n}_i(x) \hat{n}_j(y) + 3[[\Psi]] = Q_1 + Q_2. \quad (7.25)
$$

We next estimate the quantities $|Q_1|$ and $|Q_2|$. In order to estimate the term $[[|\Phi|^{-3}]]$, we use the formula

$$
a^N - b^N = (a - b) \sum_{j=1}^{3} a^{N-j} b^{j-1}, \quad (7.26)
$$

a and b being real numbers. Moreover, it is also possible to show that

$$
|z - x|^2 \ge C(|x - y| + t)^2 \quad (7.27)
$$

using the results of Appendix 7.2. From Eqs. (7.26) and (7.27), we have,

$$|Q_1| \leq C\big||\Phi^e|^2 - |\Phi^i|^2\big|(|\Phi| + t)^{-5}.$$

Finally, we use the estimate

$$\big||\Phi^e|^2 - |\Phi^i|^2\big| \leq 4t|\hat{n}(x) \cdot (x - y)| \leq Ct|x - y|^2$$

to obtain

$$|Q_1| \leq Ct(|\Phi| + t)^{-3}. \tag{7.28}$$

The term $|Q_2|$ can be estimated as follows. Using the fact that $z - y = (z - x) + (x - y) = \pm t\hat{n}(x) + (x - y)$, we obtain

$$Q_2 = \alpha \pm \Delta,$$

where $\alpha = t^2\hat{n}(x) \cdot \hat{n}(y)$ and

$$
\begin{aligned}
\pm\Delta \;=\; & \pm t(x - y) \cdot \hat{n}(y) \pm t\hat{n}(x) \cdot \hat{n}(y)[(x - y) \cdot \hat{n}(x)] \\
& + [(x - y) \cdot \hat{n}(x)][(x - y) \cdot \hat{n}(y)].
\end{aligned} \tag{7.29}
$$

The $+(-)$ sign corresponds to $x^e(x^i)$. Moreover, using Appendices A.7.2 and A.7.3 in Eq. (7.29) leads to the estimates

$$|\Delta^e| \leq \frac{1}{2}t^2 + c(t + |x - y|^2)|x - y|^2 \tag{7.30}$$

and analogously for $|\Delta^i|$. From this we have

$$\big|[[Q_2]]\big| \leq \Big|\big(|\Delta^e||\Phi^e|^{-5} - |\Delta^i||\Phi^i|^{-5}\big) + |\alpha|\big(|\Phi^e|^{-5} - |\Phi^i|^{-5}\big)\Big|\frac{3}{4\pi}. \tag{7.31}$$

Again as in the above

$$\big||\Phi^e|^{-5} - |\Phi^i|^{-5}\big| \leq Ct(|\Phi| + t)^{-5}. \tag{7.32}$$

Using Eqs. (7.28)–(7.32) finally results in

$$C\sum_{j=0}^{2} \frac{t^{j+1}}{(|x - y| + t)^{3+j}} \leq C_2 t(|x - y| + t)^{-3}.$$

Integrating the above equation over $S_\epsilon(x)$ gives

$$\int_{S_\epsilon(x)} \big|[[\partial_x \partial_y g^0]]\big| \leq C_2 t \int_{S_\epsilon(x)} (|x - y| + t)^{-3} d\Gamma(y). \tag{7.33}$$

Without any loss of generality, we can take $t = \epsilon$ and the integral in Eq. (7.34) is bounded above in the limit $t \to 0$. Finally, returning to Eq. (7.22) and using the result of Eq. (7.34) the sought-for result is obtained, namely

$$\lim_{\epsilon \to 0} \big|[[\partial_x v]]_\Gamma\big| = 0.$$

Therefore, for $\Gamma \in C^2$ and $\psi \in C(\Gamma)$,

$$\lim_{\epsilon \to 0}[\hat{n}(x) \cdot (\nabla v)(x)] = [[\partial_{n(x)}v]]_\Gamma = 0. \tag{7.34}$$

It is important to note that Eq. (7.35) does not guarantee that $\partial_x v(x^i)$ and $\partial_x v(x^e)$ go to zero individually as $t \to 0$. It only tells us that it is their difference $\partial_x(v(x^i) - v(x^e))$ that attains this limit.

Collecting these results together we have the following theorem.

Theorem 7.1 (The double-layer potential) *Consider a surface* $\Gamma \in C^2$ *and a function* $\psi \in C(\Gamma)$. *Let the double-layer potential* $v(x)$ *be defined as*

$$v(x) = \int_\Gamma \hat{n}(y) \cdot \nabla_y G^0(x, y; k)\psi(y)\mathrm{d}\Gamma(y), \ x \in R^3 \setminus \Gamma.$$

Then the restriction of $v(x)$ *to* Ω_e *and* Ω *can be continuously extended to* Γ. *That is, if* $v_+(v_-)$ *denote the restriction of* $v(x)$ *to* $\Omega_e(\Omega)$, *then*

$$v_\pm(x) = \lim_{t \to 0} v(x \pm t\hat{n}(x)), t > 0,$$

the convergence being uniform. Moreover, the traces v_+ *and* v_- *of* v *on* Γ *are given by*

$$v_+(x) = \frac{1}{2}\psi(x) + (T_K\psi)(x)$$

and

$$v_-(x) = -\frac{1}{2}\psi(x) + (T_K\psi)(x).$$

Therefore, the double-layer potential undergoes a jump discontinuity $[[v]]_\Gamma$ *across the boundary. The magnitude of the discontinuity equals* ψ, *i.e.,*

$$[[v]]_\Gamma = v_+ - v_- = \psi.$$

Moreover, the normal derivative of $v(x)$ *on* Γ *is continuous in the sense that*

$$\lim_{t \to 0^+}\left\{\left(\frac{\partial v}{\partial n}\right)(x + t\hat{n}(x)) - \left(\frac{\partial v}{\partial n}\right)(x - t\hat{n}(x))\right\} = 0,$$

the convergence being uniform.

Next we consider the single-layer potential $u(x)$.

7.5   The Single-layer Potential

This potential was defined in Eq. (7.4) $\forall x \in R^3 \setminus \Gamma$ where it is well-defined. Even when $x \in \Gamma$, the integral exists uniformly as an improper integral and hence is continuous on the surface (see Appendix A.7.6). A more detailed account of improper integrals in the context of layer potentials can be found in Hackbusch, 1995. The single-layer-potential is, therefore, well-defined in R^3. As a matter of fact $u(x)$ is continuous in the entire R^3 as demonstrated below. Consider the difference $u(x) - u(x_0)$

$$u(x) - u(x_0) = \int_\Gamma [G^0(x, y; k) - G^0(x_0, y; k)]\phi(y)\mathrm{d}\Gamma(y), \qquad (7.35)$$

$$x \in V, \; x_0 \in \Gamma, \; \phi \in C(\Gamma).$$

Define again the surface patches $\Gamma_\epsilon(x_0)$ and $\Gamma_{\epsilon c}(x_0)$. As in the case of the double-layer potential, write Eq. (7.35) as

$$|u(x) - u(x_0)| \leq \int_{\Gamma_\epsilon} \left[|g^0(x_0, y)\phi(y)| + |g^0(x, y)\phi(y)|\mathrm{d}\Gamma(y) \right]$$

$$+ \int_{\Gamma_{\epsilon c}} |g^0(x_0, y) - g^0(x, y)||\phi(y)|\mathrm{d}\Gamma(y). \qquad (7.36)$$

Again each of the first two integrals is of the order of ϵ (see Eq. (7.15)). Therefore, given a $\delta > 0$, we can choose ϵ such that each integral in Eq. (7.36) is less than $\delta\backslash 3$. Moreover, the third integral over $\Gamma_{2\epsilon c}$ tends to zero uniformly as $|x_0 - x| \to 0$ since the functions g^0 do not have any singularity there. Restricting x, x_0 such that $|x - x_0| < \epsilon$, the integral can be made to be less than $\delta\backslash 3$. Therefore, we have the classical $\epsilon - \delta$ definition of continuity: $|u(x) - u(x_0)| < \delta$ whenever $|x - x_0| < \epsilon, \delta$ depending upon ϵ only. The single-layer potential $u(x)$ is thus continuous on Γ and in its toroidal neighborhood V. Consequently, $u(x)$ is continuous in all R^3.

Next we consider the normal derivatives of u on Γ. These derivatives were defined in the previous section. If z is on Γn then

$$\partial_{z_\pm} u = (T_{K^*} u)(z)$$

in the notation of Eq. (7.6). In view of Appendix A.7.2, the operator T_K results in a uniformly existing improper integral and hence a continuous function on Γ. Being adjoint to T_K, T_{K^*} also yields a function which is continuous on Γ. Therefore, $\partial_{x_\pm} u$ can be continuously extended from $V \backslash \Gamma \to \Gamma$ from either side of the boundary. It remains to determine the behavior of $\partial_n u$ as the boundary Γ is crossed. As a matter of fact, we show below that $(\partial_n u)(x)$ has a jump discontinuity at the boundary.

Let us consider a function $Q(x)$ defined by:

$$Q(x) = v(x) + (\partial_n u)(x), \; x \in V\backslash\Gamma$$
$$= (T_K\phi)(x) + (T_{K^*}\phi)(x), x \in \Gamma.$$

The restrictions of Q in $V \backslash \Gamma$ are continuous. Let us, therefore, consider the behavior of Q as the boundary is crossed. Let $x^e = x_0 + t\hat{n}(x_0), x_0 \in \Gamma, t > 0$, be a point in Ω_e located on the normal at $x_0 \in \Gamma$. Consider the difference of Q

at x^e and x_0, namely

$$
\begin{aligned}
\Delta Q(x^e, x_0) &= Q(x^e) - Q(x_0) \\
&= \int_\Gamma \phi(y) \left[\frac{\partial g^0(x^e, y)}{\partial n(y)} - \frac{\partial g^0(x_0 y)}{\partial n(y)} \right] d\Gamma(y) \\
&+ \int_\Gamma \phi(y) \left[\frac{\partial g^0(x^e, y)}{\partial n(x^e)} - \frac{\partial g^0(x_0 y)}{\partial n(x_0)} \right] d\Gamma(y) \\
&= (\Delta Q)_{\Gamma_\epsilon} + (\Delta Q)_{\Gamma_{\epsilon c}}.
\end{aligned}
$$

As previously, $(\Delta Q)_{\Gamma_\epsilon}$ and $(\Delta Q)_{\Gamma_{\epsilon c}}$ denote integrations of ΔQ over Γ_ϵ and $\Gamma_{\epsilon c}$, respectively. Again $(\Delta Q)_{\Gamma_{\epsilon c}}$ tends to zero uniformly as $x^e \to x_0$. As to $(\Delta Q)_{\Gamma_\epsilon}$, we have

$$
|(\Delta Q)_{\Gamma_\epsilon} \leq \| \phi \|_\infty \int_{\Gamma_\epsilon} (|\frac{\partial g^0(x^e, y)}{\partial n(y)} + \frac{\partial g^0(x^e, y)}{\partial n(x^e)}|) d\Gamma(y)
$$

+ the same with x replaced by x_0.

Now consider the following.

$$
\begin{aligned}
\hat{n}(y) \cdot \nabla_y g^0(z, y) &+ \hat{n}(x) \cdot \nabla_z g^0(z, y) \\
&= \frac{e^{ik|z-y|}}{4\pi} \left\{ \frac{1}{|z-y|^3} - \frac{ik}{|z-y|^2} \right\}(z - y) \cdot \{\hat{n}(y) - \hat{n}(x)\}.
\end{aligned}
$$

Therefore,

$$
|\hat{n}(y) \cdot \nabla_y g^0(z, y) + \hat{n}(x) \cdot \nabla_x G(z, y; k)| \leq L\left\{ \frac{1}{|z-y|^3} + \frac{k}{|z-y|^2} \right\}|z - y||x - y|,
$$

where we have used the result $|\hat{n}(a) - \hat{n}(b)| \leq L|a - b|$ from Appendix A.7.2. Replacing x by x_0 and z by x^e, we have

$$
\begin{aligned}
&|\hat{n}(y) \cdot \nabla_y g^0(x^e, y) + \hat{n}(x_0) \cdot \nabla_x^e g^0(x^e, y)| \\
&\leq \quad \text{const.} \left\{ \frac{1}{|x^e - y|^3} + \frac{k}{|x^e - y|^2} \right\}|x^e - y||x_0 - y| \\
&= \quad \text{const.} \left[|x^e - y|^{-2} + k|x^e - y|^{-1} \right]|x_0 - y|.
\end{aligned}
$$

Letting x^e approach x along $\hat{n}(x)$ gives

$$
|\hat{n}(y) \cdot \nabla_y g^0(x^e, y) + \hat{n}(x_0) \cdot \nabla_{x_0} g^0(x^e, y)| \to \text{const.} \left[|x_0 - y|^{-1} + k \right]
$$

as $x^e \to x_0$. Hence

$$
|(\Delta Q)_{\Gamma_\epsilon}| \sim O\left(\int_{\Gamma_\epsilon} |x_0 - y|^{-1} d\Gamma(y) \right) \sim O(\epsilon).
$$

Therefore, as $\epsilon \to 0$, $(\Delta Q)_{\Gamma_\epsilon} \to O$. The conclusion is that $v(x) + (\partial_n u)(x)$ can be extended continuously to Γ approaching from the exterior Ω_e. The same conclusion also follows if $t < 0$, i.e., if Γ is approached from the interior Ω since all differential operations remain unchanged in V. We, therefore, conclude that the normal derivative of the single-layer potential can be continued to Γ continuously from both Ω and Ω_e.

Now by continuity of $Q(x)$,

$$v_\pm(x) + (\partial_n u)_\pm(x) = (T_K \phi)(x) + (T_{K^*}\phi)(x). \tag{7.37}$$

Moreover, from Theorem 7.1

$$v_\pm(x) = \pm\frac{1}{2}\phi(x) + (T_K\phi)(x). \tag{7.38}$$

Combining Eqs. (7.38) and (7.39) results in:

$$(\partial_n u)_\pm(x) = \mp\frac{1}{2}\phi(x) + (T_{K^*}\phi)(x)$$

from which it follows at once that

$$[[(\partial_n u)(x)]]_\Gamma = [(\partial_n u)_- - (\partial_n u)_+](x) = \phi(x).$$

All this can be collected together into the theorem below.

Theorem 7.2 (The Single-layer Potential) *Let $\Gamma \subset C^2$ and $\phi \in C(\Gamma)$. The single-layer potential $u(x)$ defined by*

$$u(x) = \int_\Gamma G^0(x,y;k)\phi(y)\mathrm{d}\Gamma(y)$$

is continuous in all R^3. Moreover, the restrictions of the normal derivative $(\partial_n u)(x)$ to Ω and Ω_e can be continuously extended to Γ. The traces of $\partial_n u$ are given by

$$(\partial_n u)_\pm(x) = \mp\frac{1}{2}\phi(x) + (T_{K^*}\phi)(x).$$

The jump condition at the boundary is described by

$$[[(\partial_n u)(x)]]_\Gamma = [(\partial_n u)_- - (\partial_n u)_+](x) = \phi(x).$$

The normal derivatives are understood in the sense that

$$(\partial_n u)_\pm(x) = \lim_{t \to 0} \hat{n}(x) \cdot \nabla u(x \pm t\hat{n}(x))$$

exists uniformly on Γ.

Remarks

The continuity and jump relations of the potentials at the boundary can be derived heuristically by appealing to the concept of *symbolic differentiation* (Friedman, 1956) of the Laplacian (see also Filippi *et al.*, 1999). Let us first recall the definition of a symbolic derivative. Consider the function $f(x) = |x|$.

It has a regular, classical derivative everywhere except at the origin where the derivative is discontinuous. Its symbolic derivative is given by

$$f'(x) = \{f'(x)\} + \text{sgn}x.$$

$f'(x)$ denotes the classical derivative of f wherever it exists (everywhere in this case except at $x = 0$), whereas the *signum* function sgn$x$ defined by sgn$x$ = sign$x$ is the symbolic derivative of the function $|x|$. To carry this a little further, let $f(x)$ be a function having a discontinuity at a point $x = a$ in its domain of definition which is some interval of the real axis. Write this function everywhere in this interval as

$$f(x) = f_-(x)H(a - x) + f_+(x)H(x - a),$$

that is, the restriction of the function is $f_-(x)$ for $x < a$ and $f_+(x)$ when $x > a$ and H is the *Heaviside function.* By direct differentiation

$$f'(x) = [f'_-(x)H(a - x) + f'_+(x)H(x - a)] + [f_+(x)\delta(x - a) - f_-(x)\delta(a - x)]$$

and remembering that $\delta"(\xi) = -\delta"(-\xi)$, we have

$$\begin{aligned} f''(x) &= [f''_-(x)H(a - x) + f''_+(x)H(x - a)] \\ &\quad + 2\delta(x - a)\{f'_+(x) - f'_-(x)\} + \delta'(x - a)\{f_+(x) - f_-(x)\}. \end{aligned}$$

The above equation for $f''(x)$ can be put into the form

$$f''(x) = \{f''(x)\} + [[f(x)]]_a \delta'(x - a) + 2[[f'(x)]]_a \delta(x - a),$$

the first term on the R.H.S. representing the classical derivative, $[[f(x)]]_a$ the jump of f at a and $[[f'(x)]]_a$ the jump of f' at a.

Now consider a function w which is smooth everywhere in R^3 except on a surface Γ across which the function has a discontinuity. Then the Laplacian of w can be written symbolically as

$$\Delta w = \{\Delta w\} + [[w]]_\Gamma \delta'(\Gamma) + [[w_n]]_\Gamma \delta(\Gamma), \tag{7.39}$$

where w_n is the normal derivative at Γ. $\delta(\Gamma)$ and $\delta'(\Gamma)$ symbolize delta distribution and its derivative supported on the boundary (for a rigorous meaning of surface delta functions and related distributions, see Jones, 1966).

Returning to layer potentials, we know that they satisfy Helmholtz's equation everywhere except on the boundary where there are discontinuities. Applying the Helmholtz operator $\Delta + k^2$ to the defining equations (7.4) and (7.3), it follows that

$$(\Delta + k^2)u(x) = \phi\delta(\Gamma) \tag{7.40}$$

for the single-layer and

$$(\Delta + k^2)v(x) = \psi\delta'(\Gamma) \tag{7.41}$$

for the double-layer potential, whereas $\{(\Delta + k^2)u(v)\} = 0$. Upon comparing Eq. (7.40) with Eq. (7.39) we obtain

$$[[u]]_\Gamma = 0; \ [[u_n]]_\Gamma = \phi$$

since there is no support of δ' on Γ in the case of the single-layer potential. Similarly, a comparison of Eqs. (7.41) and (7.39) yields the condition

$$[[v]]_\Gamma = \psi; \ [[u_n]]_\Gamma = 0.$$

These are in accordance with the results of Theorems 7.1 and 7.2.

Thus far we have considered $\phi, \psi \in C(\Gamma)$. Let us summarize briefly the results in function spaces other than C. The operators S, T_K and T_{K^*} are compact in $C(\Gamma)$. They are thus expected to produce smoother functions. As a matter of fact, these operators map $C(\Gamma) \rightarrow C^{0,\alpha}(\Gamma)$, $0 < \alpha < 1$. Moreover, S and T_K also map $C^{0,\alpha}(\Gamma) \rightarrow C^{1,\alpha}(\Gamma)$. This means that for a continuous density, the single-layer potential is Hölder continuous in R^3 and

$$||u||_{\alpha,R^3} \leq C_\alpha ||\phi||_{\infty,\Gamma}, \ 0 < \alpha < 1.$$

If the density is $C^{0,\alpha}(\Gamma)$, then the first derivative of u can be extended to the boundary in the Hölder sense and

$$||\nabla u||_{\alpha,\Omega_a} \leq C_\alpha ||\phi||_{\alpha,\Gamma}.$$

The corresponding results for the double-layer potential v are:

$$||v||_{\alpha,\Omega_a} \leq C_\alpha ||\phi||_{\alpha,\Gamma}$$

and

$$||\nabla v||_{\alpha,\Omega_a} \leq C_\alpha ||\phi||_{1,\alpha,\Gamma}.$$

In the above equations Ω_a may refer to either the interior or the exterior domain.

Next we would like to mention that Theorems 7.1 and 7.2 also apply if the densities are $L_2(\Gamma)$. The corresponding equations are:

$$\lim_{t-0^+} ||u(\cdot, \pm t\hat{n}(\cdot)) - u||_{2,\Gamma} = 0.$$

$$\lim_{t-0^+} \left\| \partial_n u(\cdot, \pm t\hat{n}(\cdot)) - \left[\int_\Gamma \phi(y)\partial_n G^0(\cdot, y, k)\mathrm{d}\Gamma(y) \mp \frac{1}{2}\phi \right] \right\|_{2,\Gamma} = 0.$$

$$\lim_{t-0^+} \left\| v(\cdot, \pm t\hat{n}(\cdot)) - \left[\int_\Gamma \psi(y)\partial_n G^0(\cdot, y, k)\mathrm{d}\Gamma(y) \pm \frac{1}{2}\psi \right] \right\|_{2,\Gamma} = 0.$$

$$\lim_{t-0^+} ||\partial_n v(\cdot, t\hat{n}(\cdot)) - \partial_n v(\cdot, -t\hat{n}(\cdot))||_{2,\Gamma} = 0.$$

For a detailed analysis with square integrable densities, the reader is referred to the text by McLean (2000).

Finally, we would like to state the following theorems without proof. The details can be found in the recent text by Doico *et al.* (2000).

Theorem 7.3 *Consider a bounded domain $\Omega \in R^3$ in C^2 with boundary Γ. Let $\phi \in L_2(\Gamma)$ be the density. Then if the single-layer potential vanishes in the interior, that is, if $u = 0$ in Ω, then $u \sim 0$ on Γ. That is, u vanishes almost everywhere on the boundary.*

Theorem 7.4 *Let Ω_e be the exterior of Ω in the above theorem and let $k \notin \sigma_D$. If $\phi \in L_2(\Gamma)$ be the density and the single-layer potential vanishes in the exterior, that is, $u = 0$ in Ω_e, then $u \sim 0$ on Γ. That is, u vanishes almost everywhere on the boundary.*

Similar results hold for the double-layer potential also. We will need the above two theorems later in the next chapter.

We are now ready to undertake the main task of this section, namely, to show that the solutions of the Helmholtz scattering problems exist. We will first consider Dirichlet and Neumann scattering leaving the case of the transmission problem for later.

7.6   The Helmholtz Scattering Problems

7.6.1   The Dirichlet and Neumann Obstacle Scattering

We already know that the layer potentials are solutions to the Helmholtz equations and satisfy Sommerfeld's radiation condition at infinity because G^0 does. However, they require the moments ϕ and ψ which are defined on the boundary. These are in turn determined by solving the boundary integral equations stated in Theorems 7.1 and 7.2. We would like to form the problem in terms of integral equations of the second kind to which the Fredholm theory can be applied. This is achieved by choosing the double-layer potential $v(x)$ as the solution of the Dirichlet and the single-layer potential $u(x)$ to solve the Neumann problem.[5]

Let us consider the Dirichlet problem first. From Theorem 7.1, $v_\pm$ are given by

$$v_\pm = \pm \frac{1}{2}\psi + T_K \psi.$$

$v_-(x)$ solves the interior Dirichlet problem if

$$v_-(x) = -\frac{1}{2}\psi(x) + (T_K\psi)(x) = f(x), x \in \Gamma, f \in C(\Gamma). \qquad (7.42)$$

Similarly, $v_+(x)$ is a solution of the exterior Dirichlet problem if

$$v_+(x) = \frac{1}{2}\psi(x) + (T_K\psi)(x) = f(x), x \in \Gamma, f \in C(\Gamma). \qquad (7.43)$$

Thus the solvability of the Dirichlet problem reduces to that of Eqs. (7.42) and (7.43). Similarly, the interior (exterior) Neumann problem is satisfied by the

---

[5]The role of the potentials can be reversed, in which case the resulting integral equations for the moments are of the first kind.

single-layer potential $u(x)$ provided that the following integral equations are satisfied, namely

$$(\partial_n u)_-(x) = \frac{1}{2}\phi(x) + (T_{K^*}\phi)(x) = g(x), x \in \Gamma, g \in C(\Gamma)$$

for the interior, and

$$(\partial_n u)_+(x) = -\frac{1}{2}\phi(x) + (T_{K^*}\phi)(x) = g(x), x \in \Gamma, g \in C(\Gamma)$$

for the exterior. We thus have the following theorems.

Theorem 7.5 (Dirichlet Problem) *The double-layer potential*

$$v(x) = \int_\Gamma \partial_y G^0(x, y; k)\psi(y)\mathrm{d}\Gamma(y)$$

is a solution to the interior Dirichlet problem if the density $\psi(x) \in C(\Gamma)$ satisfies the boundary integral equation

$$\psi(x) - 2(T_k\psi)(x) = -2f(x), x \in \Gamma, f \in C(\Gamma).$$

Similarly, $v(x)$ solves the exterior Dirichlet problem if

$$\psi(x) + 2(T_K\psi)(x) = 2f(x), x \in \Gamma, f \in C(\Gamma).$$

Theorem 7.6 (Neumann Problem) *The single-layer potential*

$$u(x) = \int_\Gamma G^0(x, y; k)\phi(y)\mathrm{d}\Gamma y$$

solves the interior Neumann problem if the density $\phi \in C(\Gamma)$ is a solution to the boundary integral equation

$$\phi(x) + 2(T_{K^*}\phi)(x) = 2g(x), x \in \Gamma, g \in C(\Gamma).$$

Similarly, $u(x)$ is a solution to the exterior Neumann problem provided that

$$\phi(x) - 2(T_{K^*}\phi)(x) = -2g(x), x \in \Gamma, g \in C(\Gamma).$$

For the sake of later convenience, the various boundary integral equations are summarized below:

$$\psi - 2T_K\psi = -2f : \text{interior Dirichlet problem}$$

$$\phi + 2T_{K^*}\phi = 2g : \text{interior Neumann problem}$$

$$\psi + 2T_K\psi = 2f : \text{exterior Dirichlet problem}$$

$$\phi - 2T_{K^*}\phi = -2g : \text{exterior Neumann problem}.$$

Moreover,

$$
\begin{aligned}
< (I + 2T_K)\psi, \phi > &= \ <\psi, \phi > +2 < T_K\psi, \phi > \\
&= \ <\psi, \phi > +2 < \psi, T_{K^*}\phi > \\
&= \ <\psi, (I + 2T_{K^*})\phi >.
\end{aligned}
$$

Thus the adjoint of the exterior Dirichlet operator $I + 2T_K$ is $I + 2T_{K^*}$, the operator for the interior Neumann problem. Similarly, the adjoint of the interior Dirichlet operator $I - 2T_K$ is $I - 2T_{K^*}$, the operator for the exterior Neumann problem. The exterior (interior) Dirichlet and interior (exterior) Neumann problem are, therefore, adjoint to each other: $I \pm 2T_K$ adjoint $I \pm 2T_{K^*}$.

The solvability of the scattering problems thus reduces to analyzing the corresponding homogeneous problems (for the density functions) for triviality or nontriviality of solutions. Let us consider the homogeneous exterior Dirichlet problem first. By Theorem 7.5, this involves solving the boundary integral equation

$$\psi(x) + 2(T_K\psi)(x) = 0, x \in \Gamma, \psi \in C(\Gamma).$$

For the homogeneous exterior Dirichlet problem, the boundary data on Γ is zero. Now we have sought the solution in the form of a double-layer potential and, consequently, by Theorem 7.1, the trace $v_+ = 0$ on Γ. Since $\psi = v_+ - v_-$, this implies that $\psi = -v_-$. It then follows that the solution to the homogeneous exterior Dirichlet problem is to be obtained by solving an interior problem for the density function $\psi(x)$. Moreover, the solution of the Dirichlet problem is unique (see Appendix A.7.6). Therefore, v must be zero in $R^3 \backslash \Omega$. Then $(\partial_n v)_+$ must also vanish on Γ. By Theorem 7.1, $\partial_n v$ is continuous across Γ implying that $(\partial_n v)_- = 0$ on Γ. ψ is, therefore, the boundary value of the solution of a homogeneous interior Neumann problem. That is, $\psi = v_-|_\Gamma$, where v_- is the solution to the problem

$$
\begin{aligned}
(\Delta + k^2)v_-(x) &= 0, \ x \in \Omega \\
\partial_n v_- &= 0, \ x \in \Gamma.
\end{aligned}
$$

Let $\tilde{V}$ be the space defined by

$$\tilde{V} = \{w_-|_\Gamma; (\Delta + k^2)w(x) = 0, \ x \in \Omega, \ \partial_n w = 0, \ x \in \Gamma\}.$$

w is in $C^2(\Omega) \cap C(\overline{\Omega})$. It also follows that $\tilde{V} = \mathcal{N}(I + 2T_K)$, $\mathcal{N}$ denoting the null space.

Let us show that $\tilde{V}$ can be nonempty. That is, the homogeneous interior Neumann problem can have nontrivial solutions. In order to keep the discussion simple, let us consider the interior Neumann problem on a circle of radius R with sources in the interior. Thus

$$
\begin{aligned}
(\Delta + k^2)v_-(x) &= \ \delta(x - x_s), \ x, x_s \in \Omega \\
\partial_n v_- &= \ 0, \ x \in \Gamma.
\end{aligned}
$$

The Helmholtz representation yields

$$v_-(x) = \frac{i}{4} H_0(k|x - x_s|) - \frac{i}{4} \int_\Gamma v_-(y)\partial_y H_0(k|y - x|) d\Gamma(y), \quad x \in \Gamma. \quad (7.44)$$

H_n and J_n stand for the cylindrical Hankel and Bessel function of order n, respectively. Now (Gradshtein and Rhyzhik, 1980)

$$H_0(|X - Y|) = \sum_{n=-\infty}^{\infty} H_n(k|X|_{>(<)}) J_n(k|Y|_{<(>)}) e^{in(\theta_X - \theta_Y)}. \quad (7.45)$$

$|X|_{<(>)}$ on X indicates that $|X|$ is greater (less) than $|Y|$, respectively, and similarly for $|Y|_{<(>)}$. θ_X, θ_Y are the angular coordinates of X and Y, respectively. Substituting Eq. (7.45) into Eq. (7.43) and simplifying yields

$$v_-(x) = \frac{i}{4} \sum_{n=-\infty}^{\infty} \{J_n(k|x_s|) H_n(k|R|) e^{-in\theta_{x_s}} - a_n kR H_n'(kR) J_n(k|x|)\} e^{in\theta_x},$$

$$(7.46)$$

where a_n is given by

$$a_n = \int_0^{2\pi} v_-(y) e^{-in\theta_y} d\theta_y. \quad (7.47)$$

Taking the normal derivative of Eq. (7.46) and setting it to zero (homogeneous boundary condition) results in the equation

$$\sum_{-\infty}^{\infty} \{J_n(k|x_s|) H_n'(kR) e^{-in\theta_{x_s}} - a_n kR H_n'(kR) J_n'(kR)\} e^{in\theta_x} = 0,$$

from which we obtain

$$a_n = \frac{J_n(k|x_s|)}{kR J_n'(kR)} e^{-in\theta_{x_s}}, \quad \forall n. \quad (7.48)$$

For real k there exist countably infinite values of $k = k_{nm}, m = 1, 2, \cdots$, (called the *interior eigenvalues*) for which $J_n'(k_{nm}R) = 0$. The subscript n on k implies that the n-th order Bessel function is being considered. Consequently, the homogeneous interior Neumann problem may have countably infinite solutions given by

$$(v_-)_{nm}(x) = J_n(k_{nm}|x|) e^{in\theta_x}.$$

The nullspace of the homogeneous interior Neumann operator $\mathcal{N}(I + 2T_{K^*})$, therefore, may not be empty. Now the operator $I + 2T_K$ of the exterior Dirichlet problem is adjoint to $I + 2T_{K^*}$ of the interior Neumann problem. It is known (see, for example, McLean, 2000) in the theory of integral equations of the second kind that if A and B are two compact operators which are adjoint to each other, then the null spaces $\mathcal{N}(I + A)$ and $\mathcal{N}(I + B)$ are of the same dimension. Thus if k is not an interior Neumann eigenvalue, then $\dim \mathcal{N}(I + 2T_{K^*}) = \dim \mathcal{N}(I + 2T_{K^*}) = \{0\}$. By the Fredholm theory, the exterior Dirichlet problem has a unique solution in this case. Hence we have the following theorem.

Theorem 7.7 (The Exterior Dirichlet Problem) *The double-layer poten-tial*

$$v(x) = \int_\Gamma \frac{\partial G^0(x, y; k)}{\partial n(y)} \psi(y) d\Gamma(y), x \in R^3 \setminus \Gamma, \psi \in C(\Gamma),$$

solves the exterior Dirichlet problem uniquely provided that k is not the eigen-value of the interior Neumann problem. That is, k is such that the boundary value problem

$$(\Delta + k^2)v_- = 0, \; x \in \Omega$$

$$\partial_n v_- = 0, \; x \in \Gamma$$

has a trivial solution.

Let us next consider the inhomogeneous Neumann problem in the interior which is

$$(\Delta + k^2)u_- = 0, \; x \in \Omega$$
$$\partial_n u_- = g \; x \in \Gamma.$$

Let $u_h(x)$ (we are using the letter u since the solution is in the form of a single-layer potential) solve the homogeneous problem, i.e., $g = 0$. Applying Green's second identity, we have

$$\int_\Omega (u_- \Delta u_h - u_h \Delta u_-) dx = \int_\Gamma (u_- \frac{\partial u_h}{\partial n} - u_h \frac{\partial u_-}{\partial n}) d\Gamma = 0$$

from which

$$\int_\Gamma u_h g d\Gamma = 0.$$

If k is not an interior eigenvalue, then this condition is satisfied automatically since the only homogeneous interior Neumann solution is the trivial solution. If, on the contrary, k is an eigenvalue then this must constitute the condition for the solvability of the interior problem. In other words, the boundary data g must be such that

$$\int_\Gamma u_h g d\Gamma = 0, \; \forall u_h$$

in order for a solution to exist.[6] We have the following theorem.

Theorem 7.8 (The Inhomogeneous Interior Neumann Problem) *The inhomogeneous interior Neumann problem is solvable if and only if*

$$\int_\Gamma u_h g d\Gamma = 0, \forall u_h,$$

where u_h is the solution of the homogeneous interior Neumann problem.

---

[6] For the Laplace problem, namely, $\Delta u = 0$, $\frac{\partial u}{\partial n} = g$ on Γ, the analogous result is $\int_\Gamma g d\Gamma = 0$.

Returning to Eq. (7.48), we see that if k is an interior eigenvalue, then a_n is unbounded and, the inhomogeneous problem has no solution. If, however, k is not an interior eigenvalue, then a_n are well-defined and are given by Eq. (7.48). The interior Neumann solution is given by

$$
\begin{aligned}
v_-(x) &= \frac{i}{4} H_0(k|x - x_s|) \\
&\quad - \sum_{-\infty}^{\infty} \frac{J_n(k|x_s|)}{J_n'(kR)} H_n'(kR) J_n(k|x|) e^{in(\theta_x - \theta_{x_s})}.
\end{aligned}
$$

The corresponding results for the exterior Neumann and the interior Dirichlet problem can be established in an analogous way and are presented in the following two theorems.

Theorem 7.9 (The Exterior Neumann Problem) *The exterior Neumann problem is uniquely solvable if and only if k is not an eigenvalue of an interior Dirichlet problem.*

Theorem 7.10 (The Inhomogeneous Interior Dirichlet Problem) *The inhomogeneous interior Dirichlet problem is solvable if and only if*

$$
\int_\Gamma f \frac{\partial v_h}{\partial n} d\Gamma = 0
$$

for all solutions v_h of the homogeneous interior Dirichlet problem.

7.7 Unconditionally Unique Solution

We note that an exterior problem must have a unique solution purely on physical grounds. The necessity of excluding the interior spectrum in order to achieve uniqueness of the exterior solution cannot, therefore, be intrinsic to the scattering problem *per se*. The difficulty must lie with the integral representations of the solutions. It is, therefore, necessary to pose the problem in such a way that unique solutions can result for any physical frequency. This is not merely of theoretical interest, but a practical necessity, both from a numerical as well as an experimental standpoint. Numerically, using a frequency from the interior spectrum will result in an ill-conditioned matrix. On the other hand, from an experimental point of view, it is difficult to choose a pulse which will selectively eliminate only those frequencies that constitute the interior spectrum if for no other reason than for the fact that for an obstacle of arbitrary shape, it is difficult to search for the interior spectrum. It goes without saying that the situation is destined to be detrimental toward an inverse solution. Fortunately, there exist methods for deriving unconditionally unique solutions to exterior problems. Several techniques are discussed in Ramm (1986). Here we limit our attention to the widely used technique (see Colton and Kress, 1983, 1992, and references therein) of using a combination of a single and a double-layer potential instead of working with them separately. Note that Helmholtz's representations are solved by any such combination. Also discussed is a technique due to Jones (1974) in which the free-space Green's function is replaced by a modified one.

7.7.1 Combining Single and Double-layer Potentials

Let us try to explain the basic motivation behind combining a single and a double-layer potential. Consider the previous example of a Neumann problem for a circular scatterer of radius R. There it was found that the root of our inability to determine the coefficients a_n was in the vanishing of the derivative of the Bessel function $J_n'(kR)$ at countably infinite values of k. For the problem in the interior, this resulted in unbounded solutions at these frequencies, whereas in the exterior, a_n simply could not be determined at the interior spectrum. The difficulty arose from the fact that k was real and, consequently, so was the function $J_n'(kR)$. The situation remains unaltered if the Neumann problem is replaced by a Dirichlet problem in which case $J_n(kR)$ would occur instead of $J_n'(kR)$. However, if a combination such as $J_n'(kR) \pm \mathrm{i}J_n(kR)$ is considered instead of $J_n'(kR)$ or $J_n(kR)$ individually, then for real k, no cancellation can take place. Indeed if we write the solution of the exterior Neumann problem of the example as

$$
u(x) = \frac{\mathrm{i}}{4}H_0(k|x - x_s|)
$$
$$
- \frac{\mathrm{i}}{4}\int_\Gamma \phi(y)[\partial_y H_0(k|x - y|) + \alpha H_0(k|x - y|)]\mathrm{d}\Gamma(y), \ \phi \in C(\Gamma)
$$

$\alpha \neq 0$, then we find that

$$
u(x) = \frac{\mathrm{i}}{4}\sum_{n=-\infty}^{\infty} J_n(k|x|)H_n(k|x_s|)e^{\mathrm{i}n(\theta_x - \theta_{x_s})}
$$
$$
- \frac{\mathrm{i}}{4}R\sum_{n=-\infty}^{\infty} a_n[kJ_n'(kR) + \alpha J_n(kR)]H_n(k|x|)e^{\mathrm{i}n\theta_x}, \ |x| < |x_s|
$$

with

$$
a_n = \int_0^{2\pi} \phi(y)e^{-\mathrm{i}n\theta_y}\mathrm{d}\theta_y.
$$

The homogeneous boundary condition, namely, $\partial_n u = 0$, results in the solution

$$
a_n = \frac{J_n'(kR)H_n(k|x_s|)}{R[kJ_n'(kR) + \alpha J_n(kR)]H_n^1(kR)}e^{-\mathrm{i}n\theta_{x_s}}.
$$

Since the denominator cannot cancel to zero for real k, the coefficients a_n remain bounded.

Let us look at the problem a little more rigorously. Consider the exterior Dirichlet case and assume that the solution can be written as

$$
w(x) = v(x) + \alpha u(x), \ x \in R^3 \setminus \Gamma,
$$

where $u(x)$ and $v(x)$ are the single and double-layer potential, respectively, with a moment function $\phi \in C(\Gamma)$. The parameter α will be discussed momentarily. The solution $w(x)$ is to be found by determining ϕ by solving the boundary integral equation

$$(I + 2T_K + 2\alpha S)\phi = 2f, \ f \in C(\Gamma). \tag{7.49}$$

In order to determine the uniqueness of the solution of Eq. (7.49), we consider, as usual, the homogeneous equation

$$(I + 2T_K + 2\alpha S)\phi = 0.$$

Using the jump relations and the homogeneous boundary data, it follows that

$$\begin{aligned} w_+ - w_- &= -w_- = (v_+ - v_-) + \alpha(u_+ - u_-) \\ &= v_+ - v_- = \phi. \end{aligned}$$

On the other hand, considering the normal derivative of w on the boundary and from the jump relations again, we have

$$\begin{aligned} \partial_n(w_+ - w_-) &= -\partial_n w_- \\ &= \partial_n[(v_+ - v_-) + \alpha(u_+ - u_-)] = -\alpha\phi. \end{aligned}$$

Using the Helmholtz equation $(\Delta + k^2)w(x) = 0$, $x \in \Omega$, and applying Green's first identity to w_-^* and Δw_- in Ω, it follows (in the bilinear form) that

$$\begin{aligned} < w_-^*, \Delta w_- >_\Omega &= -k^2 < w_-^*, w_- >_\Omega = -k^2 \|w_-\|^2_{L_2(\Omega)} \\ &= < w_-^*, \partial_n w_- >_\Gamma - < \nabla w_-^*, \nabla w_- >_\Omega, \end{aligned}$$

resulting in the identity

$$-\alpha\|\phi\|^2_{L_2(\Gamma)} = [\|\nabla w_-\|^2_{L_2(\Omega)} - k^2\|w_-\|^2_{L_2(\Omega)}]. \tag{7.50}$$

If both α and k are real, $\|\phi\|^2_{L_2(\Gamma)}$ can vanish without ϕ being zero. But if α is imaginary, then the only solution possible is the trivial solution of $\phi = 0$ and uniqueness is obtained. Next consider Eq. (7.50) $\forall k = k' + ik''$, $k'' \geq 0$. Then

$$-\alpha\|\phi\|^2_{L_2(\Gamma)} = [\|\nabla w_-\|^2_{L_2(\Omega)} - (k'^2 - k''^2)\|w_-\|^2_{L_2(\Omega)}] - 2ik'k''\|w_-\|^2_{L_2(\Omega)}.$$

There is no reason to consider real α since α must be imaginary if k is real. Therefore, if $\alpha = -i\eta$, η real, then from Eq. (7.50)

$$-\eta\|\phi\|^2_{L_2(\Gamma)} = 2k'k''\|w_-\|^2_{L_2(\Omega)}. \tag{7.51}$$

Thus if $\eta \neq 0$, η real, then the only solution of Eq. (7.51) is the trivial solution showing that the exterior Dirichlet problem is uniquely solvable $\forall k = k' + k''$, $k', k'' \geq 0$. Hence the following theorem.

Theorem 7.11 (The Exterior Dirichlet Problem) *The exterior Dirichlet problem with a continuous boundary data is uniquely solved by a combined single and double-layer potential* $\forall k = k' + ik'', k', k'' \geq 0$.

The above procedure, however, runs into difficulties for the Neumann problem because of the double differentiation of the double-layer potential in both x and y. If we write this as

$$N\phi(x) = \partial_x \int_\Gamma \partial_y G^0(x,y)\phi(y)\mathrm{d}\Gamma(y), x \in R^3, \tag{7.52}$$

we see that Eq. (7.52) has a singularity of $|x - y|^{-3}$ and is, therefore, strongly singular. The operator N is thus called *hypersingular*. N is defined only as an improper integral. A full discussion of the hypersingular operator and the hypersingular equation (7.52) will take us too far-field. We will, therefore, say only the following.

Consider a function $\overline{\psi}$ in $\mathcal{D}(R^3)$. That is, $\overline{\psi}$ is infinitely differentiable and is compactly supported in R^3. Let ψ be its trace on the boundary Γ or $\overline{\psi}|_\Gamma = \psi$. Let $B_\epsilon(x)$ be a ball of radius ϵ centered at $x \in \Gamma$ and write

$$N_\epsilon\phi(x) = \partial_x \int_{\Gamma_\epsilon} \partial_y G^0(x,y)\phi(y)\mathrm{d}\Gamma(y), \tag{7.53}$$

where $\Gamma_\epsilon = \Gamma \setminus B_\epsilon$. We can express the double-layer potential in terms of ψ in the form

$$v(x) = -(S\partial_n\psi)(x)+ < G^0, \mathcal{L}\psi >_\Omega, x \in \Omega, \tag{7.54}$$

where $\mathcal{L}$ is the Helmholtz operator $\Delta + k^2$ and $S\partial_n\psi$ is the single-layer potential with the derivative $\partial_n\psi$ as the density. Starting from the double-layer equation (7.54) and using Green's identities it is possible to show (McLean, 2000) that the hypersingular equation (7.53) can be reduced to the form

$$(N\psi)(x) = \lim_{\epsilon \to 0} \Big(N_\epsilon\phi(x)$$
$$+ \frac{1}{2}\Big[\int_- - \int_+ \Big] \partial_{nx}\partial_{ny}G^0(x,y)\psi(y)\mathrm{d}\Gamma(y)\Big), \tag{7.55}$$

where

$$\int_\pm = \int_{\Omega_\pm \cap \partial B_\epsilon(x)},$$

Ω_- and Ω_+ being the interior and exterior domain of the scatterer, respectively, and $N_\epsilon\phi(x)$ is as given by Eq. (7.53). Furthermore, Eq. (7.55) can be shown (McLean, 2000) to constitute the *finite part* of the divergent integral $(N\psi)(x)$.

Recall at this point Hadamard's regularization of a divergent integral by its finite part (Jones, 1966; Kanwal, 1997). For example, consider the function $x^{-3/2}$ over the interval $(0, \infty)$. It does not generate a regular distribution. In other words, if $\tilde{\phi}$ is a test function having a support in an interval L, then the integral

$$\int_0^\infty \tilde{\phi}(x)x^{-3/2}\mathrm{d}x \qquad (7.56)$$

is not convergent, the function $x^{-3/2}$ not being summable. However, let us consider the integral

$$I = \int_\epsilon^\infty \tilde{\phi}(x)x^{-3/2}\mathrm{d}x, \quad \epsilon > 0$$

instead of the integral (7.56). Writing $\tilde{\phi}$ as $\tilde{\phi}(x) = \tilde{\phi}(0) + x\psi(x)$, and integrating by parts yields

$$\underset{\epsilon \to 0}{\text{limit}} \left(-\frac{1}{2}\right) \int_\epsilon^L \tilde{\phi}(x)x^{-3/2}\mathrm{d}x = \underset{\epsilon \to 0}{\text{limit}} \left[\tilde{\phi}(0) \left(\frac{1}{\sqrt{\epsilon}} - \frac{1}{\sqrt{L}}\right) + \frac{1}{2}\int_\epsilon^L \psi(x)x^{-1/2}\mathrm{d}x \right].$$
$$(7.57)$$

The upper limit of integration ∞ was replaced by L because $\tilde{\phi}$ vanishes outside L. Taking the limit $L \to \infty$ in Eq. (7.57) results in the finite part

$$-\frac{1}{2} \text{ FP} \int_0^\infty \phi(x)x^{-3/2}\mathrm{d}x = \underset{\epsilon \to 0}{\text{limit}} \left[\frac{\phi(0)}{\sqrt{\epsilon}} - \frac{1}{2}\int_\epsilon^\infty \phi(x)x^{-3/2}\mathrm{d}x \right]. \qquad (7.58)$$

The terms within the braces in Eq. (7.58) are divergent individually, but taken together, the limit exists and defines the regularization of the divergent integral (7.56). Therefore, *the regularization of a divergent improper integral*

$$\int_L f(x)\phi(x)\mathrm{d}x$$

is a distribution which is identifiable with f everywhere except at the singularity where it is considered to be the finite part. With the regularized form of the integral, it is possible to establish existence results for the Neumann problem.

 There have been attempts to deal with the hypersingularity such as that of Brakehage and Werner (1965). The method of Brakehage and Werner is one of combining a single- and a double-layer potential. This technique of combining the potentials will be discussed below. Instead we present an approach introduced by Nédélec (2001). Let us return to Eq. (7.52). The equation actually involves calculating the quantity

$$(Nv)(x) = v_n(x) = (\nabla v \cdot \hat{n})(x).$$

From now on, we will drop the argument x. Let $\mathbf{g} = \nabla v$ denote the gradient vector of the double-layer potential in both Ω_i and Ω_e. Note that $\mathbf{g}$ satisfies the radiation condition at infinity. Moreover, from Theorem 7.1, $[\mathbf{g} \cdot \hat{n}]|_\Gamma = 0$. Remember, however, that the derivatives themselves may be discontinuous at the boundary. In other words, $[\mathbf{g}\cdot\hat{n}]\big|_+$ may not be the same as $[\mathbf{g}\cdot\hat{n}]\big|_-$ in general. The problem must, therefore, be understood in the sense of distributions. Thus

$$\nabla\cdot\mathbf{g} + k^2 v = 0$$

distributionally, and $\nabla\cdot\mathbf{g}\in L_2(R^3)$. Furthermore, in the notation of symbolic calculus (see Remarks in Section 7.5), we can write

$$\nabla v = \{\nabla v\} + (\nabla v)_d = \mathbf{g} + \phi\hat{n}\delta(\Gamma).$$

$\mathbf{g}$ is the distributed part, whereas $(\nabla v)_d = \phi\hat{n}\delta(\Gamma)$ is the distributional component. ϕ is a test function and $\delta(\Gamma)$ is the Dirac distribution which is concentrated on the surface Γ. Thus, we have a coupled system

$$\left.\begin{array}{l}\nabla\cdot\mathbf{g} + k^2 v = 0 \\ \nabla v - \mathbf{g} = \phi\hat{n}\delta(\Gamma).\end{array}\right\} (1)$$

in $(v,\mathbf{g})$.

The solution v is to be obtained by determining the fundamental solution of the coupled system (1), and then convolving the R.H.S., $\phi\hat{n}\delta(\Gamma)$, with the fundamental solution. Toward that end, define $\mathbf{G} = -\nabla G^0$, and write

$$\nabla\mathbf{G} = -\nabla[\nabla\cdot(G^0\mathbf{I})].$$

Note that this is the same as writing

$$\nabla\mathbf{G} = -\sum_i\sum_j\hat{e}_i\hat{e}_j\partial_{ij}G^0.$$

From vector calculus (Morse and Feshbach, 1953),

$$\Delta = \nabla(\nabla\cdot) - \nabla\times\nabla\times\cdot.$$

Then

$$\nabla\mathbf{G} = -\Delta(G^0\mathbf{I}) - \nabla\times\nabla\times(G^0\mathbf{I}) + k^2(G^0\mathbf{I}).$$

Define

$$\Sigma = -\nabla\times\nabla\times(G^0\mathbf{I}) + k^2(G^0\mathbf{I}).$$

Then

$$\nabla\mathbf{G} = -\Sigma = -(\Delta + k^2)(G^0\mathbf{I}) = \delta\mathbf{I},$$

where δ is Dirac's delta. On the other hand,

$$\nabla\cdot\Sigma = k^2\nabla\cdot(G^0\mathbf{I}),$$

leading to

$$\nabla\cdot\Sigma - k^2\nabla\cdot(G^0\mathbf{I}) = \nabla\cdot\Sigma + k^2\mathbf{G} = 0.$$

Thus

$$\nabla\cdot\Sigma + k^2\mathbf{G} = 0,$$

and

$$\nabla \mathbf{G} - \Sigma = \delta \mathbf{I}.$$

The function $\mathbf{G}$ is, therefore, the fundamental solution of the system (1).

In accordance with our earlier discussion, the double-layer potential is now given by

$$
\begin{aligned}
v(x) &= -\mathbf{G} * (\phi \hat{n} \delta(\Gamma)) \\
&= -\int_\Gamma \left\{ \nabla_y G^0(|x - y|) \cdot \hat{n}(y) \right\} \phi(y) d\Gamma(y),
\end{aligned}
$$

and

$$\mathbf{g} = \nabla[-\mathbf{G} * (\phi \hat{n} \delta(\Gamma))].$$

Using the definition of Σ given above, we obtain

$$
\begin{aligned}
\mathbf{g} &= -\mathbf{G} * (\phi \hat{n} \delta(\Gamma)) = -\Sigma * (\phi \hat{n} \delta(\Gamma)) \\
&= \nabla \times \nabla \times (G^0 \mathbf{I}) * \phi \hat{n} \delta(\Gamma) \\
&\quad - k^2 \int_\Gamma G^0(|x - y|) \phi(y) \hat{n}(y) d\Gamma(y) \\
&= \left\{ \nabla \times (G^0 \mathbf{I}) \right\} * \left\{ \nabla \times * \phi \hat{n} \delta(\Gamma) \right\} \\
&\quad - k^2 \int_\Gamma G^0(|x - y|) \phi(y) \hat{n}(y) d\Gamma(y). \quad (7.59)
\end{aligned}
$$

The last line of Eq. (7.60) followed from the fact that in a convolution, the differential operators commute.

Our purpose here is to demonstrate how the strong singularity arising in the evaluation of Nv can be avoided. It is, therefore, not necessary for us to reduce the identity (7.60) further. This is done in Nédélec (2001). The main point already appears in Eq. (7.60) itself. The primary objective is to calculate Nv, and from Eq. (7.59), this is the same as calculating $(\nabla v \cdot \hat{n})$, i.e., $\mathbf{g} \cdot \hat{n}$, $\mathbf{g}$ given by Eq. (7.60). Assuming that $\phi \in C(\Gamma)$, the first term on the R.H.S. in this equation is integrable. The second term is simply the derivative of a single-layer potential, and thus also regular.

We remark in passing that the four operators S, T_K, T_{K^*} and N are sometimes linked together in what is known as the *Calderon relations*. One defines an operator $\mathbf{C}$ by

$$\mathbf{C} = \begin{pmatrix} -T_K & S \\ -N & T_{K^*} \end{pmatrix}.$$

With the operator $\mathbf{C}$ are associated two other operators

$$\mathbf{C}\pm = \frac{I}{2} \mp \mathbf{C}.$$

It is then shown that

$$\mathbf{C}\pm^2 = I \quad \text{and} \quad \mathbf{C}_+ + \mathbf{C}_- = I.$$

The operators $\mathbf{C}_\pm$ are, therefore, projection operators (see Chapter 4), and are known as *Calderon projectors*. The following relations are found to hold among the various operators:

$$\mathbf{C}^2 = \frac{I}{4},$$
$$T_K S = S T_K,$$
$$N T_K = T_{K^*} N,$$
$$T_K^2 - SN = \frac{I}{4},$$

and

$$T_{K^*}^2 - NS = \frac{I}{4}.$$

7.8   The Transmission Problem

The transmission problem is as follows

$$(\Delta + k_+^2)u_+(x) = 0, \; x \in \Omega_+, k_+ > 0$$
$$(\Delta + k_-^2)u_-(x) = 0, \; x \in \Omega_-, k_- > 0$$
$$u_+(x) - u_-(x) = 0, \; x \in \Gamma, \qquad (7.60)$$
$$\partial_n(\mathcal{F}_+ u_+(x) - \mathcal{F}_- u_-(x)) = 0, \; x \in \Gamma. \qquad (7.61)$$

Subscripts $+, -$ refer to the exterior and interior of the scatterer, respectively. $u_+ = u^{sc} + u^{inc}$ and $\mathcal{F}_\pm = \rho_\pm^{-1}$, where ρ is the mass density. We seek the solution in the form of a combined single and double-layer potential and write

$$u_\pm = (SL\phi)_\pm + (DL\psi)_\pm = (S_\pm\phi) + (T_{K\pm}\psi), \qquad (7.62)$$

where $(SL\phi)_\pm$ denotes the single-layer potential with density ϕ. The subscripts $+(-)$ indicate that Green's function kernels in the integral operators $S_+(S_-), T_{K_+}(T_{K_-})$ have frequencies $k_+(k_-)$, respectively.[7] Substituting Eq. (7.63) in the boundary conditions (7.61) and (7.62), using the jump conditions and after minor algebra, the transmission problem is reduced to the following integral equation

$$(I - \mathbf{T}_{tr})\xi_{tr} = \mathbf{d}_t r. \qquad (7.63)$$

$\mathbf{T}_{tr}$ is given by

$$\begin{pmatrix} \mathcal{F}_+ T_{K_+} - \mathcal{F}_- T_{K_-} & S_+ - S_- \\ -(\mathcal{F}_+ R_+ - \mathcal{F}_- R_-) & -\mathcal{F}_+ T_{K_+^*} - \mathcal{F}_- T_{K_-^*} \end{pmatrix}$$

---

[7]We have introduced a slight change in notation here. Since u is being used for the fields, the single-layer potential is now denoted by SL instead of u and in order to be consistent notation wise, the double-layer potential is designated by DL.

is the transmission matrix. $\xi_{tr} = \{\psi, \phi\}^T$ is the solution vector and similarly, $\mathbf{d}_{tr} = \{-u^{inc}, -\mathcal{F}_+\partial_n u^{inc}\}^T$ is the transmission data.

We next demonstrate that the quantity $\mathcal{F}_+ R_+ - \mathcal{F}_- R_-$ is weakly singular and, therefore, compact. All other operators of the matrix $\mathbf{T}_{tr}$ are weakly singular . We follow Kittappa and Kleinman (1975) here. In reducing the $\mathcal{F}_+ R_+ - \mathcal{F}_- R_-$ element of the matrix, we need to consider integrals of the form

$$\int_\Gamma \partial_x \partial_y \left[\frac{a_+ e^{ik_+ R} - a_- e^{ik_- R}}{R}\right], \quad x, y \in \Gamma, \tag{7.64}$$

where $R = |x - y|$ and $a_\pm$ are constants. However, the constant factors do not influence the calculations and, therefore, it suffices to consider the integral (7.65) with these factors set to unity. We note that

$$\partial_y f(R) = (\hat{n}(y) \cdot \hat{R}) f_R, \tag{7.65}$$

and similarly,

$$\partial_x f(R) = -(\hat{n}(x) \cdot \hat{R}) f_R, \tag{7.66}$$

where $f_R = \partial_R f(R)$. In order to be concise, let us write

$$\xi(R) = \frac{e^{ik_+ R} - e^{ik_- R}}{R}.$$

Differentiating $\xi(R)$ in $\hat{n}(x)$ and upon using relations (7.66) and (7.67) yields

$$\begin{aligned}\partial_x \partial_y \xi(R) &= \partial_x \left[(\hat{n}(y) \cdot \hat{R})(\xi(R)_R\right] \\ &= -(\hat{n}(x) \cdot (\hat{R})(\hat{n}(y) \cdot \hat{R})(\xi(R)_{RR} \\ &\quad + (\partial_x(\hat{n}(y) \cdot \hat{R}))(\xi(R)_R.\end{aligned} \tag{7.67}$$

Now

$$\nabla_x(\hat{n}(y) \cdot \hat{R}) = \nabla_x\left(\frac{\mathbf{R} \cdot \hat{n}}{R}\right) = \frac{\hat{R} \cdot \hat{n}(y)}{R}\hat{R} - \frac{\hat{n}(x)}{R}. \tag{7.68}$$

Taking the scalar product of Eq. (7.69) with $\hat{n}(x)$ yields

$$\partial_{nx}(\hat{n}(y) \cdot \hat{R}) = -\frac{\hat{n}(x) \cdot \hat{n}(y) + (\hat{n}(y) \cdot \hat{R})(\hat{n}(x) \cdot \hat{R})}{R}. \tag{7.69}$$

Substituting Eq. (7.70) in Eq. (7.68) and then expanding the exponentials, it follows that

$$\begin{aligned}\partial_{nx}\partial_{ny}(\xi(R)) &= i\sum_{j=1}^\infty \frac{(j+1)!}{(j+2)!}(k_+^{j+2} - k_-^{j+2})(iR)^{j-1} \\ &\quad \cdot [(j-1)(\hat{n}(y) \cdot \hat{R})(\hat{n}(x) \cdot \hat{R})(\hat{n}(x) \cdot \hat{n}(y)] \\ &\quad - \frac{k_+^2 - k_-^2}{2R}[(\hat{n}(x) \cdot \hat{n}(y) - (\hat{n}(y) \cdot \hat{R})(\hat{n}(x) \cdot \hat{R})]. \end{aligned} \tag{7.70}$$

The first term of the R.H.S. in Eq. (7.71) is clearly continuous on the boundary. The second term is actually a single-layer potential with a continuous density. This, therefore, shows the weakly singular nature of the operator $\mathcal{F}_+ R_+ - \mathcal{F}_- R_-$. The operator $\mathbf{T}_{tr}$ is thus weakly singular and hence compact and Eq. (7.64) is a Fredholm equation of the second kind. It follows from the Fredholm alternative that the transmission problem has a unique solution provided that the homogeneous transmission equation

$$(I - \mathbf{T}_{tr})\xi_{tr} = 0 \qquad (7.71)$$

has only a trivial solution.

Let $\xi_h = (\psi_h, \phi_h)^T$ be the solution of the homogeneous equation (7.72). We would like to show that ξ_h vanishes identically. Let $u_{\pm h}$ be the corresponding solution of Eq. (7.72). That is,

$$u_{\pm h} = (SL_\pm)\phi_h + (DL_\pm)\psi_h$$

are solutions to the transmission problem with boundary data

$$u_{+h} - u_h = 0 \qquad (7.72)$$

and

$$\partial_n(\mathcal{F}_+ u_{+h} - \mathcal{F}_- u_h) = 0. \qquad (7.73)$$

It is, therefore, sufficient to show that $u_{\pm h}$ vanishes for homogeneous boundary conditions. Toward that we consider the following.

Let B_R be a ball of radius R (boundary Γ_R) enclosing the scatterer Ω (with boundary Γ), its origin being inside Ω. Consider the annular region Ω_R between Γ and Γ_R in which region u_{+h} satisfies the Helmholtz equation $(\Delta + k_+^2)u_{+h} = 0$ and so does its complex conjugate $\bar{u}_{+h}$. Now apply Green's second identity to u_{+h} and $\bar{u}_{+h}$ in Ω_R and obtain

$$0 = \int_{\Gamma_R} [u_{+h}\partial_R \bar{u}_{+h} - \bar{u}_{+h}\partial_R u_{+h}]d\Gamma_R - \int_{\Gamma} [u_{+h}\partial_n \bar{u}_{+h} - \bar{u}_{+h}\partial_n u_{+h}]d\Gamma, \qquad (7.74)$$

where ∂_R is the radial derivative on Γ_R. Equation (7.75) can be recast in the form

$$\text{Im} \int_{\Gamma_R} u_{+h}\partial_R \bar{u}_{+h}d\Gamma_R = \text{Im} \int_{\Gamma} u_{+h}\partial_n \bar{u}_{+h}d\Gamma. \qquad (7.75)$$

We can further rewrite Eq. (7.76) as

$$\text{Im} \int_{\Gamma_R} [u_{+h}(\partial_R \bar{u}_{+h} - ik_+ \bar{u}_{+h}) + ik_+ |u_{+h}|^2]d\Gamma_R = \text{Im} \int_{\Gamma} u_{+h}\partial_n \bar{u}_{+h}d\Gamma. \qquad (7.76)$$

Since u_{+h} satisfies the radiation condition at infinity, the first term of the integral on the L.H.S. of Eq. (7.77) vanishes as $R \to \infty$ and the L.H.S. reduces to

$$\mathrm{Im} \int_{\Gamma_R} |u|^2 \mathrm{d}\Gamma_R \sim O(1).$$

Therefore,

$$0 = \mathrm{Im} \int_\Gamma u_{+h} \partial_n \overline{u}_{+h} \mathrm{d}\Gamma \geq 0, \qquad (7.77)$$

and by Rellich's lemma $u_{+h} = 0$ in $R^3 \setminus \Omega$. Since $u_{\pm h}$ satisfies the homogeneous boundary conditions (7.73) and (7.74), we have $u_h = 0$ on Γ. Therefore,

$$0 = \mathrm{Im} \int_\Gamma u_h \partial_n \overline{u}_h \mathrm{d}\Gamma.$$

From the Helmholtz interior representation (Theorem 6.1) it follows that $u_h = 0$ in Ω. Hence the following theorem.

Theorem 7.12 *The transmission problem with boundary conditions (7.61) and (7.62) has a unique solution.*

7.9 Jones' Method

There is a method due to Jones (1974) for dealing with exterior problems. Jones' method applies for all positive wavenumbers and avoids the need to regularize. Instead of a combined single- and double-layer potential approach, Jones uses a modified Green's function $\tilde{G}(x, y; k)$. The corresponding modified layer potential operators $\tilde{S}$ and $\tilde{T}_{K(K^*)}$ (that is, the operators S and $T_{K(K^*)}$ having the modified Green's function as kernel) are compact on $C(\Gamma) \times C(\Gamma)$. The newly formed Green's function $\tilde{G}(x, y, k)$ is defined as

$$\tilde{G}(x, y, k) = G^0(x, y, k) + \tilde{g}(x, y, k), \qquad (7.78)$$

where $G^0(x, y, k)$ is the usual free-space Green's function. The extra term $\tilde{g}(x, y, k)$ in Eq. (7.79) consists of a series of outgoing scattering solutions and is constructed as

$$\tilde{g}(x, y, k) = ik \sum_{\ell=0}^\infty \sum_{m=-\ell}^\ell \tilde{g}_{\ell m} h_\ell^{(1)}(k|x|) h_\ell^{(1)}(k|y|) Y_{\ell m}(\hat{x}) Y_{\ell m}(\hat{y}), \qquad (7.79)$$

where, as usual, $\hat{z} = z \setminus |z|$. The free-space Green's function $G^0(x, y, k)$, on the other hand, has the following expansion:

$$G^0(x, y, k) = \frac{e^{ik|x-y|}}{4\pi|x-y|}$$

$$= ik \sum_{\ell=0}^\infty \sum_{m=-\ell}^\ell j_\ell(k|x|) h_\ell^{(1)}(k|y|) Y_{\ell m}(\hat{x}) Y_{\ell m}^*(\hat{y}), \qquad (7.80)$$

where $|x| < |y|$. The modified single and double layer potentials are given by

$$\tilde{u}(x) = (\tilde{S}\phi)(x) = \int_\Gamma \tilde{G}(x, y, k)\tilde{\phi}(y)\mathrm{d}\Gamma(y), \qquad (7.81)$$

and

$$\tilde{v}(x) = (\tilde{T}_K\psi)(x) = \int_\Gamma \partial_y\tilde{G}(x, y, k)\tilde{\phi}(y)\mathrm{d}\Gamma(y), \qquad (7.82)$$

$\forall x \in R^3 \setminus \Gamma \setminus \overline{B}_R$, where B_R is a ball of radius R such that $B_R \subset \Omega$. In other words, B_R is strictly contained in the scattering region Ω, the origin being in B_R.

Consider Dirichlet scattering. If we substitute the expansions (7.80) and (7.81) into Eq. (7.83) and do the y-integration, we obtain for $\tilde{v}$ the following series

$$\tilde{v}(x) = \sum_{\ell=0}^\infty \sum_{m=-\ell}^\ell \tilde{v}_{\ell m}\left[j_\ell(k|x|) + \tilde{g}_{\ell m}h_\ell^{(1)}(k|x|)\right]Y_{\ell m}(\hat{x}), \qquad (7.83)$$

with coefficients $\tilde{v}_{\ell m}$. From the asymptotic (in order) behavior of the spherical Hankel functions, namely

$$h_n^{(1)}(k|z|) \sim O(\frac{(2n)^n}{(ek|z|)^n}), n \to \infty,$$

it follows that

$$\left|\tilde{g}_{\ell m}h_\ell^{(1)}(k|x|)h_\ell^{(1)}(k|y|)\right| \sim O\left(R^{2n}|x|^{-n}|y|^{-n}\right).$$

Therefore, for $|x|, |y| > R + \epsilon$, $\epsilon > 0$, the series in Eq. (7.80) is uniformly convergent and can, therefore, be differentiated in x and y term-by-term as many times as desired. The $\tilde{g}(x, y, k)$ part of the double-layer potential, therefore, satisfies the vanishing jump condition in the normal derivative across Γ as in Theorem 7.1.

Suppose that $\tilde{\psi}_h$ solves the homogeneous boundary integral equation

$$\tilde{\psi}_h + \tilde{T}_K\tilde{\psi}_h = 0.$$

Then $\tilde{v}_h$ with $\tilde{\psi}_h$ as density is a solution of the exterior Dirichlet problem. Since the exterior Dirichlet problem is uniquely solvable, it follows that $\tilde{v}_h \equiv 0$ in $R^3 \setminus \Omega$. Therefore,

$$\partial_{n+}\tilde{v}_h = \partial_{n-}\tilde{v}_h = 0$$

on Γ by the jump condition and the fact that both T_K and $g(x, y, k)$ part do so.

Next consider a sphere Γ_{R_1} of radius R_1 enclosing the obstacle. Applying Green's second identity to $\tilde{v}$ and its complex conjugate $\overline{\tilde{v}}$ in the annular region between Γ_{R_1} and Γ, we have

$$\int_{\Gamma_{R_1}}[\tilde{v}\partial_{R_1}\overline{\tilde{v}} - \overline{\tilde{v}}\partial_{R_1}\tilde{v}] = \int_\Gamma[\tilde{v}_-\partial_n\overline{\tilde{v}} - \overline{\tilde{v}}\partial_n\tilde{v}_-] = 0. \qquad (7.84)$$

The appearance of $\tilde{v}_-$ in Eq. (7.85) is due to the fact that Green's identity requires that the normal $\hat{n}$ be pointing inward into the scattering region. Substituting Eq. (7.84) for $\tilde{v}$ into Eq. (7.85) and upon simplification results in the identity

$$\frac{i}{2k}\sum_{\ell=0}^{\infty}\sum_{m=-\ell}^{\ell}|\tilde{v}_{\ell m}|^2\left[1-|1+2\tilde{g}_{\ell m}|^2\right].\tag{7.85}$$

It readily follows that a unique solution exists for the exterior Dirichlet problem, that is, $\tilde{v}$ in Eq. (7.84) is zero if and only if the coefficients $\tilde{v}_{\ell m}$ are zero, and we must impose the condition

$$|1+2\tilde{g}_{\ell m}|\neq 1,\forall\ell\in Z^+,\forall m\in[-\ell,\ell].$$

The exterior Dirichlet problem can then be solved uniquely for all positive wavenumbers in terms of only the double-layer potential $\tilde{v}$ given by Eq. (7.83) with the modified Green's function (7.79) as the kernel of the integral operator $\tilde{T}_K$. As a matter of fact, as demonstrated by Jones, the number of the coefficients $\tilde{v}_{\ell m}$ in Eq. (7.84) equals the number of the interior Neumann eigenvalues removed. Similar considerations also apply to the exterior Neumann problem. We will say no more about Jones' method. Details appear in Colton and Kress (1983) (the present discussion was patterned after them), Ursell (1978) and Kleinman and Roach (1982).

Finally, we would like to mention two important numerically oriented studies in solving exterior problems. The first is by Schenk (1967) regarding the solvability of the exterior problems for all wavenumbers. The second is the more theoretical work of Burton and Miller (1971) dealing with the hypersingular integral equations in Helmholtz scattering. However, we do not discuss them here and refer the reader instead to the original papers cited.

7.10   Appendix A.7.1

Lemma

$$\int_{\Gamma}\hat{n}(y)\cdot\nabla_y G^0(x,y;k)\mathrm{d}\Gamma(y)\;=\;-1,\;x\in\Omega$$
$$=\;0,\;x\in\Omega_e$$
$$=\;-\frac{1}{2},\;x\in\Gamma.$$

The free-space Green's function G^0 on the L.H.S. can be replaced by the potential theoretical fundamental solution g^0.

Proof: Consider Theorem 6.1 for the Helmholtz representation in the interior. Note that the function u in the theorem is arbitrary subject only to the smoothness conditions necessary for the theorem to hold. We can, therefore, set

u equal to unity in the theorem and the results in Ω and Ω_e follow at once. The evaluation of the integral on the boundary is more involved and can be done as follows.

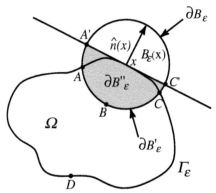

Figure A.7.1 – A schematic illustration for the proof of the lemma. $B_\epsilon(x)$ is a ball of radius ϵ at $x \in \Gamma$. ∂B_ϵ is the boundary surface of B_ϵ. $\partial B_\epsilon' = \partial B_\epsilon \cap \Omega$ equals the portion ABC of ∂B_ϵ. $\partial B_\epsilon'' = \{y \in \partial B_\epsilon | y \cdot \hat{n}(x) < 0\}$ equals the shaded hemisphere $A'ABCC'$. $\tilde{\Gamma}_\epsilon = \Gamma \setminus (B_\epsilon \cap \Gamma)$ equals the surface ADC.

Consider a point $x \in \Gamma$ and a ball B_ϵ of radius ϵ centered at x. Referring to Figure A.7.1, let us define the following: $\tilde{\Gamma}_\epsilon = \Gamma \setminus (\Gamma \cap B_\epsilon)$, $\partial B_\epsilon' = \partial B_\epsilon \cap \Omega$ and $\partial B_\epsilon'' = \{y \in \partial B_\epsilon : \hat{n}(x) \cdot y < 0\}$. In other words, $\partial B_\epsilon''$ is the hemisphere of the ball B_ϵ which is on the same side of the tangent plane through x as is Ω. According to the second part of the Lemma, we have

$$\int_{\tilde{\Gamma}_\epsilon \cup \partial B_\epsilon'} \hat{n}(y) \cdot \nabla_y G^0(x, y; k) d\Gamma(y) = 0, \ x \in B_\epsilon \cap \Gamma.$$

Therefore,

$$\int_{\tilde{\Gamma}_\epsilon} \hat{n}(y) \cdot \nabla_y G^0(x, y; k) d\Gamma(y) = -\int_{\partial B_\epsilon'} \hat{n}(y) \cdot \nabla_y G^0(x, y; k) d\Gamma(y).$$

Now

$$\int_\Gamma = \lim_{\epsilon \to 0} \int_{\tilde{\Gamma}_\epsilon}.$$

Then

$$\int_\Gamma \hat{n}(y) \cdot \nabla_y G^0(x, y; k) d\Gamma(y) = -\lim_{\epsilon \to 0} \int_{\partial B_\epsilon'} \hat{n}(y) \cdot \nabla_y G^0(x, y; k) d\Gamma(y). \qquad (1)$$

Now

$$\hat{n}(y) \cdot \nabla_y G^0(x, y; k) = \frac{e^{ik|x-y|}}{4\pi} \left[\frac{1}{|x-y|^3} - \frac{ik}{|x-y|^2} \right] \hat{n}(y) \cdot (x - y).$$

As $\epsilon \to 0$, it is sufficient to replace G^0 by g^0 and consider only the integral

$$\frac{1}{4\pi} \int_{\partial B'_\epsilon} \frac{\hat{n}(y) \cdot (x-y)}{|x-y|^3} d\Gamma(y) = \frac{1}{4\pi} \int_{\partial B'_\epsilon} \frac{\hat{n}(y) \cdot \widehat{(x-y)}}{|x-y|^2} d\Gamma(y), \qquad (2)$$

where $\widehat{(x-y)} = (x-y)\backslash|x-y|$ is the unit vector along $x-y$. This is again due to the fact that the term involving ik results in a uniformly existing improper integral which vanishes as $\epsilon \to 0$. Now the integration in Eq. (2) equals integration over the hemisphere $\partial B''_\epsilon$ minus that over the area $\partial B''_\epsilon \backslash \partial B'_\epsilon$. The integral over $\partial B''_\epsilon \backslash \partial B'_\epsilon$ goes to zero as $\epsilon \to 0$. In this limit the integral over the hemisphere is simply the solid angle subtended by the hemisphere at x which is 2π. Substituting in Eq. (1), the third line of the Lemma on the boundary follows.

7.11   Appendix A.7.2

Lemma. Let $\Gamma \in C^2$. Then there exists a constant $C > 0$ depending upon the structure of Γ such that

$$|\hat{n}(y) \cdot (x-y)| \le C|x-y|^2, \quad \forall x, y \in \Gamma.$$

Proof: Let x be a point on Γ and $\Gamma_\epsilon(x)$ a surface patch such that $\Gamma_\epsilon(x) = \{y \in \Gamma; |x-y| \le \epsilon\}$. Pass a tangent plane T to Γ through x. For a sufficiently small ϵ, the patch $\Gamma_\epsilon(x)$ can be described in terms of the local coordinates $\xi = (\xi_\perp, \xi_3)$ with x as the origin. The coordinate ξ_3 is along the normal $\hat{n}(x)$ at x and $\xi_\perp$ spans the tangent plane T. In other words, $y = \eta(\xi_\perp)$, $y \in \Gamma_\epsilon(x)$, η describing the functional form of the surface in terms of ξ. Moreover, $\eta(0) = (\nabla\eta)(0) = 0$. Since $\Gamma \in C^2$, $\eta(\xi_\perp)$ can be expanded in a Taylor series around $\xi = 0$ yielding

$$\begin{aligned}(x-y) \cdot \hat{n}(x) &= \xi_3 \\ &= \eta(0) + (\nabla\eta)_0 \xi_\perp + \frac{1}{2}(\nabla^2\eta)_0 \xi_\perp^2 + \cdots.\end{aligned}$$

By virtue of the fact that $\eta(0) = (\nabla\eta)(0) = 0$, we have

$$|(x-y) \cdot \hat{n}(x)| \le \frac{1}{2}|(\nabla^2\eta)_0||x-y|^2 \le C|x-y|^2,$$

the constant C depending on the surface Γ. Hence is the Lemma. In general, if $\Gamma \in C^{1,\alpha}$, $0 < \alpha \le 1$, then $|(x-y) \cdot \hat{n}(x)| \le C|x-y|^{1+\alpha}$.

Next let x_0 and x_1 be two points on the boundary and consider the difference $\hat{n}(x_0) - \hat{n}(x_1)$. Applying the mean value theorem to $\hat{n}(x_0) - \hat{n}(x_1)$ gives

$$|\hat{n}(x_0) - \hat{n}(x_1)| \le |(\hat{n})_x(z)||x_0 - x_1|,$$

where $(\hat{n})_x(z)$ is the derivative of the normal with respect to $x \in \Gamma$ at a point $z \in \Gamma$ intermediate between x_0 and x_1. Now $(\hat{n})_x(z)$ is a vector on the tangent plane at z and since $\Gamma \in C^2$, this is a vector of finite length. Therefore,

$$|\hat{n}(x_0) - \hat{n}(x_1)| \le C_1|x_0 - x_1|,$$

where C_1 is some constant. If $L = \max(C, C_1)$, then we have

$$|(x - y) \cdot \hat{n}(x)| \leq L|x - y|^2$$

and

$$|\hat{n}(x_0) - \hat{n}(x_1)| \leq L|x_0 - x_1|.$$

We also have

$$|1 - \hat{n}(x_0) \cdot \hat{n}(x_1)| \leq L|x_0 - x_1|$$

and a simple geometric argument shows that for x_0 close to x_1,

$$\hat{n}(x_0) \cdot \hat{n}(x_1) > \frac{1}{2}.$$

7.12   Appendix A.7.3

Lemma. *Let $x \in R^3 \setminus \Gamma$. Then $\exists$ a constant C such that*

$$I = \int_\Gamma |\hat{n}(y) \cdot \nabla_y G^0(x, y; k)| d\Gamma(y) \leq C < \infty, \ \forall x.$$

Moreover, the free-space Green's function G^0 can be replaced by the potential theoretic fundamental solution g^0.

Proof: Let $x = x_0 + t\hat{n}(x_0)$, $x_0 \in \Gamma$, $t > 0$. Then

$$I = I_{\Gamma_\epsilon} + I_{\Gamma_{\epsilon c}}.$$

For $y \in \Gamma_{\epsilon c}$, the integral over $\Gamma_{\epsilon c}$ is bounded. It is then sufficient to consider the integral over Γ_ϵ only. For $y \in \Gamma_\epsilon$, $x - y = (x - x_0) + (x_0 - y)$ and we have

$$|x - y|^2 = |x - x_0|^2 + |x_0 - y|^2 + 2(x - x_0) \cdot (x_0 - y). \tag{3}$$

Now since $x - x_0 = |x - x_0|\hat{n}(x_0)$, then

$$2(x - x_0) \cdot (x_0 - y) = 2|x - x_0|\{\hat{n}(x_0) \cdot (x_0 - y)\}.$$

By the Lemma in Appendix A.7.2, it follows that

$$|2(x - x_0) \cdot (x_0 - y)| \leq 2C|x - x_0||x - y|^2.$$

Choose $\epsilon < \frac{C}{2}$. Then

$$2(x - x_0) \cdot (x - y)| < |x - x_0||x_0 - y|.$$

Therefore,

$$\min |x - y|^2 = |x - x_0|^2 + |x_0 - y|^2 - |x - x_0||x_0 - y|. \tag{4}$$

Now let $|x - x_0| < \frac{\epsilon}{2}$ and $|x_0 - y| < \epsilon$. Then $|x - x_0||x_0 - y| < \frac{\epsilon^2}{2}$. Using the bound in Eq. (4) in Eq. (3), we obtain the relation

$$|x - y|^2 \geq \frac{1}{2}(|x - x_0|^2 + |x_0 - y||^2).$$

Now

$$
\begin{aligned}
|\hat{n}(y) \quad \cdot \quad \nabla y G^0(x, y; k)| &\leq \frac{1}{4\pi}|\hat{n}(y) \cdot (x - y)|\left[\frac{1}{|x - y|^3} + \frac{k}{|x - y|^2}\right] \\
&= \frac{1}{4\pi}|\hat{n}(y) \cdot (x - x_0) + \hat{n}(y) \cdot (x_0 - y)|\left[\frac{1}{|x - y|^3} + \frac{k}{|x - y|^2}\right] \\
&\leq \frac{1}{4\pi}[|||x - x_0|| + C|x_0 - y|^2]\left[\frac{1}{|x - y|^3} + \frac{k}{|x - y|^2}\right].
\end{aligned}
$$

It is sufficient to show that the $|x - y|^{-3}$ part of the integral remains bounded as this part is most singular for small values of ϵ. Now

$$
\begin{aligned}
&\frac{1}{4\pi}\frac{|x - x_0|}{\left[|x - x_0|^2 + |x_0 - y|^2\right]^{3/2}} + \frac{C}{4\pi}\frac{|x_0 - y|^2}{\left[|x - x_0|^2 + |x_0 - y|^2\right]^{3/2}} \\
&\leq \frac{1}{4\pi}\frac{|x - x_0|}{\left[|x - x_0|^2 + |x_0 - y|^2\right]^{3/2}} + \frac{C}{4\pi}|x_0 - y|^{-1}.
\end{aligned}
$$

Let $x - x_0 = \xi$ and $x_0 - y = \eta$. Then

$$
\begin{aligned}
\int_{\Gamma_\epsilon} \frac{|x - x_0|}{\left[|x - x_0|^2 + |x_0 - y|^2\right]^{3/2}} d\Gamma(y) \\
\leq 2\int_0^{2\pi} d\theta \int_0^\epsilon \frac{\eta\xi}{\xi^2 + \eta^2}^{3/2} d\eta \\
= 4\pi \int_0^\lambda \frac{z\, dz}{(i + z^2)^{3/2}},
\end{aligned}
$$

where $\lambda = \frac{\epsilon}{|x - x_0|}$. As $z \to \infty$, the integral remains bounded since the integrand falls off as z^{-3}. The integral, therefore, converges. Furthermore, the integral of $|x_0 - y|^{-1} \sim O(\epsilon)$ over the same limits. The lemma is thus established.

7.13 Appendix A.7.4

The Fredholm Alternative
Let A be a bounded linear operator between two Banach spaces X and Y. Assume that the range, $\mathcal{R}(A)$, of A is a closed subspace in Y and the subspaces $\mathcal{N}(A)$ and $Y \setminus \mathcal{R}(A)$ are finite-dimensional. An operator A satisfying these

conditions is called an operator of the *Fredholm type*. Moreover, such operators are characterized by a quantity $I(A)$ defined by

$$I(A) = \dim \mathcal{N}(A) - \dim Y \setminus \mathcal{R}(A).$$

The number $I(A)$ is known as the *index* of A. For operators of the class $A = I + K, K$ being compact, it is known (McLean, 2000) that the index is zero. The operators of the Fredholm type, therefore, include the important integral operators of the layer potentials. Also, if A is Fredholm, then so is its adjoint A^* and furthermore, $I(A^*) = -I(A)$. It is also known (see Chapter 3) from functional analysis that $\overline{\mathcal{R}(A)} = \mathcal{N}(A^*)^{\perp}$.

Consider now the inhomogeneous equation $Af = g$ and assume that the homogeneous equation $Af = 0$ has only a trivial solution $f = 0$. The nullspace $\mathcal{N}(A)$ is, therefore, empty and consequently, A is one-to-one. By virtue of the fact that the index $I(A)$ is zero and the nullspace is empty, $\dim(Y \setminus \mathcal{R}(A)) = 0$. It follows at once that A maps X onto Y. Therefore, A is an injective and onto map and, consequently, has a bounded inverse and the inhomogeneous equation $Af = g$ has a unique solution $\forall g \in Y$. Analogous results follow for the inhomogeneous adjoint problem $A^* \tilde{f} = \tilde{g}$.

Next let $\mathcal{N}(A)$ be nonempty having a dimension, say, p. Since the index is zero, it follows that $\dim(Y \setminus \mathcal{R}(A)) = p$. Both $\mathcal{N}(A)$ and $Y \setminus \mathcal{R}(A)$ are spanned by p linearly independent solutions $\{u_i\}_1^p$ of the homogeneous equation. Since $\overline{\mathcal{R}(A)} = \mathcal{N}(A^*)^{\perp}, f$ must be perpendicular to the subspace $Y \setminus \mathcal{R}(A)$. This implies that the scaler product $< f, u_i > = 0, \forall i = 1, 2, \cdots, p$. Again, an identical conclusion holds for the adjoint A^*.

In analysis, it is usual to collect the above results into the celebrated *Fredholm alternative* stated below.

Theorem 7.13 (The Fredholm Alternative) *Let* $A : X \rightarrow Y, X, Y$ *Banach spaces, be a bounded linear operator of the Fredholm type with index zero. The Fredholm alternative states that either*

(I) *The nonhomogeneous problems* $Af = g$ *and* $A^* \tilde{f} = \tilde{g}$ *have unique solutions* f *and* $\tilde{f}$, *respectively, for every given* $g \in Y$ *and* $\tilde{g} \in X$ *in which case the homogeneous problems* $Af = 0$ *and* $A^* \tilde{f} = 0$ *must have only trivial solutions;*

Or

(II) *The homogeneous equations* $Af = 0$ *and* $A^* \tilde{f} = 0$ *have the same number of linearly independent solutions* $\{u_i\}_1^p$ *and* $\{v_i\}_1^p, p \geq 1$, *respectively. The nonhomogeneous problems* $Af = g$ *and* $A^* \tilde{f} = \tilde{g}$ *are not solvable for all* g *and* $\tilde{g}$. *Nonunique solutions exist if and only if* $< f, u_i > = 0$ *and* $< \tilde{g}, v_i > = 0, \forall i = 1, 2, \cdots p$.

It is interesting to point out that originally, Fredholm's theory and the alternative bearing his name arose out of reducing the integral equation to a discretized, finite-dimensional form similar to that encountered in Chapter 4. The analysis, however, was aimed at treating the case in which the dimension n of the finite-dimensional space tended to infinity and the resultant determinant of the system came to be known as the *Fredholm determinant*. The dichotomy seen in the theorem in the behavior of the two integrals (the one involving A and the other involving A^*) above reflects that between the zero and nonzero values of the determinant. The modern formulation of the alternative (as in the theorem above) reflects the geometric language of functional analysis rather than the classical analysis of Fredholm. For more on this point and also for a more rigorous analysis of the alternative using the theory of Fredholm operators, the reader should consult the recent text by McLean (2000).

7.14   Appendix A.7.5

Uniqueness of the Dirichlet Solution

In this Appendix we demonstrate that the exterior Dirichlet problem has at most one solution. In other words, if a solution exists then it is the only solution. Consider a Dirichlet problem with homogeneous boundary data:

$$(\Delta + k^2)u = 0, \text{ in } \Omega_e, \ u = 0 \text{ on } \Gamma.$$

Let Γ_R be a sphere of radius R enclosing Γ. Apply Green's second identity to u and $\bar{u}$ in the annular region between Γ_R and Γ. This leads to the following relation:

$$\int_{\Gamma_R} (u\partial_R u^* - u^* \partial_R u) d\Gamma_R = \int_\Gamma (u\partial_n u^* - u^* \partial_n u) d\Gamma = 0.$$

Now from Chapter 6, we have

$$ik \int_{\Gamma_R} (u\partial_R u^* - u^* \partial_R u) d\Gamma_R = \left\{ \int_{\Gamma_R} (|\partial_R u|^2 + k^2 |u|^2) \right\} d\Gamma_R.$$

This implies that

$$\lim_{R\to\infty} \left\{ \int_{\Gamma_R} (|\partial_R u|^2 + k^2 |u|^2) \right\} d\Gamma_R = 0.$$

This implies that $u = 0$.

7.15   Appendix A.7.6

Improper Integrals

The *improper integrals* are discussed in any advanced text on calculus, for example, Buck (1956). A number of important characterizing relations for the potentials, their gradients and directional derivatives, happen to involve improper integrals. It is, therefore, worthwhile to briefly review the definition and the main properties of such integrals. An integral can be improper if either the limit(s) of integration is (are) infinite and/or the integrand is unbounded. Here we are concerned with the latter. Let $f(x)$ be a function defined on a region

$D \in R^m$. f is assumed to be continuous in D except in a set $S \in D$ of measure zero where the function has singularities. The Riemann sums of f then do not exist on S and the Riemann integral of f

$$J(f) = \int_D f(x)\mathrm{d}x$$

does not exist. Let $\mathcal{S}$ be a neighborhood containing S. Then

$$\tilde{J}(f) = \int_{D\backslash\mathcal{S}} f(x)\mathrm{d}x$$

is well-defined. Let $\{\mathcal{S}_k\}, \mathcal{S}_k \in \mathcal{S}_{k-1}$ be a sequence of neighborhoods of S such that $S = \cup\mathcal{S}_k$ and $\mathcal{S}_k \to S, k \to \infty$. Define

$$\tilde{J}(\mathcal{S}_k) = \int_{D\backslash\mathcal{S}_k} f(x)\mathrm{d}x.$$

If $\mathrm{limit}_{\mathcal{S}_k \to S}\, \tilde{J}(\mathcal{S}_k)$ exists for any sequence $\{\mathcal{S}_k\}$ independently of the sequence chosen, then the integral $J(f)$ of f over D is defined by this limit and is said to exist as an improper integral. The definition given above implies absolute integrability, that is,

$$|J|(f) = \int_D |f(x)|\mathrm{d}x$$

exists (improperly) and is finite. The improper integral is then said to be *convergent*. If $J(f)$ exists improperly, but $|J|(f)$ is not convergent, then the improper integral is called *conditional*.

Frequently, the function f contains a parameter, say, ξ so that $f = f(x, \xi), \xi$ itself belonging to a set, say, S_ξ. This is the case with the integral operators defining the potentials in which the parameterization takes the form $f(x, \xi) = f(x - \xi)$ (cf. the free-space Green's function and its derivatives). Consider again the quantity $\tilde{J}(\mathcal{S}_k)$. If $\mathrm{limit}_{\mathcal{S}_k \to S}\tilde{J}(\mathcal{S}_k) \to 0$, then the improper integral of f is said to exist *uniformly*. In this case, the integral $J(f)$ is considered to be continuous in ξ. That is, $|f(p) - f(q)| \leq \epsilon$ whenever $|p - q| \leq \delta, p, q \in S_\xi$ (see Hackbush, 1995). The limit $\mathcal{S}_k \to S$ is the same as the limit $\mu_k = \mu(\mathcal{S}_k) \to 0, \mu_k$ being the measure

$$\mu_k = \int_{\mathcal{S}_k} \mathrm{d}x.$$

Chapter 8

Uniqueness Theorems in Inverse Problems

We now consider the uniqueness of inverse scattering problems of Helmholtz's equation. To determine whether and under what conditions a scatterer can be reconstructed uniquely from the knowledge of the fields it scatters is of paramount importance in inverse problems. We begin with the obstacle problem and then move on to the inhomogeneity. It also stands to reason that we consider the Dirichlet and Neumann problems first before taking up the more difficult transmission boundary condition. Since the fields cannot penetrate an impedance boundary, the inverse problem in this case consists entirely of recovering the shape of the obstacle. For a penetrable body, on the other hand, refraction takes place at the boundary and, consequently, a full inverse problem must involve not only the reconstruction of the scatterer's boundary, but its interior as well.

Before delving into the details, let us briefly present some mathematical preliminaries relevant to the purpose here including short derivations of the spectrum of the Dirichlet and Neumann Laplacians. This is as much for the completeness of the presentation as for the reason that the knowledge of these spectra is essential for establishing uniqueness results. The contents of Sections 8.1 through 8.3 can be found in Mikhlin (1970), Davies (1995) and DiBenedetto (1995). For the uniqueness results for obstacle problems in Sections 8.4 through 8.6, we draw heavily from Colton and Kress (1992) and Kirsch and Kress (1993). The results for the inhomogeneity problem are given in Section 8.7 which is based on the work of Nachmann (1989), Ramm (1992) and Colton and Kress (1992).

8.1   Some Definitions

Let us introduce some definitions which will be needed in the sequel. Let K be an operator acting in an infinite-dimensional Hilbert space H and $< Kf, g >_H$ the inner product, $f, g \in H$.

Definition 8.1 (Symmetric Operator). *An operator K acting in a Hilbert space H is called symmetric if its domain $D(K)$ is dense in H and if for any $f, g \in D(K)$, the relation $< Kf, g >_H = < f, Kg >_H$ holds.*

We would like to point out that there is a distinction between an operator being symmetric and being self-adjoint. For a self-adjoint operator, it is also true that $< Kf, g >_H = < f, K^*g >_H = < f, Kg >_H$. However, if K is self-adjoint, then $D(K) \neq D(K^*)$. In other words, K^* may be a proper extension of K. Thus every self-adjoint operator is symmetric, but the converse is not true.

Let $Q(f, g)$ denote the inner product $< Kf, g >_H$. Recall from Chapter 4 that if f and g are real, then Q is linear in both arguments and is called a *bilinear form*. If, on the other hand, f and g are complex, then $Q(f, g)$ is linear in f and *antilinear* or *conjugate linear* in g, i.e., $Q(\alpha f, g) = \alpha Q(f, g)$ and $Q(f, \alpha g) = \alpha^* Q(f, g)$, α^* meaning the complex conjugate of α. An inner product Q in which both these properties combine is called *sesquilinear* meaning $1\frac{1}{2}$ - *times linear* (Kreyszig, 1978). If $f = g$, then $Q(f, f)$ becomes a *quadratic form*.

Definition 8.2 (Positive Operator) *A symmetric operator K is called positive if the quadratic form $Q(f, f) \geq 0$, and $Q(f, f) = 0$ if and only if f is zero. If $Q(f, f) \geq 0, \forall f \in D(K)$, then the operator is non-negative.*

Definition 8.3 (Positive Definite Operator) *A symmetric operator K is called positive-definite if the quadratic form $Q(f, f) \geq c^2 \|f\|^2, c^2 > 0$. Equivalently,*

$$\inf_{f \in D(K), f \neq 0} \frac{Q(f, f)}{\|f\|^2} > 0.$$

It should be pointed out that an operator can be positive without being positive-definite. An example is the operator $K = -\frac{d^2}{dx^2}, Ku = f, f \in C^2[0, \infty), u(0) = 0, u(x) = 0$ for $x > a$ (see Mikhlin, 1970).

Definition 8.4 (Resolvent Operator) *Let K be a non-negative self-adjoint operator in a Hilbert space H. Then the operator $(K+I)^{-1}$ is called the resolvent operator of K.*

Definition 8.5 (Discrete Spectrum) *Let H be an infinite-dimensional Hilbert space and A a symmetric operator on H. A is said to have a discrete spectrum if it has an infinite sequence of eigenvalues $\{\lambda_n\}_{n=1}^{\infty}$ with a unique limit point at infinity and the sequence $\{u_n\}_{n=1}^{\infty}$ of the corresponding eigenfunctions is complete in H.*

The spectrum of a positive operator K lies on the positive half-axis of the real line, i.e., $\text{Spec}(K) \in [0, \infty)$. On the other hand, if K is positive-definite such that $Q(f, f) \geq c^2 \|f\|^2, c^2 > 0$, then $\text{Spec}(K) \in [c^2, \infty)$.

Next we state the well-known *Friedrich's inequality* which is essential in establishing the spectral properties of the Dirichlet and Neumann Laplacians. The inequality is useful in establishing positive-definiteness of solutions of Laplace's equation for the Dirichlet problem and draws attention to the relation between the solution and its gradient.

Friedrich's inequality. *Let Ω be a bounded domain in an n-dimensional Euclidean space R^n with a piecewise smooth boundary. Let $u \in C^1(\Omega)$ satisfy* *the condition* $u|_\Gamma = 0$ *on the boundary. Then*

$$\|u\|_\Omega^2 \le C\|\nabla u\|_\Omega^2.$$

The constant C depends only on Ω and not on u.

Furthermore, for the sake of convenience of the discussions below, we reiterate the uniqueness and analyticity properties of the forward solution in the following two Lemmas.

Lemma 8.1. *Let $\Omega \subset R^n$ be an obstacle and $\psi \in C^2(R^3 \setminus \overline{\Omega})$ be a scattering solution for which the scattering amplitude vanishes identically. Then the corresponding scattered field vanishes identically everywhere in $R^3 \setminus \overline{\Omega}$.*

The lemma guarantees the uniqueness of the forward solution. Let F_∞ be the far-field map, that is, $F_\infty \psi^{rmsc} = \psi_\infty$. Then the lemma tells us that a scattering solution ψ^{rmsc} of the Helmholtz equation vanishes everywhere if $\psi^\infty \equiv 0$.

Lemma 8.2. *The scattering solution of the Helmholtz equation is analytic in all its variables in any compact subset of $\Omega_e = R^3 \setminus \overline{\Omega}$.*

The importance of Lemma 8.2 is two-fold. First, it shows that in order to obtain the scattering amplitude ψ_∞ over the entire unit sphere $\hat{\Omega}_3$, it is necessary to know it only over some open subset $\tilde{\Omega}_3$ of $\hat{\Omega}_3$. Secondly, it allows us to analytically continue the scattered field in compact subsets of the exterior domain. This is an important consideration for uniqueness questions.

Next we demonstrate the linear independence of the total fields for linearly independent incident fields.

8.2 Properties of the Total Fields

8.2.1 Obstacle Scattering: Linear Independence of Total Fields

The objective is to show that the total fields in an obstacle scattering corresponding to incident waves of a fixed frequency $k_0 > 0$, but with distinctly different orientations are linearly independent of each other. Let $\hat{k}_j$, $j = 1, 2, \cdots$ be an enumerable set of unit vectors in $\hat{\Omega}_3$, and $\psi(x, k_j)$, $k_j = k_0 \hat{k}_j$, $j = 1, 2, \cdots$ be the corresponding total fields. Let

$$\sum_j a_j \psi_j = 0, \tag{8.1}$$

where ψ_j is $\psi(x, k_j)$. We would like to prove that a_j is identically zero, $\forall j$.

Consider a ball B_R of radius $R > 0$ which contains the scatterer inside it. Moreover, let R be sufficiently large so that the field is asymptotic on its surface Γ_R. Now for $x \in \Gamma_R$

$$\begin{aligned} \psi_j(x) &= \psi_j^{\text{inc}}(x) + \psi_j^{\text{sc}}(x) \\ &= e^{ik_j \cdot x} + \psi_j^{\text{sc}}(x). \end{aligned}$$

Recall from Chapter 6 that for sufficiently large R, $\psi_j^{sc} \sim O(\frac{1}{R})$. Multiplying both sides of Eq. (8.1) by the complex conjugated m-th incident field, $\overline{\psi}_m^{inc}$, we have

$$a_m + \sum_{j \neq m} a_j \psi_j^{inc} \overline{\psi}_m^{inc} \sim O(R^{-1}). \tag{8.2}$$

Next integrate Eq. (8.2) over Γ_R to obtain

$$a_m + \frac{1}{4\pi R^2} \sum_{j \neq m} a_j \int_{|x|=R} e^{i(k_j - k_m)\cdot x} d\Gamma(x) \sim O(R^{-1}). \tag{8.3}$$

Upon performing the integration in Eq. (8.3) yields

$$a_m + \frac{1}{R} \sum_{j \neq m} a_j \mathrm{Sinc}(\omega) \sim O(R^{-1}), \tag{8.4}$$

where the Sinc function $\mathrm{Sinc}(\omega)$ is given by

$$\mathrm{Sinc}(\omega) = \frac{\sin(R k_0 b_{mj})}{k_0 b_{mj}},$$

$$k_0 b_{mj} = |k_m - k_j|.$$

From Eq. (8.4) it is clear that as $R \to \infty$, the coefficient a_m vanishes. Hence,

Theorem 8.1 *Let $\hat{k}_n \in \hat{\Omega}_3, n = 1, 2, \cdots$ be an infinite sequence of distinct unit vectors specifying an infinite sequence of incident fields, the magnitude $k_0 > 0$ of the wave vectors $k_j = k_0 \hat{k}_j$ being fixed. Then the corresponding infinite sequence $\psi_n = \psi(x, k_n), n = 1, 2, \cdots$, of the total fields form a linearly independent set.*

8.2.2   Inhomogeneity Scattering

A somewhat analogous result can be obtained for scattering from an inhomogeneity. Let $m(x) = 1 - \epsilon(x), x \in R^3$, and consider a ball $B(R)$ of radius R (with center inside the inhomogeneity Ω) outside which $m = 0$. Let $u(\cdot; \hat{\theta}^i) \in C^2(B(R))$ be a scattering solution of the Helmholtz equation $(\Delta + k^2 \epsilon)u = 0$ in $B(R)$ corresponding to an incident plane wave propagating along the unit vector $\hat{\theta}^i$. That is, $u^{inc}(x, \hat{\theta}^i) = e^{ikx \cdot \hat{\theta}^i}$. Also let $v(x) \in C^2(B(R_1)), R_1 > R$, denote any solution of the Helmholtz equation $(\Delta + k^2 \epsilon)v = 0$ in $B(R_1)$. Assume that

$$< v, u(\cdot; \hat{\theta}^i) >_{L_2(B(R))} = 0, \ \forall \hat{\theta}^i \in \hat{\Omega}_3. \tag{8.5}$$

$< f, g >_{L_2(B(R))}$ is the usual L_2-inner product defined by

$$< f, g >_{L_2(B(R))} = \int_{B(R)} f(y) g^*(y) dy,$$

where $*$ denotes complex conjugate. For the sake of notational simplicity, we will use the inner product symbol instead of the integral. In addition, the argument $B(R)$ in $L_2(B(R))$ will be dropped. We would like to show that if Eq. (8.5) holds, then $v \equiv 0$ or v vanishes identically. More precisely, we will show that $\|v\|_{L_2}^2 = 0$. Moreover, if we consider a sequence of solutions $v_j \to v$ in $L_2(B(R))$, then it suffices to show that $< v_j, v >_{L_2} = 0$. In the derivation below, we follow Kirsch (1996).

Now $u(\cdot; \hat{\theta}^i)$ is a solution of the Lippmann–Schwinger equation (see Chapter 6) and is given by

$$u(\cdot; \hat{\theta}^i) = (I + T)^{-1} u^{\text{inc}}(\cdot; \hat{\theta}^i), \tag{8.6}$$

where

$$T(\cdot) = -k^2 \int_{B(R)} (1 - \epsilon(y)) G^0(\cdot, y) dy.$$

From Eqs. (8,5) and (8.6) we obtain

$$
\begin{aligned}
< v, (I + T)^{-1} u^{\text{inc}}(\cdot; \hat{\theta}^i) >_{L_2} &= < (I + T^*)^{-1} v, u^{\text{inc}}(\cdot; \hat{\theta}^i) >_{L_2} \\
&= < w, u^{\text{inc}}(\cdot; \hat{\theta}^i) >_{L_2} = 0,
\end{aligned} \tag{8.7}
$$

where $w = (I + T^*)^{-1} v$. Recalling that the adjoint of an integral operator with kernel $K(x, y)$ is an integral operator with kernel $K^*(y, x)$, it follows that

$$v(x) = (1 + T^*) w = w(x) + k^2 (1 - \epsilon^*(x)) \tilde{w}, \tag{8.8}$$

where

$$\tilde{w} = < w, G^0 >_{L_2}. \tag{8.9}$$

Using Eq. (8.8) we obtain

$$
\begin{aligned}
< v_j, v >_{L_2} &= < v_j, w >_{L_2} + < v_j, k^2 (1 - \epsilon^*) \tilde{w} >_{L_2} \\
&= < v_j, w >_{L_2} + < k^2 (1 - \epsilon) v_j, \tilde{w} >_{L_2} \\
&= < v_j, w >_{L_2} + < \Delta v_j + k^2 v_j, < w, G^0 >_{L_2} >_{L_2} \\
&= < v_j + < \Delta v_j + k^2 v_j, G^{0*} >_{L_2}, w >_{L_2}.
\end{aligned} \tag{8.10}
$$

By the interior Helmholtz representation (see Chapter 6),

$$
[v_j + < \Delta v_j + k^2 v_j, G^{0*} >_{L_2}](x)
$$
$$
= \int_{|y|=R} \left[G^0(x, y) \partial_R v_j(y) - v_j(y) \partial_R G^0(x, y) \right] d\Gamma(y). \tag{8.11}
$$

Replacing Eq. (8.11) in Eq. (8.10) results in

$$
\begin{aligned}
< v_j, v >_{L_2} &= \int_{B(R)} w^*(y) \Big\{ \int_{|y|=R} \Big[G^0(x, y) \partial_R v_j(y) \\
&\quad - v_j(y) \partial_R G^0(x, y) \Big] d\Gamma(y) \Big\} dy.
\end{aligned} \tag{8.12}
$$

Next consider the interior Helmholtz representation in $B(R_1)$ for which $|y| = R_1$ is the boundary. This leads to (for $z \in B(R_1)$)

$$
\begin{aligned}
v_j(z) &= \int_{|y|=R_1} [G^0(z,y)\partial_{R_1}v_j(y) - v_j(y)\partial_{R_1}G^0(z,y)]d\Gamma(y) \\
&\quad - \int_{B(R_1)} [\Delta v_j(y) + k^2 v_j(y)]G^0(z,y)dy \\
&= \int_{|y|=R_1} [G^0(z,y)\partial_{R_1}v_j(y) - v_j(y)\partial_{R_1}G^0(z,y)]d\Gamma(y) \\
&\quad - \int_{B(R_1)\backslash B(R)} [\Delta v_j(y) + k^2 v_j(y)]G^0(z,y)dy \\
&\quad - \int_{B(R)} [\Delta v_j(y) + k^2 v_j(y)]G^0(z,y)dy \\
&= \int_{|y|=R_1} [G^0(z,y)\partial_{R_1}v_j(y) - v_j(y)\partial_{R_1}G^0(z,y)]d\Gamma(y) \\
&\quad - \int_{B(R)} [\Delta v_j(y) + k^2 v_j(y)]G^0(z,y)dy
\end{aligned}
\tag{8.13}
$$

since $\Delta v_j(y) + k^2 v_j(y) = 0$ in $B(R_1)\backslash B(R)$. In view of Eq. (8.13), it follows that $|y| = R$ in Eq. (8.11) can be replaced by $|y| = R_1$. Using this fact in Eq. (8.12) and interchanging the order of integration yields

$$
< v_j, v >_{L_2} = \int_{|y|=R_1} [\tilde{w}^*(y)\partial_{R_1}v_j(y) - v_j(y)\partial_{R_1}\tilde{w}^*(y)]d\Gamma(y).
\tag{8.14}
$$

The proof is completed by showing that $\tilde{w}$ vanishes identically outside $B(R)$. Toward this, we use Eq. (8.9) and note that

$$
\lim_{|x|\to\infty} \tilde{w}^*(x) = \lim_{|x|\to\infty} < w, G^0 >_{L_2}^* = \frac{1}{4\pi} < w, u^{\text{inc}}(\cdot; -\hat{\theta}^i) >_{L_2} = 0
$$

by virtue of Eq. (8.7). The proposition then follows from Eq. (8.14).

All this can be summed up into the following theorem.

Theorem 8.2 *Let $\epsilon \in C^2(R^3)$ and let $m = 1 - \epsilon$ vanish outside a ball $B(R)$ of radius R with center inside the inhomogeneity. Let $u(\cdot; \hat{\theta}^i)$ be the total field corresponding to an incident wave $e^{ik\hat{\theta}^i \cdot x}$, $\hat{\theta}^i \in \hat{\Omega}_3$. If $v(x) \in C^2(R^3)$ be any solution of the Helmholtz equation $(\Delta + k^2\epsilon)v = 0$ in a ball $B(R_1)$ of radius $R_1 > R$, then*

$$
< v, u(\cdot; \hat{\theta}^i) >_{L_2(B(R))} = 0 \ \forall \ \hat{\theta}^i \in \hat{\Omega}_3.
$$

8.3 The Dirichlet and Neumann Spectrum

8.3.1 The Spectrum of the Negative of the Dirichlet Laplacian in a Bounded Domain

The next to discuss is the spectrum of the negative of the Dirichlet Laplacian $-\Delta_D$. That is, we would like to solve the eigenvalue problem

$$-\Delta u(x) = \lambda u \tag{8.15}$$

subject to the boundary condition: $u = 0$ on Γ, where $u \in C^2(\Omega)$ is a classical solution that satisfies the above boundary condition. Establishing the spectrum of the Dirichlet Laplacian in a bounded domain is a nontrivial task in classical potential theory. Our aim is to give only the essential outlines of the solution leaving the rigorous details to be found in Mikhlin (1970).

Let us first show that $-\Delta_D$ is self-adjoint and positive-definite. Let $u, v \in D(-\Delta_D) = \{u : -\Delta u = \lambda u, u|_\Gamma = 0$. It follows from Green's second identity and Eq. (8.15) that

$$\int_\Omega (u\Delta v - v\Delta u)\mathrm{d}x = \int_\Gamma \left(-v\frac{\partial u}{\partial n} + u\frac{\partial v}{\partial n}\right)\mathrm{d}\Gamma = 0.$$

$-\Delta_D$ is, therefore, self-adjoint. In order to show that the Laplacian is also positive-definite, we apply Green's first identity to the inner product $< -\Delta u, u >_\Omega$ and obtain

$$< -\Delta u, u >_D = \|\nabla u\|_2^2.$$

From Friedrichs's inequality (Section 8.1), it follows that $\|u\|_2^2 \leq c\|\nabla u\|_2^2$, $c > 0$ constant. Hence $< -\Delta u, u >_\Omega \geq C\|u\|_2^2$ proving the positive-definiteness of $-\Delta_D$.

The basis for the determination of the spectrum is contained in the following theorem which we state without proof.

Theorem 8.3 (Mikhlin, 1970) *Let A be a positive-definite self-adjoint operator on an infinite-dimensional Hilbert space H such that any set bounded in the norm in H^1 is also compact in the norm of H. Then the spectrum is discrete, i.e., it consists of an infinite sequence of eigenvalues $\{\lambda_i\}_1^\infty$ with a unique limit point at infinity, the sequence of the corresponding eigenfunctions $\{\psi_i\}_1^\infty$ being complete in H.*

By the theorem then, the eigenspace is of finite multiplicity.

In order to apply Theorem (8.3) to the Dirichlet Laplacian, we need an integral representation of its solution. Toward this, consider a point x in $\Omega \in R^m, m \geq 3$, and a ball $B_\epsilon(x)$ of radius ϵ with x as center. ϵ is small enough so that the ball is entirely contained in Ω. Let ζ be any point in the complement $\Omega \setminus B_\epsilon(x)$ (see Figure 8.1). Denote the distance $|x - \zeta|$ by r and the function r^{2-m} by $v(\zeta)$. Then applying Green's first identity for the Laplace equation for the Dirichlet problem yields (u assumed to be in $C^1(\overline{\Omega})$)

$$\int_{\Omega \backslash B_\epsilon} u \Delta v d\zeta = -\int_{\Omega \backslash B_\epsilon} \nabla_\zeta u \cdot \nabla_\zeta v d\zeta + \int_\Gamma u \frac{\partial v}{\partial n} d\Gamma + \int_{\Gamma_\epsilon} u \frac{\partial v}{\partial n} d\Gamma_\epsilon. \qquad (8.16)$$

The L.H.S. of Eq. (8.16) is zero since v is harmonic in $\Omega \backslash B_\epsilon$. Using the Dirichlet boundary condition and letting $\epsilon \to 0$ in (8.16) results in the integral representation

$$u(x) = \frac{1}{(m-2)|\hat{\Omega}_m|} \int_\Omega \nabla_\zeta u \cdot \nabla_\zeta v d\zeta, \qquad (8.17)$$

where $|\hat{\Omega}_m|$ is the area of the unit sphere in R^m. In obtaining Eq. (8.17), we noted that $d\Gamma_\epsilon = \epsilon^{m-1} dS_\epsilon$, where dS_ϵ is the angular part,

$$\partial_r v \big|_{\Gamma_\epsilon} = -\partial_r r^{2-m} \big|_{r=\epsilon} = (m-2)\epsilon^{1-m}$$

and

$$\int_{\Gamma_\epsilon} u \frac{\partial v}{\partial n} d\Gamma_\epsilon = (m-2) \underset{\epsilon \to 0}{\text{limit}} \int_{\hat{\Omega}_m} u(x + \epsilon\theta) dS_\epsilon.$$

Moreover,

$$|\nabla_\zeta v| \le \frac{m-2}{r^{m-1}}.$$

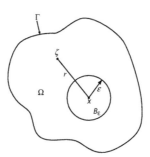

Figure 8.1. The geometry for the integral representation (8.17). $r = |x - \zeta|$.

From the above inequality, it is seen that the integral operator in Eq. (8.17) is weakly singular and hence is compact in $L_2(\Omega)$. Now for a positive-definite operator A in a Hilbert space H, the quadratic form $Q(f, g)$ is a norm functional. It is known (Davies, 1995) that the closure of the domain of this quadratic form (for the Dirichlet problem with homogeneous boundary condition) is the Sobolev space H_0^1. By this is meant that if we consider a space consisting of the elements from the domain $D(A)$ of A with Q as the norm and complete the space, then we would obtain the space H_0^1. Alternatively, this says that any classical solution of the Dirichlet Laplacian is also in $H_0^1(\Omega)$, a well-established result in functional analysis.

Next consider a set $M \in H_0^1(\Omega)$ and let u constitute a bounded set in M:

$$\|u\|_{H_0^1(\Omega)} \leq C = \text{const.}, \ \forall u \in M.$$

Then $\nabla u \in L_2(\Omega)$. Since u is given by the integral representation (8.17) and the integral operator is compact, it transforms the set ∇u bounded in $L_2(\Omega)$ to a set u which is compact in $L_2(\Omega)$. But since Eq. (8.17) holds for functions in $H_0^1(\Omega)$ (it simply recovers any function in M from its gradient), the set M must be compact in $L_2(\Omega)$. Theorem 8.3 is then applicable and shows that the spectrum of the negative of the Dirichlet Laplacian must be discrete. To each eigenvalue there corresponds only a finite number of linearly independent eigenfunctions.

It can also be established in a different way by considering the inverse operator A^{-1}. Since Friedrich's inequality gives $\|u\|_2 \leq c\|\nabla u\|_2$, it follows that $\|A^{-1}\| \leq \|u\|_2 \backslash \|\nabla u\|_2$. A^{-1} is bounded in $L_2(\Omega)$ and is symmetric. Moreover, the map $A^{-1} : L_2(\Omega) \to H_0^1$ is continuous and $A^{-1} : L_2(\Omega) \to L_2(\Omega)$ is a composition of mappings from $L_2(\Omega) \to H_0^1$ and then from $H_0^1 \to L_2(\Omega)$. Now the latter is compact from Sobolev's embedding theorem. Therefore, $A^{-1} : L_2(\Omega) \to L_2(\Omega)$ is compact and has eigenvalues $\{\mu_j\}_1^\infty$ with eigenfunctions $\{\psi_j\}_1^\infty$. Consequently, the operator A has the spectrum $\{\lambda_j\}_1^\infty$ with eigenfunctions $\{\psi_j\}_1^\infty$, where $\lambda_j = \mu_j^{-1}$. We, therefore, have the following theorem.

Theorem 8.4 *The spectrum of the negative of the Dirichlet Laplacian in a bounded region has only a discrete spectrum, the eigenvalues being of finite multiplicity.*

We would like to mention a point regarding the smoothness of the boundary for the Dirichlet problem which will prove to be important in later discussions. Let $\Omega \in R^n$ be a bounded domain and consider the Dirichlet problem

$$\Delta u = 0 \quad x \in \Omega$$
$$u|_\Gamma = \phi \quad x \in \Gamma,$$

where $u \in C^2(\Omega) \cap C(\overline{\Omega})$, and $\phi \in C(\Gamma)$. A simple sufficient condition in order for the problem to have a unique solution is that Ω should satisfy the so-called *exterior sphere* condition, namely

$\forall\, x^* \in \Gamma$, there exists a sphere $B_R(x_0)$ of radius R centered at

$$x_0 \in R^n \setminus \overline{\Omega} \text{ such that } \partial B_R(x_0) \cap \Gamma = \{x^*\}.$$

In other words, the sphere touches the surface at x^* only. This property is shared by domains whose boundaries could be irregular such as those having edges, corners and even cusps pointing outward. This condition follows from Perron's treatment of the problem in which the domain of u is divided into *subharmonic* and *superharmonic* functions. u is sub(super)harmonic if $\forall x \in \Omega, u(x) \leq (\geq)$ its spherical mean at x. The sets $\sigma(\phi;\Omega)$ of the subharmonic and $\Sigma(\phi;\Omega)$ of the superharmonic functions are nonempty. The unique solution to the problem is the unique element of separation between these two sets. The determination of this element calls for a function, known as the *barrier function H*, which has the following characteristics.

$$
\begin{aligned}
H(x^*, \cdot) \quad & \in C(\overline{\Omega}) \\
& \in \Sigma(\phi; \Omega) \text{ in a neighborhood of } \Omega \\
H(x^*, x) > 0 \quad & \forall x \in \overline{\Omega} \setminus x^* \\
H(x^*, x^*) = 0. &
\end{aligned}
$$

It is precisely the construction of the barrier function H through which the surface regularity enters. As a matter of fact, this is the only place where the structure of the boundary enters. The existence of the barrier H requires that the boundary satisfies the exterior sphere criterion.

The last item to be discussed in this section is the theorem stated below.

Theorem 8.5 (Davies, 1995) *Let $\Omega \in R^m$ be a bounded region. Then the spectrum $\{\lambda_n(\Omega)\}_1^\infty$ of the negative of the Dirichlet Laplacian is a monotonically decreasing function of the domain. That is, if Ω_n is an increasing sequence of domains such that their union is Ω, then*

$$
\lim_{j \to \infty} \lambda_n(\Omega_j) = \lambda_n(\Omega), \ \forall \ n \geq 1.
$$

This concludes our cursory presentation of the Dirichlet Laplacian.

We consider the Neumann Laplacian next.

8.3.2   The Analysis of the Neumann Laplacian

Next we consider the Neumann problem. In comparison with the Dirichlet case, establishing the existence of a discrete spectrum for the negative of the Neumann Laplacian in a bounded domain is more involved. There are some characteristic differences in the behavior of the two Laplacians. The Dirichlet Laplacian is positive-definite, while the Neumann Laplacian is not even positive (see Definitions 8.2 and 8.3). Consider, for example, the Neumann problem:

$$
\begin{aligned}
-\Delta u = 0, \quad & x \in \Omega \\
\tfrac{\partial u}{\partial n}|_\Gamma = 0 &
\end{aligned}
$$

and form the inner product

$$
< -\Delta u, u > = \int_\Omega ||\nabla u||^2 \mathrm{d}x.
$$

Now a function $u(x) = 1$ is certainly a legitimate solution. It satisfies Laplace's equation and the boundary condition. However, for such a solution, the inner product $< -\Delta u, u >$ is identically zero even when the norm of the function $||u||_2^2$ is greater than zero. According to Definition 8.2, the Neumann operator is clearly not positive. Let us next replace the homogeneous Laplace equation above by $-\Delta u = f(x)$. In a Neumann problem, the function f cannot be specified arbitrarily, because if we integrate the Laplacian over Ω by parts and use the Neumann boundary condition, we have

$$\int_\Omega f\mathrm{d}x = 0,$$

implying that f must be orthogonal to unity. Therefore, only those functions in $L_2(\Omega)$ qualify for f which occupy a subspace $\tilde{L}_2(\Omega) \subset L_2(\Omega)$, all the elements of this subspace being orthogonal to unity. Thus the Neumann Laplacian transforms any function in its domain to another function in $\tilde{L}_2(\Omega)$. A similar conclusion follows if the equation is homogeneous, but the boundary condition is inhomogeneous. Moreover, the Neumann problem cannot have a unique solution. That is, if u is a solution then so is $u + c$, where c is a constant. Also the monotonicity property of Theorem 8.5 for the Dirichlet problem does not extend to the Neumann case. Finally, we point out another important difference between the behaviors of the two Laplacians. The compactness of the Neumann resolvent (Definition 8.5) depends upon the regularity of the boundary. Thus depending upon the peculiarity of the boundary, the Neumann resolvent may fail to be compact and Theorem 8.3 cannot be invoked. The spectrum may be changed as much as one likes by perturbing the boundary in as small a neighborhood as only a single point. Fortunately, it is possible to avoid these pathologies by imposing regularity conditions on the boundary (Davies, 1995).

The Neumann Laplacian can be made positive and its solution unique by restricting the domain of the Laplacian to $\tilde{L}_2(\Omega)$. The range is, of course, $\tilde{L}_2(\Omega)$. In this newly defined space, the nonuniqueness of the Neumann problem disappears. That is, if $u + c \in \tilde{L}_2(\Omega)$ is a solution then the constant c must be identically zero. This follows because

$$0 =< c, 1 >= \int_\Omega c \, \mathrm{d}x$$

implies that c is zero. Moreover, if the inner product

$$< -\Delta u, u >= \int_\Omega |\nabla u|^2\mathrm{d}x,$$

vanishes, then ∇u must also vanish indicating that the solution is a constant which must be zero by virtue of what was just said above. Thus the Neumann operator in $\tilde{L}_2$ is positive and unique. $< -\Delta u, u >\geq 0$, the equality holding only if u is zero.

The positivity of $-\Delta_N$ is thus established. But to determine that it is also positive-definite and hence to conclude the existence of a discrete spectrum *via* Theorem 8.3 requires an integral representation for the Neumann solution similar to Eq. (8.17) for the Dirichlet problem. In the case of the Neumann problem, this calls for some restrictions on the smoothness of the boundary. Essentially, the requirement is that the region Ω should satisfy an *interior cone condition* in place of the exterior sphere condition of the Dirichlet case. Every point of Ω including points on the boundary is required to be the vertex of a cone of radius R and semi-angle α such that the cone lies entirely within Ω. This is depicted in Figure 8.2(a). The cone condition immediately excludes domains which are cuspidal and non-Lipschitzian, whereas domains which are $C^m, m \geq 0$, piecewise smooth and Lipschitz are allowed.

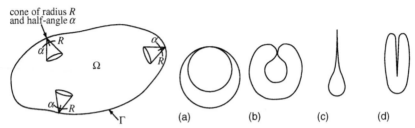

Figure 8.2. A schematic illustration of the interior cone condition and some non-Lipschitzian domains. In (a) is shown a domain for which the cone condition is clearly violated. (b) illustrates the case where the domain lies on both sides of the boundary while (c) and (d) show cuspidal regions, the cusp being exterior in (c) and interior in (d).

For the Neumann problem, an integral representation of the solution can be written as

$$u(x) = \int_\Omega F_0(x, \xi) v \mathrm{d}\xi + \int_\Omega F_1(x, \xi, \xi_j) v_{\xi_j} \mathrm{d}\xi. \qquad (8.18)$$

Equation (8.18) is the analog of Eq. (8.17) for the Dirichlet problem and is actually a variant of the well-known *integral identity of Sobolev* which is used in proving various embedding theorems in Sobolev space theory (An excellent discussion appears in Mikhailov, 1978). The function v in Eq. (8.18) is as in Eq. (8.16). The functions $F_0(x, \xi)$ and $F_1(x, \xi, \xi_j)$ are bounded and the second integral is integrable, possessing only a weak singularity.

Finally, we summarize the procedures for establishing the positive-definiteness and the existence of a discrete spectrum for the Neumann–Laplacian. For further details on these issues, the reader is referred to Mikhlin (1970). We begin by considering the operator $\Delta_1 = -\Delta + I$ which is positive-definite since

$$Q_{\Delta_1}(u, u) \geq \|u\|^2.$$

Let us denote this norm by $\|\cdot\|_1$. This is none other than the norm $\|\cdot\|_{H^1}$ associated with the positive-definite operator Δ_1. Let us call this space $H^1_{\Delta_1}$. Again, as in the Dirichlet problem, using the compactness of the integral operator in the Sobolev representation (8.18), we find that the bounded set M in $H^1_{\Delta_1}$ is also compact in $L_2(\Omega)$. Therefore, by Theorem 8.3, the existence of a discrete spectrum for Δ_1 is established.

Let $\{u_i\}_1^\infty$ be the eigenvectors with the spectra $\{\nu_i\}_1^\infty$ which can be arranged in an increasing order

$$\nu_1 \leq \nu_2 \leq \nu_3 \cdots .$$

It is known (Collatz, 1966) that any eigenvalue of a positive-definite operator cannot be less than the norm of the operator itself. This leads to the conclusion that the lowest eigenvalue ν_1 must strictly be greater than unity. This follows because by definition

$$\nu_1 = \frac{Q_{\Delta_1}(u_1, u_1)}{\|u_1\|_2^2} = \frac{\|u_1\|_{H_{\Delta_1}^1}^2}{\|u_1\|_2^2} \geq 1.$$

Now if strict equality holds, that is, if $\nu_1 = 1$, then $\|u_1\|_{H_{\Delta_1}^1}^2 - \|u_1\|_2^2 = 0$ implying that ∇u is identically zero and, consequently, u is constant. Since $u \in \tilde{L}_2(\Omega)$, u vanishes identically. This contradicts the definition of u being an eigenfunction. Hence ν_1 must be strictly greater than unity. An immediate consequence is that $-\Delta_N$ is positive-definite as shown below.

Now

$$\nu_1 = \inf_{u \in H_{\Delta_1}^1} \frac{\|u\|_{H_{\Delta_1}^1}^2}{\|u\|_2^2}.$$

It follows that

$$\|u\|_{H_{\Delta_1}^1}^2 \geq \nu_1 \|u\|_2^2.$$

From this

$$Q(-\Delta u, u) \geq (\nu_1 - 1)\|u_1\|_2^2.$$

Since $\nu_1 > 1$, the positive-definiteness of $-\Delta_N$ is established.

Let $\phi \in H_{\Delta_1}^1$. Then

$$\nu_k < u_k, \phi > = < \Delta_1 u_k, \phi > = < -\Delta u_k, \phi > + < u_k, \phi >$$

from which

$$< -\Delta u_k, \phi > = < (\nu_k - 1)u_k, \phi > .$$

This shows at once that $-\Delta_N$ has countable eigenvalues $(\nu_k - 1)$ which increase without limit. The corresponding eigenfunctions $\{u_k\}$ form a complete orthonormal system in $\tilde{L}_2(\Omega)$. We have, therefore, the following theorem

Theorem 8.6 *The negative Laplacian $-\Delta_N$ of a Neumann problem has a discrete spectrum. The eigenfunctions and eigenvalues are $\{u_k\}$ and $\{\nu_k - 1\}$, respectively, where $\{u_k\}, \{\nu_k\}$ are the eigenfunctions and eigenvalues of the extended operator $-\Delta + I$.*

8.4 The Uniqueness of the Inverse Dirichlet Obstacle Problem

Let us now consider the uniqueness for the Dirichlet scattering following arguments that are due to Schiffer (Colton and Kress, 1992). We would like to prove that if there are two distinctly different scatterers both of which give rise to identical scattering amplitudes for the same incidence (to be discussed momentarily), then the scatterers must be identical. Let Ω_1 and Ω_2 be two scatterers and ψ_1^∞ and ψ_2^∞ their scattering amplitudes, respectively. By the condition of the problem, $\Delta_{12}^\infty = \psi_1^\infty - \psi_2^\infty = 0$. Now consider a ball B_R of radius $R > 0$ and which is sufficiently large so as to enclose both obstacles within it. Thus $\overline{\Omega}_1 \cup \overline{\Omega}_2 \subset B_R$. Now Lemma 1 says that the scattering amplitude determines the scattered field uniquely in $R^n \setminus B_R$. On the other hand, according to Lemma 2, the scattered field can be analytically continued between any two compact subsets of the domain $R^n \setminus (\overline{\Omega}_1 \cup \overline{\Omega}_2)$. Taken together, they imply that the difference $\Delta_{12}^{sc} = \psi_1^{sc} - \psi_2^{sc}$ of the scattered fields vanishes identically in $B_R \setminus (\overline{\Omega}_1 \cup \overline{\Omega}_2)$.

First assume that the scatterers are disjoint, i.e., $\overline{\Omega}_1 \cap \overline{\Omega}_2 = \emptyset$ and consider the scattering by Ω_1 in the region occupied by the second scatterer Ω_2. We can write

$$(\Delta + k_0^2)\psi_1(x) = 0, \quad x \in \Omega_2$$
$$\psi_1(x) = 0, \qquad\qquad x \in \Gamma_2.$$

The above boundary condition follows from the fact that $\psi_1^{sc} = \psi_2^{sc}$ anywhere in $B_R \setminus (\overline{\Omega}_1 \cup \overline{\Omega}_2)$ and by the Dirichlet condition on the boundary Γ_2 of Ω_2. ψ_1^{sc} is, therefore, a solution to the interior Dirichlet problem in Ω_2 with homogeneous boundary condition for the total field ψ_1.

Now consider that the incidence consists of plane waves propagating along a fixed direction, but having a band of frequencies in an interval I over which $k_i \leq k_0 \leq k_f, k_f > k_i, k_0 > 0$. The data consists of the scattered fields over an open set of the unit sphere $\tilde{\Omega}_n$ for a fixed direction of incidence $\hat{k}_0$, but for a range of frequencies of the incident plane waves, i.e, the data is in $\tilde{\Omega}_n \times I$. Note that $\psi_1(x)$ cannot vanish in Ω_2 or it will vanish everywhere in $R^n \setminus \Omega_1$. This follows from its being an entire function everywhere in the exterior (see Theorem 6.4). The consequence of all this is that $-\Delta\psi_1(x) = k_0^2\psi_1(x)$ in Ω_2 for any wavenumber in the band. This implies that $-\Delta$ has a continuous spectrum which is in direct contradiction to the conclusion of Theorem 8.4. Hence the scatterers cannot be disjoint and, therefore, $\overline{\Omega}_1 \cap \overline{\Omega}_2 \neq \emptyset$.

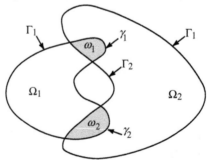

Figure 8.3. An illustration of Schiffer's arguments for establishing uniqueness for Dirichlet scattering. ω_1 and ω_2 are the connected component parts of the intersection $\Omega_1 \cap \Omega_2$ of $\overline{\Omega}_1$ and $\overline{\Omega}_2$.

Let us next consider scatterers which are not disjoint as shown in Figure 8.3. Let ω_i denote the i-th connected component of $\overline{\Omega}_1 \cap \overline{\Omega}_2$ as shown in the figure. Consider the scattering between the components ω_1 and ω_2. By virtue of the arguments above, ω_1 and ω_2 cannot be disjoint. The conclusion, therefore, is that the scatterers must be identical and $\overline{\Omega}_1 = \overline{\Omega}_2$.

We, therefore, have the following uniqueness theorem.

Theorem 8.7 *The scattering data for $k_0 > 0$ fixed, $k_i \le k_0 \le k_f, k_f > k_i$, and $\hat{k}_s \in \tilde{\Omega}_n$, determine a Dirichlet scatterer uniquely.*

Next let us consider the data set: $\hat{k}_n \in \tilde{\Omega}_n, \hat{k}_{sn} \in \tilde{\Omega}_n$, and $k_0 > 0$ is fixed. $\hat{k}_n$ and $\hat{k}_{sn}$ are unit vectors along the incident and the scattered direction, respectively. The data set is now in $\tilde{\Omega}_n \times \tilde{\Omega}_n$ with a single given k_0. In other words, we are considering data for a fixed frequency, but for the directions of incidence and scattering over some open sets of the unit sphere. These sets do not necessarily have to be identical. The proof is similar to that of Theorem 8.7 above. Again consider the connected region ω_1. The j-th field, $\psi_j = \psi(x, k_j)$ which corresponds to the j-th incidence, solves the problem

$$(\Delta + k_0^2)\psi_j = 0, \ x \in \omega_1,$$

$$\psi_j = 0, \ x \in \gamma_1,$$

γ_1 being the boundary of ω_1. By Theorem 8.1, the solutions ψ_js are linearly independent. It then follows that k_0^2 is again an eigenvalue of the negative of the Dirichlet Laplacian, but of infinite multiplicity. As before, this is in violation of Theorem 8.4 according to which the degeneracy must be finite. The scatterers must, therefore, be identical.

Theorem 8.8 *The scattering data: $\hat{k}_n \in \tilde{\Omega}_n, \hat{k}_{sn} \in \tilde{\Omega}_n, k_0 > 0$ fixed, determine a Dirichlet scatterer uniquely.*

8.5 The Uniqueness of Inverse Neumann Obstacle Scattering

We now return to the inverse obstacle problem with Neumann boundary conditions. From what was said in Section 8.3.2, a discrete spectrum of finite multiplicity exists for the Neumann problem provided that the domain satisfies the interior core condition as, for example, when no cusps are present. Although the domains Ω_1 and Ω_2 may separately fit the category, there is no guarantee that the connected parts ω_i of their intersection will do so. An example is given in Figure 8.4 where the intersection contains a cusp. Therefore, Schiffer's arguments that served for the Dirichlet case cannot be directly extended to the Neumann problem or for that matter, to transmission scattering also since the latter involves Neumann boundary conditions (more on the transmission problem later).

It is, therefore, necessary to consider the problem from a different perspective. This was initiated by Isakov (1990) and was later simplified by Kirsch and Kress (1993). (See also Hettlich, 1994 and Potthast, 1998). The approach which can be called the method of Green's function uses point source incidence, the sources being located in $R^3\backslash(\overline{\Omega}_1 \cup \overline{\Omega}_2)$, i.e., outside the closure of the union of the scatterer domains. Assume without a loss of generality that the intersection of $\overline{\Omega}_1$ and $\overline{\Omega}_2$ is nonempty: $\overline{\Omega}_1 \cup \overline{\Omega}_2 \neq \emptyset$. A contradiction is shown to arise if the source is allowed to approach that part of the boundary of one of the scatterers which is not in the closure of the other scatterer's domain, the argument being symmetric in either obstacle. From this contradiction, the uniqueness is concluded.

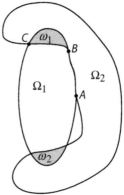

Figure 8.4 – Two scatterers Ω_1 and Ω_2 and their intersection $\overline{\Omega}_1 \cap \overline{\Omega}_2 = \overline{\omega}_1 \cup \overline{\omega}_2$. ω_1 and ω_2 are unconnected. $\omega_1 = ABC$ is cuspidal.

It is important to note that invoking point source incidence is entirely mathematical. The actual enunciation of the problem is in terms of the plane wave incidence. One then shows that the fundamental assumption that both Ω_1 and Ω_2 give rise to identical scattering amplitudes for incident plane waves of all orientations, but for a fixed wavenumber, carries over to the point sources. This in turn calls for certain completeness relations involving plane waves, the main result being the following.

Let V be the span of the plane waves, that is,

$$V = \text{Span}\{u^i(\cdot, k_0\hat{k}_j), \hat{k}_j \in \hat{\Omega}_3, k_0 > 0 \text{ fixed}, \ j = 1, 2, \cdots, n\}$$

and distributed uniformly on the unit sphere. An element $v_n \in V$ can be expressed as

$$v_n(x) = \Sigma_{j=1}^{n} v_{nj} e^{ik_0(\hat{k}_j \cdot x)}, x \in R^3, v_{nj} \in \mathcal{C}, j = 1, 2, \cdots n.$$

Then it is shown that

$$\lim_{n \to \infty} v_n \to u \text{ and } \lim_{n \to \infty} \nabla v_n \to \nabla u,$$

where u is a solution of the Helmholtz equation in a domain Ω with a boundary in C^2.

Let us show that the plane waves $e^{ik(\hat{k}_j \cdot x)}$ form a complete set in $L_2(\Gamma)$. First consider the regular wavefunctions

$$\chi_{mn}(x) = j_n(k|x|)Y_{nm}(\hat{\theta}_x)$$

and demonstrate that these constitute a complete set in $L_2(\Gamma)$. Toward that consider the single-layer potential

$$u(x) = \int_\Gamma \phi(y)G^0(x,y)\mathrm{d}\Gamma(y), \quad \phi \in L_2(\Gamma), \quad x \in R^3 \setminus \Gamma. \tag{8.19}$$

Let B_R be a ball of radius R containing Ω and let R be large enough so that the sphere B_c circumscribing Ω is also contained in B_R. Let $x \in B_R \setminus B_c$. Using the spherical harmonic expansion of Green's function (Chapter 7) in Eq. (8.19), we have

$$u(x) = \sum_{n=0}^{\infty} \sum_{m=-n}^{n} \alpha_{nm} h_n^{(1)}(k|x|)Y_{nm}(\hat{\theta}_x) \int_\Gamma \phi(y)j_n(k|y|)Y_{nm}^*(\hat{\theta}_y)\mathrm{d}\Gamma(y),$$

where α_{nm} are constants. Assume that the integral over Γ vanishes. Then $u = 0, \forall x \in B_R \setminus B_c$. From the analyticity of the single-layer potential, then $u \equiv 0$ everywhere in $B_R \setminus \overline{\Omega}$ implying that the density ϕ must vanish identically. The completeness of the set $\chi_{mn}(x)$ in $L_2(\Gamma)$ is, therefore, demonstrated.

Next consider that

$$\int_\Gamma \phi(y)e^{ik(\hat{k}_j \cdot y)}\mathrm{d}\Gamma(y) = 0. \quad j = 1, 2 \cdots . \tag{8.20}$$

Moreover, let $\hat{k}_j$ be dense in $\hat{\Omega}_3$. This means that the above integral vanishes for any $\hat{k}_j \in \hat{\Omega}_3$. Now invoking the spherical harmonic expansion of a plane wave, namely

$$e^{ik(\hat{k}_j \cdot y)} = 4\pi \sum_{n=0}^{\infty} \sum_{m=-n}^{n} i^n j_n(k|x|)Y_{nm}(\hat{\theta}_y)Y_{nm}^*(\hat{\theta}_k)$$

we have from Eq. (8.20)

$$0 = 4\pi \sum_{n=0}^{\infty} \sum_{m=-n}^{n} i^n Y_{nm}^*(\hat{\theta}_k) \int_\Gamma j_n(k|y|)Y_{nm}(\hat{\theta}_y)\mathrm{d}\Gamma(y).$$

Since $\chi_{mn}(x)$ is complete in $L_2(\Gamma)$, it follows that ϕ must vanish.

Exactly analogously, one can consider a double-layer potential instead of a single-layer potential, invoke analyticity of the double-layer potential and show that the normal derivatives $\partial_n \chi_{mn}$ and $\partial_n e^{ik(\hat{k}_j \cdot y)}$ are also complete in $L_2(\Gamma)$. We have, therefore, the following theorem.

Theorem 8.9 *Let* $\{\hat{k}_n\}_{n=1}^{\infty}$ *be dense in the unit sphere* $\hat{\Omega}_3$. *Then the set of plane waves* $\{e^{ik(\hat{k}_n \cdot y)}\}$ *and the wavefunctions* $\chi_{mn}(x) = j_n(k|x|)Y_{nm}(\hat{\theta}_x)$, $n = 0, 1, \cdots$, $m = -n, \cdots n$, *constitute complete sets in* $L_2(\Gamma)$, $\Gamma \in C^2$. *In addition, the normal derivatives* $\partial_n \chi_{mn}$ *and* $\partial_n e^{ik(\hat{k}_j \cdot y)}$ *are also complete in* $L_2(\Gamma)$.

Now consider a domain Ω with boundary $\Gamma \in C^2$ and let x be a point in Ω. Consider a single-layer potential u with density $\phi \in L_2(\Gamma)$. $u(x)$ is a solution of the Helmholtz's equation. Let ϕ_n be in the span of the plane waves. Then

$$\int_{\Gamma} (\phi - \phi_n)(y) G^0(x, y) \mathrm{d}\Gamma(y) = u(x) - u_n(x), \qquad (8.21)$$

where u and u_n correspond to ϕ and ϕ_n, respectively. Taking absolute values of both sides of Eq. (8.21) yields

$$|u(x) - u_n(x)| \le C \|\phi - \phi_n\|_{L_2(\Gamma)},$$

where

$$C = \sup_{x \in \Omega, y \in \Gamma} .$$

By construction, u_n is a combination of plane waves. Then letting $n \to \infty$ and from Theorem 8.9 of L_2-convergence of the boundary value of ϕ_n, it follows that $u_n \to u$ uniformly in Ω. A similar conclusion can be arrived at for ∇u by considering a double-layer potential. Therefore, we have

$$v_n \to u \text{ and } \nabla v_n \to \nabla u$$

as $n \to \infty$ on every compact sets of Ω. Note that the convergence holds for a regular solution of the Helmholtz equation such as $u \in C^2(\Omega_e)$. Therefore, the convergence should also apply to a point source. In other words

$$v_n \to G^0 \text{ and } \nabla v_n \to \nabla G^0. \qquad (8.22)$$

We now return to the Neumann problem:

$$\left(\Delta + k_0^2\right) \psi_j^{\mathrm{sc}} = 0, \ x \in \Omega_{ej} = R^3 \backslash \overline{\Omega}_j$$

$$\frac{\partial \psi_j^{\mathrm{sc}}}{\partial n} = -\frac{\partial G^0(\cdot, x_s)}{\partial n} \text{ on } \Gamma_j,$$

where $j = 1, 2$. We are told that the scattering amplitudes are identical for both scatterers for all orientations of the incident plane waves. Let us first show that this also holds for ψ_1^{sc} and ψ_2^{sc} above. Toward this, consider that the incident field is a linear combination of m number of plane waves. That is,

$$\psi^{\mathrm{inc}}(x) = v_m(x) = \sum_{j=1}^{m} \alpha_j e^{ik_0(\hat{k}_j \cdot x)}.$$

Denoting the corresponding scattering solutions by ψ_{mj}^{sc}, we have

$$(\Delta + k_0^2)\psi_{mj}^{\mathrm{sc}} = 0, \ x \in \Omega_{ej}$$

$$\frac{\partial \psi_{mj}^{\mathrm{sc}}}{\partial n} = -\frac{\partial v_{mj}}{\partial n}, j = 1, 2, \ \text{on } \Gamma_j.$$

By linearity of the Helmholtz equation, $\psi_{m1}^{\mathrm{sc}} = \psi_{m2}^{\mathrm{sc}} = \psi_m^{\mathrm{sc}}$ in Ω_e as this holds for each individual wave. Since $\nabla v_m \to \nabla G^0(\cdot, x_s), m \to \infty$, by Eq. (8.22), it follows at once from the well-posedness of the problem that $\psi_1^{\mathrm{sc}} = \psi_2^{\mathrm{sc}}$ in Ω_e for a point source if the scattered fields are identical for the plane-wave incidence. Hence is the following Lemma.

Lemma 8.3. *Let $\Omega_1, \Omega_2, \Omega_1 \cap \Omega_2 \neq 0$ be two sound-hard scatterers such that the scattered fields coincide for incident plane waves of all orientations, but of a fixed frequency. Then this also holds if the incident plane waves are replaced by a point source located in $\Omega_e = R^3 \backslash (\overline{\Omega}_1 \cup \overline{\Omega}_2)$.*

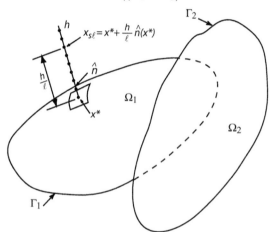

Figure 8.5 – Illustration of the proof of uniqueness for inverse Neumann obstacle scattering.

Refer to Figure 8.5 and consider a point $x^* \in \Gamma_1$, $x^* \in \partial \Omega_e \backslash \overline{\Omega}_2$. In other words, x^* is on that part of the boundary Γ_1 of scatterer Ω_1 which is not in the closure of the domain Ω_2. Let there be a sequence of point sources located at

$$x_{s\ell} = x^* + \frac{h}{\ell}\hat{n}(x^*), h > 0, l = 1, 2, \cdots,$$

where $x_{s\ell} \in \Omega_e$, for a constant h. Consider the scatterer Ω_2. By assumption, $\psi_1^{\mathrm{sc}} = \psi_2^{\mathrm{sc}} = \psi^s$ in Ω_e. Now the solution of the Helmholtz equation in the exterior domain is analytic in all its variables (Chapter 6) and has no poles in $R^n \backslash \overline{\Omega}_j$. Therefore, both the field as well as its gradient are bounded in every compact subset of Ω_e. In particular, considering scattering from Ω_2, we have

$$\left|\frac{\partial \psi^s}{\partial n_1}(x^*)\right| \leq C,$$

where C is a constant. The situation is, however, dramatically different if scatterer Ω_1 is considered. In that case

$$\left|\frac{\partial \psi^s}{\partial n_1}(x^*)\right| = \left|\frac{\partial G^0(x^*; x_{s\ell})}{\partial n_1}\right| \sim \frac{1}{4\pi} \frac{|1 - ik_0||x^* - x_s||}{|x^* - x_{s\ell}|^2} \to \infty, \quad \ell \to \infty.$$

This is a contradiction. Therefore, Ω_1 must be identical to Ω_2. Hence

Theorem 8.10 *Let Ω_1, Ω_2 be two Neumann obstacles the scattering amplitudes of which coincide for incident plane waves of all directions $\hat{k}_i \in \hat{\Omega}_3$, the wavenumber $k_0 > 0$ being fixed. Then $\Omega_1 \equiv \Omega_2$.*

Following Isakov (1990) and Potthast (1998), the essence of the method can also be described thus. Let the boundary and the source point $x^*, x_{s\ell}$, respectively, be as in the above. Assume that there exist two nonidentical obstacles Ω_1 and Ω_2 such that they have identical scattering amplitudes for plane waves of a fixed wavenumber $k > 0$, but with all directions of incidence. As already demonstrated, the scattering amplitudes are also identical if a point source incidence is considered. Let $u^s_{j\ell}$, $j = 1, 2$, be the scattered fields of obstacles 1 and 2 when the source is at $x_{s\ell}$. Then from the singularity of the point source and the boundary condition, it follows that

$$|\partial_{n_1} u^s_{1,\ell}(x^*)| = |(\partial_{n_1} G^0(|x_{s\ell}|)(x^*))| \sim O\left(\frac{h}{\ell}\right)^{-2} \geq C_1 \left(\frac{h}{\ell}\right)^{-1}. \quad (8.23)$$

for small h and large ℓ. The constant C_1 depends on the boundary.

On the other hand, replacing $u^s_{1,\ell}$ by $u^s_{2,\ell}$ (since the scattered fields are the same for both obstacles by assumption), we have the following. Let $F^{sc} : \partial_n u^{inc} \to u^{sc}$ be the scattering operator for scatterer 2. The operator F^{sc} can be a regularized version of the combined single and double-layer potentials as in Chapter 7. The important point is that F^{sc} is bounded in some suitable function space and let the bound be denoted by C^{sc}. Now $u^s_{2,\ell} = F^{sc}\partial_n G^0(|x_{s\ell} - y|), y \in \Gamma_2$. Then

$$|(\partial_{n_1} u^s_{2,\ell})(x^*)| \leq C^{sc}|(\partial_{n_1}\partial_{n_2} G^0(|x_{s\ell} - y|))(x^*) \leq C_2 d(x^*, \Gamma_2)^{-3}, \quad (8.24)$$

where $d(x^*, \Gamma_2)$ is the set-theoretical distance between x^* and the boundary Γ_2 of the second scatterer. Comparing Eqs. (8.23) and (8.24) yields

$$d(x^*, \Gamma_2) \leq \text{const.} \left(\frac{h}{\ell}\right)^{\frac{1}{3}}. \quad (8.25)$$

This shows that there is a direct relation between the location of the point source and the set theoretical distance between the two supposedly different obstacles Ω_1 and Ω_2 ($\Omega_1 \cap \Omega_2 \neq \emptyset$). Importantly, the result is a consequence of the assumption that the scattering amplitudes are identical for both scatterers. Moreover, Eq. (8.25) shows that as the distance between the point source and x^* vanishes, so does the distance $d(x^*, \Gamma_2)$ between the obstacles. For further details along this line of reasoning, see Potthast (1998).

8.6 Uniqueness in Inverse Transmission Obstacle Scattering

Next we discuss the uniqueness in inverse transmission obstacle scattering. As before, the problem is to demonstrate that if two different transmission obstacles give rise to identical scattering amplitudes for plane waves of a fixed frequency, but of all directions of incidence, then the obstacles must be identical themselves. There are, of course, two problems involved in this case, the uniqueness of the boundary and the material parameters. We confine our attention only to the problem of recovering the boundary assuming that the wavenumbers of the interior are known. Thus

$$\left(\Delta + k_+^2\right) u_{j+} \;=\; 0 \text{ in } \Omega_{je} = R^3 \setminus \overline{\Omega}_j \tag{8.26}$$

$$\left(\Delta + k_-^2\right) u_{j-} \;=\; 0 \text{ in } \Omega_j \tag{8.27}$$

$$u_{j+} - u_{j-} \;=\; -u_j^{\text{inc}} \text{ on } \Gamma_j \tag{8.28}$$

$$\partial_n(\mathcal{F}_+ u_{j+} - \mathcal{F}_- u_{j-}) \;=\; -\partial_n u_j^{\text{inc}} \text{ on } \Gamma_j, \tag{8.29}$$

$j = 1, 2$, $k_\pm > 0$. $\mathcal{F}_\pm$ were defined in Chapter 7.

Isakov (1990) initiated the proof of uniqueness by considering special singular solutions of the Helmholtz equation. His arguments were later simplified by Kirsch and Kress (1993) and by Hettlich (1994). An interesting proof based on the concept of an *approximate, ϵ-uniqueness* was given by Potthast (1998). The Kirsch–Kress approach for the transmission problem follows basically the same line of reasoning as that for the inverse Neumann problem discussed in Section 8.5, but with important modifications necessitated by the transmission boundary condition. The proof by Hettlich (1994) which is closer to that of Isakov is based on a variational approach and consists of two steps. In the first step, a variational relation is derived between the solutions corresponding to the two scattering domains Ω_j, $j = 1, 2$, exploiting the condition of the problem, namely, that the scattering amplitudes be identical for both scatterers for incident waves of a fixed frequency, but of all directions of incidence. The relation involves the products of the solutions as well as the product of their gradients. In the second step, the solutions are replaced by special singular solutions[1] by considering a sequence of point sources as shown in Figure 8.5. These special singular solutions are then used in the variational relation just mentioned. A comparison of the two reveals a contradiction if the source is allowed to approach the boundary in the same manner as in the previous Neumann case. The uniqueness result follows from this contradiction. Comparatively to the variational proof, the arguments of Kirsch and Kress (1993) are simpler and more transparent. We, therefore, sketch the outline of their proof below.

---

[1] The idea of using special solutions of a scattering problem in order to solve an associated inverse problem has a long history much of which is associated with multidimensional inversions in the Schrödinger equation. This will be discussed in the following Section 8.7 where the inverse inhomogeneity problem is discussed.

Consider a point source at $x_s \in \Omega_{12}$, Ω_{12} being the exterior of the closure of the union $\Omega_1 \cup \Omega_2$. Equations (8.26)–(8.29) become

$$
\begin{aligned}
(\Delta + k_+^2)u_{j+} &= 0 \text{ in } \Omega_{je} \\
(\Delta + k_-^2)u_{j-} &= 0 \text{ in } \Omega_j \\
u_{j+} - u_{j-} &= -G_+^0(\cdot, x_s) \text{ on } \Gamma_j & (8.30) \\
\partial_n(\mathcal{F}_+ u_{j+} - \mathcal{F}_- u_{j-}) &= -\mathcal{F}_+ \partial_n G_+^0(\cdot, x_s) \text{ on } \Gamma_j, & (8.31)
\end{aligned}
$$

As before, $j = 1, 2$, $k_\pm > 0$, and $G_\pm^0$ indicates free-space Green's functions with wavenumbers $k_\pm$. Again by the assumption of the problem and following the same line of arguments as in the Neumann case of the previous section, we obtain

$$u_{1+} = u_{2+} \text{ in } \Omega_{12}.$$

Next assume that the scatterers are not identical and consider the point x^* and a sequence of sources at $x_{j\ell}^s$ as shown in Figure 8.5. Let $u_{j+}^{(\ell)}$ $j = 1, 2$, be the scattered fields when the source at the ℓ-th point x_ℓ^s. Then

$$u_{1+}^{(\ell)} = u_{2+}^{(\ell)} \text{ in } \Omega_{12}. \tag{8.32}$$

Let us consider the scattered field at x^* due to the second scatterer. Now the forward transmission problem is well-posed. Then at x^* which is outside the closure of Ω_2, we have

$$||u_{2+}^{(\ell)}(x^*)||_\infty + ||\partial_n u_{2+}^{(\ell)}(x^*)||_\infty \le C_1, \ \forall \ell.$$

Now by definition

$$\partial_n f(x^*) = \lim_{t \to 0} \frac{1}{t}[f(x^* + t\hat{n}(x^*)) - f(x^*)],$$

where the limit is assumed to be uniform. Since by the assumption of the problem, $u_{1+} = u_{2+} \in \Omega_{12}$, it follows that

$$||u_{1+}^{(\ell)}(x^*)||_\infty + ||\partial_n u_{1+}^{(\ell)}(x^*)||_\infty \le C_1 \tag{8.33}$$

as $\ell \to 0$.

In the next step, let us add $G_-^0(x^*, x_\ell)$ to both sides of Eq. (8.30) and $\partial_n(\mathcal{F}_- G_-^0(x^*, x_\ell))$ to both sides of Eq. (8.31). Adding the two resulting boundary conditions using Eq. (8.25), and after a straightforward manipulation yields

$$
\begin{aligned}
||(\partial_n(\mathcal{F}_+ G_+^0 &- \mathcal{F}_- G_-^0))(x^*)||_\infty \le ||G_+^0 - G_-^0||_\infty \\
&+ ||u_{1+}^{(\ell)}(x^*)||_\infty + ||\partial_n u_{1+}^{(\ell)}(x^*)||_\infty + ||u_{1-}^{(\ell)} - G_-^0||_\infty \\
&+ \mathcal{F}_- ||\partial_n(u_{1-}^{(\ell)} - G_-^0)||_\infty. \tag{8.34}
\end{aligned}
$$

In the limit $\ell \to 0$, we have

$$||G_+^0 - G_-^0||_\infty \to 0$$

and by Eq. (8.33),

$$||u_{1+}^{(\ell)}(x^*)||_\infty + ||\partial_n u_{1+}^{(\ell)}(x^*)||_\infty$$

remains bounded. The boundedness of the norms $||u_{1-}^{(\ell)} - G_-^0||_\infty$ and $||\partial_n(u_{1-}^{(\ell)} - G_-^0)||_\infty$ is proved by Kirsch and Kress. Therefore, the R.H.S. of Eq. (8.34) remains bounded as the source approaches the boundary point x^*. This implies that

$$||(\partial_n(\mathcal{F}_+ G_+^0 - \mathcal{F}_- G_-^0))(x^*)||_\infty$$

is bounded in the same limit. This is, however, impossible since $\mathcal{F}_+ \neq \mathcal{F}_-$. Ω_1 cannot, therefore, be different from Ω_2. Note that the origin of the contradiction is in Eq. (8.32) which is a consequence of assuming that the domains Ω_1 and Ω_2 are different and yet give rise to identical scattering amplitudes. Hence is the following theorem.

Theorem 8.11 *Let Ω_1 and Ω_2 be two transmission obstacles. Assume that for a fixed frequency, the scattering amplitudes for both scatterers coincide for all directions of incidence. Then the scatterers must be identical.*

8.7 Uniqueness of Inverse Inhomogeneity Scattering

We conclude this chapter with a short discussion of the uniqueness of inverse inhomogeneity scattering. The corresponding direct problem was introduced in the final section of Chapter 6 where its two basic varieties; the plasma and the variable velocity wave equation were discussed. Here we consider the general equation of the form

$$[\Delta + k^2 + k^2 \epsilon(x)]\psi = 0 \tag{8.35}$$

in R^3, $v \in L_2(\Omega)$. The problem is, as before, to demonstrate that if there exist two different potentials ϵ_1 and ϵ_2 such that their corresponding scattering amplitudes $u_1^\infty(\hat{x}, \hat{k})$ and $u_2^\infty(\hat{x}, \hat{k})$ are identical over $\hat{\Omega}_3 \times \hat{\Omega}_3$, i.e., $u_1^\infty(\hat{x}, \hat{k}) = u_2^\infty(\hat{x}, \hat{k})$, $\forall \hat{x}, \hat{k} \in \hat{\Omega}_3$, $k > 0$ fixed, then ϵ_1 is identical to ϵ_2. We are, therefore, considering the problem at a fixed energy.

Consider a ball $B_R(0)$ of radius R with the origin as the center and which contains both scatterers in its interior. Let v_1 and v_2 be any two solutions of the Helmholtz equation corresponding to ϵ_1 and ϵ_2, respectively. Thus

$$[\Delta + k^2 + k^2 \epsilon_1(x)]v_1 = 0 \ x \in \Omega_1, \tag{8.36}$$

and

$$[\Delta + k^2 + k^2 \epsilon_2(x)]v_2 = 0 \ x \in \Omega_2. \tag{8.37}$$

Let v denote the difference $v_1 - v_2$. Then from Eqs. (8.36) and (8.37), we obtain

$$[\Delta + k^2 + k^2 \epsilon_1(x)]v = k^2(\epsilon_2 - \epsilon_1)v_2. \tag{8.38}$$

Multiplying both sides of Eq. (8.38) by v_1 and upon using Eq. (8.36), yields

$$v_1 \Delta v - v \Delta v_1 = k^2 (\epsilon_1 - \epsilon_2) v_1 v_2. \tag{8.39}$$

Let $B_{R_1}(0)$, $R_1 > R$, be another ball such that $B_R(0) \subset B_{R_1}(0)$. Next apply Green's second identity to both sides of Eq. (8.34) in the annular region Ω_R between the balls $B_{R_1}(0)$ and $B_R(0)$. Since the scattered fields are identical in the unbounded connected complement of the union $\overline{\Omega}_1 \cup \overline{\Omega}_2$, it follows that

$$\int_{\Omega_R} (\epsilon_1 - \epsilon_2) v_1 v_2 = 0. \tag{8.40}$$

Equation (8.40) shows that the desired uniqueness result requires that the product $v_1 v_2$ of the solutions be complete in L_2. It should be pointed out that by the density result of Theorem 8.2, the product $v_1 v_2$ can be replaced by the product $u_1 u_2$ of the total fields in $B_R(0)$.

The proof of completeness is furnished in terms of a certain special solution of the Helmholtz equation (8.35). Let us briefly describe the background of the problem (see also Nachman, 1989). Consider the equation

$$\nabla(\gamma \nabla v) = 0. \tag{8.41}$$

If γ is identically equal to unity, then Eq. (8.41) is just Laplace's equation. Let $\zeta = \xi + i\eta \in \mathcal{C}^3$, ξ, $\eta \in R^3$, be a complex frequency such that $\zeta \cdot \zeta = \zeta^2 = 0$ and $\zeta_1 + \zeta_2 = -iz$, $z \in R^3$. Then the functions $e^{\zeta \cdot x}$ are harmonic and solve Laplace's equation. Moreover, let $\gamma \in L_2(\Omega)$ and assume that

$$0 = \int_\Omega \gamma(x) e^{\zeta_1 \cdot x} e^{\zeta_2 \cdot x} dx = \int_{R^3} \gamma(x) e^{-iz \cdot x}, \tag{8.42}$$

where γ has been extended to all R^3 by zero outside Ω. This holds for every z and hence the Fourier transform vanishes identically in R^3. It follows at once that γ is identically zero. Thus the product functions such as $e^{\zeta_1 \cdot x} e^{\zeta_2 \cdot x}$ are complete in $L_2(\Omega)$.

If γ is close to a constant, then Calderón (1980) suggested that an approximation of the Fourier transform of γ, $\hat{\gamma}$, be obtained with the family of harmonic functions $e^{\zeta \cdot x}$. However, if the assumption that γ is close to a constant is discarded, then demonstrating that the product functions are complete becomes difficult. This is primarily due to two factors. First, each function in the product is now a solution of a different equation, i.e., with a different potential. Secondly, the potentials are variables instead of constants. However, completeness can still be obtained provided that one works with the solutions $u(x, \zeta)$ which behave like $e^{\zeta \cdot x}$ for large values of $|\zeta|$ (Sylvester and Uhlmann, 1987; Nachman, 1989). The uniqueness problem then reduces to demonstrating that such solutions to Eq. (8.35) exist. We now turn to this.

The proof of uniqueness is obtained essentially by modifying the arguments which were developed for solving the multidimensional inverse Schrödinger equation (Fadeev, 1956, 1965; Newton, 1989; Chadan and Sabatier, 1989. See also

Ablowitz and Clarkson, 1990; Henkin and Novikov, 1987). As a matter of fact, the Fourier-transformed plasma wave equation is the Schrödinger equation, whereas, the variable velocity equation, when Fourier transformed, is also Schrödinger's equation, but in which the potential $k^2 q(x)$ depends upon the wavenumber k. The theory of the multidimensional inverse Schrödinger equation applies also to these equations. In the case of the plasma wave equation, the corresponding Schrödinger's equation has no bound states, whereas the analysis of the variable velocity equation fails for the large $|k|$ asymptotics (Newton, 1989). Excellent discussions regarding various interconnections between the Schrödinger, plasma wave and the variable velocity equations can be found in Cheney (1984), Cheney and Rose (1988), Cheney, et. al (1987), Rose, et al. (1985) and Newton (1985). The analysis of inverse scattering for the inhomogeneity problem is carried out mostly within the framework of the Lippmann-Schwinger equation, although exceptions exist (see Tabbara, et al. 1988; Wang and Chew, 1989 and Gutman and Klibanov, 1994). There exist several proofs of uniqueness for the inverse reconstruction of the refractive index (or rather the dielectric constant, see comments in Chapter 6) from the scattering data (Hähner, 1996; Kirsch, 1996, Colton and Kress, 1992). However, in the discussions below we follow closely somewhat simpler arguments put forth by Ramm (1992).

Let us recapitulate our arguments. We would like to show that the potential $\epsilon(x)$ in Eq. (8.35) can be reconstructed uniquely from the scattering data.. This is to be done by showing that there exist special solutions of this equation such that the product solutions $v_1 v_2$ of Eq. (8.39) or Eq. (8.40) are complete in $L_2(\Omega)$. This is in turn to be accomplished by showing that solutions exist such that they are of the form $e^{i\zeta \cdot x}$, $\zeta \in \mathcal{C}^3$ in the asymptotic limit $|\zeta| \to \infty$. Moreover, $\{\zeta + \eta\}$, ζ, $\eta \in \mathcal{C}^3$ contains R^3. We first demonstrate that the Helmholtz equation (8.35) admits a special solution which behaves asymptotically as $e^{ik\zeta \cdot x}$. Consider the equation

$$[\Delta + k^2 + k^2 \epsilon(x)]u = [\Delta + k^2 + q(x)]u = 0 \tag{8.43}$$

for a fixed $k > 0$ and $q(x) \in L_2(\Omega)$, q being zero outside $\Omega \subset R^3$. Thus q has a compact support in Ω. Let $\zeta = a + ib \in \mathcal{C}^3$, $a, b \in R^3$ be such that $\zeta \cdot \zeta = k^2$. This means that $a^2 - b^2 = k^2$, and $a \cdot b = 0$. Let the solution to Eq. (8.42) be written as

$$u(x) = e^{ik\zeta \cdot x}(1 + r(x, \zeta)). \tag{8.44}$$

Substituting Eq. (8.43) in Eq. (8.42) results in

$$[\Delta + 2ik\zeta \cdot \nabla]r = -q(x)(1 + r(x, \zeta)). \tag{8.45}$$

Consider next the equation

$$[\Delta + 2ik\zeta \cdot \nabla]w = f, \quad f \in L_2(\Omega). \tag{8.46}$$

Let $\hat{w}(\lambda)$, $\lambda = (\lambda_1, \lambda_2, \lambda_3)^T \in R^3$, be the Fourier transform of u. Then

$$\hat{w}(\lambda) = (2\pi)^{-3} \int_{R^3} e^{-ik\lambda \cdot x} w(x) dx.$$

Equation (8.45) is solved by

$$w(x) = - \int_{R^3} \tilde{f}(\lambda)(\lambda^2 + i\zeta \cdot \lambda)^{-1} e^{ik\lambda \cdot x} d\lambda. \tag{8.47}$$

Since $a \cdot b = 0$, we can write $a = \hat{e}_3 t$ and $b = \hat{e}_1 t$, where $\hat{e}_j$, $j = 1, 2, 3$, are the orthonormal basis vectors of R^3. By direct calculation, we have

$$\lambda^2 + i\zeta \cdot \lambda = \lambda_1^2 + i\lambda_1 t + \Lambda,$$

where $\Lambda = \lambda_3^2 + (\lambda_2 + t\backslash 2)^2 - t^2 \backslash 4$. It then follows that $(\lambda^2 + i\zeta \cdot \lambda)$ vanishes if $\lambda_1 = 0$ and $\Lambda = 0$. The last condition gives a circle in the (λ^2, λ^3)-plane with center at $(0, -t\backslash 2, 0)$, the radius being $t\backslash 2$.

Following Ramm, we next consider a torus T_δ of width and depth 2δ centered around the circle $\Lambda = 0$ and write

$$\int_{R^3} = \int_{T_\delta} + \int_{R^3 \backslash T_\delta}.$$

Then from Eq.(8.46)

$$|w(x)| \leq \left| \int_{T_\delta} \tilde{f}(\lambda)(\lambda^2 + i\zeta \cdot \lambda)^{-1} e^{ik\lambda \cdot x} d\lambda \right|$$
$$+ \left| \int_{R^3 \backslash T_\delta} \tilde{f}(\lambda)(\lambda^2 + i\zeta \cdot \lambda)^{-1} e^{ik\lambda \cdot x} d\lambda \right|. \tag{8.48}$$

It is not difficult to show that

$$\|w\|_{L_2(\Omega)} \leq \text{const.} \|f\|_{L_2(\Omega)} \{ (\delta t)^{-1} + \delta \}.$$

For a fixed $t > 0$, it is easily seen that

$$\min_{\delta}[(\delta t)^{-1}, \delta] = 2t^{-\frac{1}{2}}.$$

Therefore, Eq. (8.47) yields

$$\|w\|_{L_2(\Omega)} \leq \text{const.} \|f\|_{L_2(\Omega)} t^{-\frac{1}{2}}.$$

From the definition of ζ in the above paragraphs, it follows at once that

$$\|w\|_{L_2(\Omega)} \leq \text{const.} \|f\|_{L_2(\Omega)} |\zeta|^{-\frac{1}{2}}.$$

This leads to the following lemma.

Lemma 8.3. *Let $\zeta = a + ib \in C^3$, $a, b \in R^3$, $|\zeta| \geq 1, \operatorname{Im}\zeta \neq 0$, $\zeta \cdot \zeta = k^2$, be a complex vector. Let $f \in L_2(\Omega)$, $\Omega \in R^3$ a bounded domain. Consider the equation*

$$[\Delta + i\zeta \cdot \nabla]u(x, \zeta) = f$$

in R^3. Then the following estimate holds, namely

$$\|u\|_{L_2(\Omega)} \leq \text{const.}\|f\|_{L_2(\Omega)}|\zeta|^{-\frac{1}{2}}.$$

Finally, we consider the equation

$$Lu = -qu + f, \tag{8.49}$$

where $L = \Delta + i\zeta \cdot \nabla$. Now since q has a compact support and in view of Lemma 8.3, it follows that $qu \in L_2(\Omega)$. Then taking the L_2-norm on both sides of Eq. (8.48) yields

$$\|u\|_{L_2(\Omega)} \leq \text{const.}|\zeta|^{-\frac{1}{2}}[\|f\|_{L_2(\Omega)} + \|qu\|_{L_2(\Omega)}.$$

This shows that

$$\|u\|_{L_2(\Omega)} \sim O(|\zeta|^{-\frac{1}{2}}).$$

Returning to Eq. (8.44), it follows that

$$\|r(x, \zeta)\|_{L_2(\Omega)} \sim O(|\zeta|^{-\frac{1}{2}})$$

for any compact subset of $R^3 \setminus \Omega$. This then demonstrates that the equation

$$[\Delta + k^2 + q(x)]u(x) = 0$$

has special solutions of the form

$$u(x) = e^{ik\zeta \cdot x}(1 + r(x, \zeta))$$

which behaves exponentially as $|\zeta| \to \infty$.

The proof is completed by demonstrating that there exist elements $\zeta, \eta \in C^3$ such that $\zeta + \eta = p, p \in R^3$. Without any loss of generality, p is assumed to be along the x_3-axis, that is, $p = |p|\hat{e}^3$. Write $\zeta = \zeta_1 + i\zeta_2$, $\eta = \eta_1 + i\eta_2$, $\zeta_1, \zeta_2, \eta_1, \eta_2 \in R^3$. Since $\zeta + \eta = p$, it follows that $\zeta_2 = -\eta_2$, $\zeta_{1i} = -\eta_{1i}$, $i = 1, 2$, and $\zeta_{13} = |p| - \eta_{13}$. Moreover, $\zeta \cdot \zeta = \eta \cdot \eta = k^2$ leading to the identities $\zeta_1^2 - \zeta_2^2 = \eta_1^2 - \eta_2^2 = k^2$, $\zeta_1 \cdot \zeta_2 = \eta_1 \cdot \eta_2 = 0$. Altogether, these identities give rise to 10 relations. It is to be shown that there exist $\zeta, \eta \in C^3$ such that all these 10 relations are satisfied. For the sake of convenience, let us arrange these equations systematically.

$$\zeta_1^2 - \zeta_2^2 = \eta_1^2 - \eta_2^2 = k^2. \tag{8.50}$$

$$\zeta_1 \cdot \zeta_2 = \eta_1 \cdot \eta_2 = 0. \tag{8.51}$$

$$\zeta_2 + \eta_2 = 0. \tag{8.52}$$

$$\zeta_{1i} + \eta_{1i} = 0, \ i = 1, 2. \tag{8.53}$$

$$\zeta_{13} + \eta_{13} = |p|. \tag{8.54}$$

Eliminating η between the identities given in Eq. (8.50) and using Eqs. (8.51) and (8.52) results in $\zeta_{23} = 0$. Let $\zeta_2 = \hat{e}_1|a|$. Moreover, choose

$$\zeta_1 = \hat{e}_2 b + \hat{e}_3 \frac{|p|}{2}.$$

Therefore,

$$\zeta = \hat{e}_2 b + \hat{e}_3 \frac{|p|}{2} + i\hat{e}_1|a|. \tag{8.55}$$

Replacing Eq. (8.54) in Eq. (8.49), we obtain

$$b^2 - a^2 = k^2 - \frac{|p|^2}{4}. \tag{8.56}$$

Let $a = \gamma b$ and choose

$$\gamma = \pm \left[1 + \frac{k^2 - \frac{|p|^2}{4}}{b^2}\right]^{\frac{1}{2}}. \tag{8.57}$$

Take $\eta = \mathcal{R}_\pi \zeta$, where $\mathcal{R}_\pi$ is a rotation matrix in the plane through π. Then from Eqs. (8.54)–(8.56), it is seen that all ten conditions given in Eqs. (8.49)–(8.53) are satisfied. Therefore, the set

$$\{\zeta + \eta\}, \zeta, \eta \in \mathcal{M},$$

$$\mathcal{M} = \{\zeta, \eta \in C^3, \zeta \cdot \zeta = \eta \cdot \eta = k^2\},$$

contains R^3. The uniqueness is thus proved.

The above uniqueness analysis was carried out at a fixed energy, that is, for a fixed $k > 0$. A somewhat simpler proof appears in Colton and Kress (1992) in which the frequency ranges over an interval. Appendix A.8.1 and A.8.2 contains the background material necessary for the discussions below. Consider the homogeneous harmonic polynomial

$$H_{\ell m} = r^\ell Y_{\ell m}(\hat{\theta})$$

defined in Eq. (1) of the Appendix. $H_{\ell m}$ is harmonic in every finite domain. Let us next consider the expansion

$$u(r, \hat{\theta}) = \sum_{\ell=0}^{\infty} \sum_{m=-\ell}^{\ell} u_{\ell m} H_{\ell m}. \tag{8.58}$$

From Eq. (3) of the Appendix, we know that

$$\left| \sum_{m=-\ell}^{\ell} u_{\ell m} Y_{\ell m} \right| \leq \left[\frac{2\ell+1}{4\pi} \right]^{\frac{1}{2}} \sum_{m=-\ell}^{\ell} |u_{\ell m}|.$$

Let Ω be a bounded domain in R^3 contained within a sphere S_R of radius R. Moreover, assume that $u \in H^s(S_R)$, $s \in R^1$. Now we also know (*the maximum-minimum principle*) (see any text on complex variables, Churchill, 1960, for example) that a harmonic function in a bounded domain cannot take its maximum or minimum value in that domain unless it is a constant. The series in Eq. (8.57), therefore, converges for any real s.

Next let us consider harmonic functions of the form $h = e^{i\zeta \cdot x}$ of complex frequency $\zeta = \xi + i\eta$, $\xi, \eta \in R^3$, $\zeta \cdot \zeta = 0$. Let

$$h_1 = e^{i\zeta \cdot x} \quad \text{and} \quad h_2 = e^{i\zeta^* \cdot x}$$

be two such functions. Now we can represent h in the series

$$h = \sum_{\ell=0}^{\infty} \sum_{m=-\ell}^{\ell} h_{\ell m} H_{\ell m}. \tag{8.59}$$

Moreover, from earlier discussions, it is known that the product of functions h are complete in $L_2(\Omega)$. Consequently, expressing h_1 and h_2 in the series form of Eq. (8.58), it follows that the product functions $H_{\ell m} H_{\ell' m'}$ are also complete in $L_2(\Omega)$, $\forall \ell = 0, 1, \cdots$, $-\ell \leq m \leq \ell$. From now on, we will consider Y_ℓ instead of $Y_{\ell m}$ and the polynomials

$$H_\ell = r^\ell Y_\ell.$$

We now return to the problem of inhomogeneity scattering. As before, let $m_i = 1 - \epsilon_i, i = 1, 2$, be two inhomogeneities which give rise to identical scattering amplitudes for all incident waves $e^{ik\hat{k} \cdot x}, \hat{k} \in \hat{\Omega}_3, k \in I$. Therefore,

$$u_\infty^1(\hat{x}, \hat{k}) = u_\infty^2(\hat{x}, \hat{k}), \tag{8.60}$$

where u_∞^1 and u_∞^2 are the scattering amplitudes corresponding to inhomogeneity 1 and 2, respectively. Moreover, let B be a ball in R^3 which encloses the supports of both m_1 and m_2. Using the Lippmann–Schwinger equation (see Chapter 6), we have, from Eq. (8.59)

$$0 = \int_B e^{-ik\hat{x} \cdot y}[m_1(y)u_1(y; \hat{k}, k) - m_2(y)u_2(y; \hat{k}, k)]dy, \tag{8.61}$$

where u_j is the total field. We multiply Eq. (8.60) by $Y_\ell(\hat{x})$, integrate over the unit sphere and use the Funk–Hecke formula (Eq. (10), Appendix A.8.1) to obtain

$$0 = \int_B j_\ell(k|y|)Y_\ell(\hat{y})[m_1(y)u_1(y; \hat{k}, k) - m_2(y)u_2(y; \hat{k}, k)]dy, \tag{8.62}$$

Multiplying Eq. (8.61) in turn by $Y_\ell(\hat{k})$ and integrating over the unit sphere results in

$$0 = \int_B j_\ell(k|y|)Y_\ell(\hat{y})[m_1(y)v_{\ell'1}(y;k) - m_2(y)v_{\ell'2}(y;k)]dy, \qquad (8.63)$$

where

$$v_{\ell'j}(y;k) = \int_{\hat{\Omega}_3} u_j(y;\hat{k},k)Y_\ell(\hat{k})d\hat{k}.$$

Now by the Lippmann–Schwinger equation

$$u_j(y;\hat{k},k) = e^{ik\hat{k}\cdot y} - k^2 T_j u_j,$$

where

$$T_j(\cdot) = \int_B G^0(\cdot,y',k)m_j(y')u_j(y')dy'.$$

Therefore,

$$v_{\ell'j}(y;k) = u_{\ell'j}(y;k) - k^2(T_j v_{\ell'j})(y;k),$$

where

$$u_{\ell'}(y;k) = \int_{\hat{\Omega}_3} e^{ik\hat{k}\cdot y}Y_{\ell'}(\hat{k})d\hat{k}. \qquad (8.64)$$

Thus

$$v_{\ell'j} = [I + k^2 T_j]^{-1}. \qquad (8.65)$$

Replacing Eq. (8.64) in Eq. (8.62) gives

$$0 = \int_B j_\ell(k|y|)Y_\ell(\hat{y})[m_1[I + k^2T_1]^{-1} - [I + k^2T_2]^{-1}u_{\ell'}(y)dy. \qquad (8.66)$$

Now the operator $[I + k^2T_j]^{-1}$ is invertible $\forall k > 0$ as shown in Appendix A.8.2. Taking limit $k \to 0$ in the above Eq. (8.65), we obtain

$$0 \sim \int_B j_{\ell\downarrow0}(k|y|)Y_\ell(\hat{y})[m_1(y) - m_2(y)]u_{\ell\downarrow0}dy, \qquad (8.67)$$

where $f_{\ell\downarrow0}$ indicates the value of f in the limit $k \to 0$. Using the Funk–Hecke formula in Eq. (8.66) gives

$$u_{\ell'j}(z) = 4\pi i^\ell j_{\ell'}(k|z|)Y_{\ell'}(\hat{z}).$$

Moreover,

$$j_{\ell\downarrow0}(z) \sim \frac{z^\ell}{(2\ell+1)!!}.$$

Therefore

$$u_{\ell\downarrow0}(z) \sim 4\pi i^\ell \frac{(k|z|)^\ell}{(2\ell+1)!!}Y_\ell(\hat{z}).$$

Combining all this in Eq. (8.66), we obtain

$$0 \sim \int_B [m_1(y) - m_2(y)]|y|^\ell Y_\ell(\hat{y})|y|^{\ell'} Y_{\ell'}(\hat{y})dy$$

$$= \int_B [m_1(y) - m_2(y)]H_\ell(|y|, \hat{y})H_{\ell'}(|y|, \hat{y}).$$

Since the products $H_\ell H_{\ell'}$ are complete in $L_2(\Omega)$, we finally obtain $m_1 = m_2$, and the scatterers are identical.

The following theorem can thus be stated. The dielectric constant ϵ can be uniquely determined from the scattering amplitudes $u_\infty(\hat{x}, \hat{k}), \hat{x}, \hat{k} \in \hat{\Omega}_3$, for either at a fixed energy $k > 0$, or for energies in an interval.

8.8 Appendix A.8.1

Spherical Harmonics

In this Appendix we briefly review the spherical harmonics. Exhaustive discussions are given in Muller (1997) and Groemer (1996) (see also Nedelec, 2001, for a highly useful detailed review). We begin by recalling that by a polynomial $P_\ell(x)$ of degree ℓ in the single variable x is meant a function of the form

$$P_\ell(x) = a_0 + a_1 x + a_2 x^2 + \cdots + a_\ell x^\ell, \quad a_0 \neq 0.$$

The dimension of the space $\mathcal{P}_\ell$ of such polynomials is simply $(\ell + 1)/1!$. If, on the other hand, $x = (x_1, x_2, x_3)$, that is, the polynomial P_ℓ is a function of three variables, then

$$\dim\{\mathcal{P}_\ell\} = \frac{(\ell + 3)(\ell + 2)(\ell + 1)}{3!}.$$

A polynomial $P_\ell(x)$ is called *homogeneous* if

$$P_\ell(tx) = t^\ell P_\ell(x),$$

where t is a parameter. In addition, if P_ℓ satisfies Laplace's equation $\Delta P_\ell = 0$, then it is called *harmonic*. Let $\mathcal{H}_\ell$ denote the space of all homogeneous harmonic polynomials of degree ℓ.

Consider a spherical polar coordinate system in which $(r, \hat{\theta})$, $\hat{\theta} = (\theta, \phi) \in \hat{\Omega}_3$, replaces the triple (x_1, x_2, x_3). An element $H_\ell(r, \hat{\theta})$ of $\mathcal{H}_\ell$ is of the form

$$H_\ell(r, \hat{\theta}) = r^\ell Y_\ell(\hat{\theta}). \tag{1}$$

H_ℓ is also known (MacRobert, 1967) as a *solid spherical harmonic of order* ℓ, whereas the angular part $Y_\ell(\hat{\theta})$ is called the *surface spherical harmonic of order* ℓ. The nomenclature is reasonable since $Y_\ell(\hat{\theta}) = H_\ell(r = 1, \hat{\theta})$. Therefore, a solid spherical harmonic of order ℓ, when restricted to the unit sphere, is called a *spherical harmonic of order* ℓ.

Since $\Delta H_\ell = 0$, it is seen that the polynomial $Y_\ell(\hat{\theta})$ must solve the equation

$$(\Delta_S + \ell(\ell+1))Y_\ell(\hat{\theta}) = 0,$$

where

$$\Delta_S = \frac{1}{\sin\theta}\partial_\theta(\sin\theta\partial_\theta) + \frac{1}{\sin^2\theta}\partial_\phi^2$$

is the so-called *Laplace–Beltrami operator* on the unit sphere. The subscript S on Δ indicates sphere. The *surface gradient* ∇_S is given by

$$\nabla_S = \hat{e}_\theta\partial_\theta + \hat{e}_\phi\frac{1}{\sin\theta}\partial_\phi.$$

Y_ℓ also satisfies the orthogonality relation, namely

$$\int_{\hat{\Omega}_3} Y_\ell(\hat{y})Y_m^*(\hat{y})d\Omega(\hat{y}) = \delta_{\ell m}.$$

$d\Omega(\hat{y}) = \sin\theta d\theta d\phi$ is the surface measure on the unit sphere. It is also not difficult to show that Δ_S is self-adjoint in $L_2(\hat{\Omega}_3)$. That is,

$$< \Delta_S u, v >_{L_2(\hat{\Omega}_3)} = < u, \Delta_S v >_{L_2(\hat{\Omega}_3)} . \tag{2}$$

Moreover,

$$\int_{\hat{\Omega}_3} (\Delta_S u \cdot \Delta_S v)d\Omega(\hat{y}) = -\int_{\hat{\Omega}_3} v^*\Delta_S u d\Omega(\hat{y}).$$

Being self-adjoint, the eigenfunctions of Δ_S (that is, the spherical harmonics) constitute an orthogonal basis in $L_2(\hat{\Omega}_3)$.

Now $\mathcal{H}_\ell$ is a subspace spanned by the eigenfunctions of Δ_S with $-\ell(\ell+1)$ as the eigenvalue. However, it is necessary to specify a basis for each finite eigenspace $\mathcal{H}_\ell$ which is of multiplicity $2\ell + 1$. Such a basis is provided by the well-known classical spherical harmonics

$$Y_{\ell m}(\hat{\theta}) = \left[\frac{2\ell+1}{4\pi}\frac{(\ell-|m|)!}{(\ell+|m|)!}\right]^{\frac{1}{2}} P_\ell^m(\cos\theta)e^{im\phi},$$

$\ell = 0, 1, 2, \cdots$, $-\ell \le m \le \ell$, and P_ℓ^m are the *associated Legendre polynomials*. The functions $Y_{\ell m}$, $\forall \ell \ge 0$, $-\ell \le m \le \ell$, constitute an orthonormal basis for $L_2(\hat{\Omega}_3)$. Some properties of $Y_{\ell m}$ are:

$$\int_{\hat{\Omega}_3} Y_{\ell m}(\hat{y})Y_{\ell'm'}^*(\hat{y})d\Omega(\hat{y}) = \delta_{\ell\ell'}\delta_{mm'}.$$

$$\sum_{m=-\ell}^{\ell} Y_{\ell m}(\hat{x})Y_{\ell'm'}^*(\hat{y}) = \frac{2\ell+1}{4\pi}P_\ell(\cos\theta), \angle(\hat{x}, \hat{y})$$

$$Y_{\ell m}(\mathcal{R}\hat{y}) = \sum_{k=-\ell}^{\ell} \alpha_{mk}Y_{\ell k}(\hat{x}),$$

where $\mathcal{R}$ is the rotation matrix in R^3. The above relation follows because the rotation matrix $\mathcal{R}$ transform a homogeneous polynomial of degree ℓ to another homogeneous polynomial of the same degree. Moreover, remembering that $P_\ell(0) = 1$, one obtains the following relation

$$\sum_{m=-\ell}^{\ell} |Y_{\ell m}(\hat{x})|^2 = \frac{2\ell + 1}{4\pi}$$

and hence

$$|Y_{\ell m}(\hat{x})| \leq \left[\frac{2\ell + 1}{4\pi}\right]^{\frac{1}{2}}. \tag{3}$$

This implies that $|P_\ell(\cos\theta)| \leq 1$.

There exist operators for which the spherical harmonics are the eigenfunctions. There are also the so-called *ladder operators* which increase or decrease the index m. Some of these operators are

$$\begin{aligned}
L_\phi &= -i\partial_\phi, \\
L_\pm &= e^{\pm i\phi}(\pm\partial_\theta - \cot\theta L_\phi), \\
L_+ L_- &= -(\partial_\theta^2 + \cot^2\theta\partial_\phi^2 - L_\phi + \cot\theta\partial_\theta).
\end{aligned}$$

The corresponding operations on $Y_{\ell m}$ are given by

$$\begin{aligned}
L_\phi Y_{\ell m} &= m Y_{\ell m}, \\
L_\pm Y_{\ell m} &= \sqrt{(\ell \mp m)(\ell \pm m + 1)} Y_{\ell m\pm 1}, \\
L_\pm L_\mp Y_{\ell m} &= (\ell \pm m)(\ell \mp m + 1) Y_{\ell m}.
\end{aligned}$$

Next we present a brief derivation of the important *Jacobi–Anger* expansion of a plane wave

$$e^{ik\hat{k}\cdot x} = e^{ik|x|\cos\theta} = e^{ik|x|z},$$

where $\theta = \angle(\hat{k}, \hat{x})$ and $z = \cos\theta$. Now the Legendre polynomial $P_\ell(z)$ is described by the equation

$$(1 - z^2)P_\ell''(z) - 2zP_\ell'(z) = -\ell(\ell + 1)P_\ell(z),$$

where $'$ denotes differentiation with respect to z. The above equation can be rewritten as

$$((1 - z^2)P_\ell'(z))' = -\ell(\ell + 1)P_\ell(z). \tag{4}$$

Multiplying Eq. (4) on both sides by $e^{ik|x|z}$, integrating over the interval $[-1, +1]$, and using the well-known relation

$$(1 - z^2)P_\ell'(z) = \frac{\ell(\ell + 1)}{2\ell + 1}(P_{\ell-1}(z) - P_{\ell+1}(z)),$$

we obtain

$$g_{\ell+1}(k|x|) + g_{\ell-1}(k|x|) = \frac{2\ell+1}{k|x|} g_\ell(k|x|). \tag{5}$$

In Eq. (4),

$$g_\ell(k|x|) = (-i)^\ell \int_{-1}^{+1} e^{ik|x|z} P_\ell(z) dz.$$

We note that the spherical Bessel (j_ℓ), Neumann (η_ℓ),and Hankel (h_ℓ) functions all satisfy the recursion relation given by Eq. (5). However, by setting $\ell = 0, 1$, it is seen that $g_\ell = 2j_\ell$. Therefore,

$$(-i)^\ell \int_{-1}^{+1} e^{ik|x|z} P_\ell(z) dz = 2j_\ell(k|x|). \tag{6}$$

Now the basis functions given by

$$\left[\frac{2\ell+1}{2\pi}\right] P_\ell$$

are orthogonal. Exploiting this in Eq. (6) immediately leads to the Jacobi–Anger formula, namely

$$e^{ik|x|z} = \sum_{\ell=0}^{\infty} i^\ell (2\ell+1) j_\ell(k|x|) P_\ell(z). \tag{7}$$

Next we discuss the important *Funk–Hecke* formula. Toward that we multiply both sides of the eigenvalue equation for the spherical harmonics, namely

$$\Delta_S Y_{\ell m}(\hat{z}) = -\ell(\ell+1) Y_{\ell m}(\hat{z})$$

by $e^{-ik|x|\hat{x}\cdot\hat{z}}$ and integrate over $\hat{\Omega}_3$. Noting from Eq. (2) that Δ_S is self-adjoint in $L_2(\hat{\Omega}_3)$, we have

$$\int_{\hat{\Omega}_3} Y_{\ell m}(\hat{z}) \Delta_S e^{-ik|x|\hat{x}\cdot\hat{z}} d\Omega(\hat{z}) = -\ell(\ell+1) \int_{\hat{\Omega}_3} e^{-ik|x|\hat{x}\cdot\hat{z}} Y_{\ell m}(\hat{z}) d\Omega(\hat{z}). \tag{8}$$

Next we replace $e^{-ik|x|\hat{x}\cdot\hat{z}}$ in Eq. (8) by its Jacobi–Anger expansion given by Eq. (7). Let $\hat{x}$ be along the x_3-axis and note that

$$\Delta_S Y_{\ell 0}(\theta) = -\ell(\ell+1) Y_{\ell 0}(\theta). \tag{9}$$

Replacing $P_\ell(z)$ by

$$P_\ell(z) = \left[\frac{4\pi}{2(\ell+1)}\right]^{\frac{1}{2}} Y_{\ell 0}(\theta),$$

and after a straightforward algebra, the Funk–Hecke formula is obtained which is

$$\int_{\hat{\Omega}_3} e^{-ik|x|\hat{x}\cdot\hat{z}} Y_{\ell m}(\hat{z}) d\Omega(\hat{z}) = \frac{4\pi}{i^\ell} j_\ell(k|x|) Y_{\ell m}(0,0)$$

$$= \frac{4\pi}{i^\ell} j_\ell(k|x|) Y_{\ell m}(\hat{x}). \tag{10}$$

8.9   Appendix A.8.2

Solution of Direct Inhomogeneity Scattering

In this Appendix we present a short discussion of the nature of the solution of a direct inhomogeneity scattering problem. The governing equations are

$$(\Delta + k^2\epsilon(x))u(x) = 0, \; x \in R^3 \setminus \Omega \tag{11}$$

$$u = u^{\text{inc}} + u^{\text{sc}}$$

$$\frac{\partial u^{\text{sc}}}{\partial n} - iku^{\text{sc}} = O\left(\frac{1}{|x|^2}\right). \tag{12}$$

It is assumed that the dielectric constant $\epsilon \in C^2(R^3)$ is compactly supported in a ball $B_R \in R^3$ of radius R outside which ϵ is unity. Therefore, if we define $m(x) = 1 - \epsilon(x)$, then m is identically zero outside B_R. As we already know from Chapter 6, the integral representation of the solution u is the Lippmann–Schwinger equation which, in this case, takes the form

$$u(x) = u^{\text{inc}}(x) - k^2 \int_{B_R} G^0(x, y, k)m(y)u(y)dy. \tag{13}$$

Let us rewrite the Lippmann–Schwinger equation (13) in terms of the integral operator $T : C[B_R] \to C[B_R]$ as

$$u(x) = u^{\text{inc}}(x) - k^2 T u, \tag{14}$$

where

$$Tu(x) = \int_{B_R} G^0(x, y, k)m(y)u(y)dy.$$

T is weakly singular and hence compact. The solution is then given by

$$u = [I + k^2 T]^{-1} u^{\text{inc}}.$$

Let us first show that Eq. (13) has at most one solution. The standard technique consists of two parts: Rellich's lemma and the so-called *principle of unique continuation*. Of the two, Rellich's lemma was already discussed in Chapter 6. The unique continuation principle has two equivalent enunciations which we state without proof below. In version one, the principle (in the present context) runs thus.

Let $\epsilon \in C^2(R^3)$, $\epsilon(x) = 1 \; \forall \; x \notin B_R$. Let B_{R_1}, $R_1 \geq R$. Consider a solution $u \in C^2(R^3)$ of the Helmholtz equation $(\Delta + k^2\epsilon(x))u(x) = 0$ in R^3 such that $u = 0$ for all $|x| \geq R_1$ for some $R_1 \geq R$. Then u vanishes in all R^3.

In the second version which belongs to the long tradition of classical potential theory, the unique continuation principle has the following formulation.

Let Ω be a domain, i.e., an open and connected bounded region in R^3. Consider that $u \in C^2(R^3)$ is a solution of the Helmholtz equation $(\Delta + k^2\epsilon(x))u(x) =$

0 in Ω such that $u = 0$ in some neighborhood of a point $x_0 \in \Omega$. Then u vanishes in all Ω.

Next consider the following integral of the radiation condition, namely

$$I = \int_{|x|=R_0} \left| \frac{\partial u^{\mathrm{sc}}}{\partial n} - iku^{\mathrm{sc}} \right|^2 d\Gamma. \tag{15}$$

If the incident field is zero, then Eq. (15) can also be written as

$$I = \int_{|x|=R_0} \left| \frac{\partial u}{\partial n} - iku \right|^2 d\Gamma. \tag{16}$$

By Eq. (12), $I \sim O(1/(|R_0|^2))$. Next we reduce I in a manner already discussed in Chapter 6 in order to show that

$$\lim_{R_0 \to \infty} \|u\|_2^2 \to 0$$

for ϵ real or complex. By Rellich's lemma, it follows that $u = 0$ everywhere outside B_R. Then the unique continuation principle (either version) tells us that $u = 0$ in all of R^3. The homogeneous problem, therefore, has only the trivial solution. This proves the uniqueness of the direct problem of inhomogeneity scattering. The uniqueness thus established means that $[I + k^2 T]u = 0$ has only the trivial solution. By Riesz' theorem, it follows that the inverse operator $[I + k^2 T]^{-1}$ is bounded in $C(B_R)$.

Chapter 9

Some Algorithms for Obstacle Reconstructions

9.1 Introduction

In this final chapter, we present some numerical reconstructions of obstacles from the scattering data for plane-wave incidence. Many of the results in inverse obstacle scattering are based on optimizational methods. The inverse problem is transformed into one in nonlinear optimization, the nonlinearity arising from the integral operator of scattering being nonlinear in scatterer characteristics although the Helmholtz equation itself is linear. In most optimizational methods, the forward problem has to be solved repeatedly. In addition, they suffer from the occurrence of local minima and, therefore, do not have unique solutions. Most of this chapter will discuss the optimizational methods because of their important role in generating approximate inversions. Nonoptimizational techniques will be reviewed at the end.

Consider the scattering problem:

$$(\Delta + k^2)u(x) = 0, \ x \in \Omega_e$$

$$\mathcal{B}u(x) = g, \ x \in \Gamma \in C^2. \tag{9.1}$$

$\mathcal{B}$ is a boundary operator and g the boundary data. $\mathcal{B}$ is the identity operator I for the Dirichlet and the normal derivative operator ∂_n for the Neumann problem. Let us focus attention for the moment on the inverse problem of reconstructing the boundary Γ from the scattering amplitude u_∞. Consider a star-shaped scatterer with the boundary parameterization

$$r(\hat{y}) = \hat{y}f(\hat{y}) \ \hat{y} \in \hat{\Omega}_n, n = 2, 3,$$

where, as usual, $\hat{\Omega}_n$ is the unit sphere in $R^n \in V$, V being an appropriate function space to be defined. An optimizational inversion consists in determining

$f(\hat{y})$ from the scattering data by minimizing an objective functional χ which is defined by

$$\chi = \chi(f, u_\infty(f)) = \chi_1(u_\infty(f)) + \chi_2(f). \qquad (9.2)$$

The first part of the functional $\chi_1 : L_2(\hat{\Omega}_n) \to R$ and the second part $\chi_2 : V \to R$. The inverse solution can now be formulated as:

$$\text{find } \tilde{f}_0 \in U \text{ such that } \chi(\tilde{f}_0) \leq \chi(f), \ \forall f \in U \subset V.$$

The functionals χ_1 and χ_2 are assumed to be continuous so that when f is restricted to a compact subset $U \subset V$, a solution would exist in U. Since $\chi_1(u_\infty(f))$ is a composite functional, the continuity of the map $f \to u_\infty(f)$ is to be demonstrated. For Γ in C^2, the intersection $C^{1,\alpha}(\hat{\Omega}_n) \cap C^2(\hat{\Omega}_n)$ is compact (Colton and Kress, 1983), where the Hölder space $C^{1,\alpha}(D), 0 < \alpha \leq 1$, is defined by the norm

$$
\begin{aligned}
||\phi||_{1,\alpha} \ &= \ ||\phi||_\infty + ||\nabla\phi||_\infty \\
&+ \ \sup_{x,y \in D, x \neq y} \frac{|\nabla\phi(x) - \nabla\phi(x)|}{|x-y|^\alpha}.
\end{aligned}
$$

Therefore,

$$U = \{f \in C^{1,\alpha}(\hat{\Omega}_n), 0 < \alpha \leq 1, n = 2, 3; ||f||_{1,\alpha} \leq a, f \geq b\}.$$

Let $u(\overline{\Omega}_e)$ be a predicted scattering solution of the Helmholtz problem (9.1) which depends on the parameterization $f(\hat{y})$. The optimizational inverse problem of recovering the boundary is:
find a parameterization $\tilde{f}(\hat{y})$ such that the functional

$$
\begin{aligned}
\chi \ &= \ ||A_\infty u(\tilde{f}(\hat{y})) - u_\infty||^2_{L_2(\hat{\Omega}_n)} \\
&+ \alpha ||u(\tilde{f}(\hat{y}))||^2_{L_2(\tilde{\Gamma})} + \beta ||g - \mathcal{B}(\tilde{f}(\hat{y}))||^2_{L_2(\hat{\Omega}_n)} \qquad (9.3)
\end{aligned}
$$

is minimum, where u_∞ is the *measured data*. A_∞ is the far-field (i.e., $|x| \to \infty$) operator of Eq. (9.1) and $\tilde{\Gamma}$ is the boundary described by $\tilde{f}(\hat{y})$. Note that χ in Eq. (9.3) is essentially the Tikhonov functional (3.16). Here α is the regularization parameter and β is a weight. The operator A_∞ depends on Γ which is what is to be reconstructed indicating once again the nonlinear nature of the optimization. $||A_\infty \tilde{u}(\tilde{f}(\hat{y})) - u_\infty||^2_{L_2(\hat{\Omega}_n)}$ in Eq. (9.3) which corresponds to χ_1 of Eq. (9.2) is the square of the residual and is designed to prevent the predicted data to deviate too far from the actual data, whereas the last term on the R.H.S. of Eq. (9.3) with weight β does the same for the boundary condition and corresponds to the χ_2 part of Eq. (9.2). The proof that the scattering amplitude varies continuously with the boundary can be quite involved. However, for the important Dirichlet, Neumann and transmission boundary conditions, the continuity results exist (Colton and Kress, 1992; Kittappa and Kleinman, 1975).

Therefore, continuous dependence of the scattering amplitude on the boundary will be assumed below.

The conditions (either experimental or numerical) under which the reconstructions are performed cannot possibly satisfy the requirements of the uniqueness theorems of Chapter 8. For example, it is not possible in practice to either excite the object and collect data continuously, without interruption, on the unit sphere. One must necessarily be satisfied with a finite number of incident angles, frequencies and data points. It may be conjectured that since the scattering solutions are analytic in all of their arguments, the data can be analytically continued from an arbitrary patch to the entire unit sphere. In reality, however, attempts at such analytic continuation are foiled by the presence of noise. A further complication arises from the fact that the minimization of the objective functional χ seldom results in a global minimum even if one exists. Instead, it is the occurrence of the local minima that is more than likely. In other words, the optimization problem does not have a unique solution. Under these conditions what is known to hold in general is: (a) the space of solutions of the optimization problem is not empty, that is, the minimization problem (9.3) has a solution in U; (b) there exists a sequence of parameterizations $\{f_n\}$ containing a convergent subsequence with a limit point in U, but the limit is not guaranteed to coincide with the true parameterization; and (c) as the regularization parameter α goes to zero, the infimum $m(\alpha)$ of χ vanishes. In practice, the sequence $\{f_n\}$ is generated through iterations and as discussed in Chapter 3, this is a regularization procedure. However, in accordance with the conclusions there, as $\alpha \to 0$, the original ill-posed problem is obtained.

The important task is, however, the calculation of the predicted solution $u(f(\hat{y}))$ and it is in here that various inversion methods within the optimizational framework differ from one another. There exist several approaches to the problem. Of these, the well-known are the methods of:

a. the potential over known boundaries as *ansatz* for the scattered field (Kirsch and Kress, 1987; Zinn, 1989, 1991);

b. the superposition of incident fields[1]

c. the spherical wavefunction expansion (Angell *et al.*, 1987; Jones and Mao, 1989, 1990; see also Buchanan *et al.*, 2000).

A fourth method which was reported recently is one of boundary variation and rational extrapolation (Ghosh Roy *et al.*, 1997, 1998; Warner *et al.*, 2000) based on the seminal work of Bruno and Reitich (1992, 1995) on diffraction grating problems. Since methods (a) and (b) (and to some extent (c)) have already been discussed in detail in the monograph of Colton and Kress (1992), we summarize them here while discussing the boundary variation technique in some detail.

---

[1]This method was developed in a series of papers by Colton and Monk. However, since the results are presented in a unified manner in the 1992 monograph of Colton and Kress, we cite only the latter reference. The references of the original papers can be found in their text.

9.2   The Method of Potentials over Known Boundaries

The method is originally due to Kirsch and Kress (1987) and we illustrate it here by taking a transmission problem following Zinn (1991). Zinn (1989) has also discussed the Dirichlet condition. It is assumed (a relatively mild restriction) that the boundary Γ lies within an annular region bounded by an internal surface Γ_i and an outer boundary Γ_e and $f(\hat{y}) \in U$ as described above. The scattered field u^{sc} and the interior field u_i are represented by single-layer potentials

$$u^{\text{sc}}(x) = \int_{\Gamma_i} \phi(y) G^0(x, y; k_0) \mathrm{d}\Gamma_i(y) = (S_i \phi)(x) \tag{9.4}$$

$$u_i(x) = \int_{\Gamma_e} \psi(y) G^0(x, y; k_0) \mathrm{d}\Gamma_i(y) = (S_e \psi)(x). \tag{9.5}$$

The densities $\phi \in L_2(\Gamma_i)$ and $\psi \in L_2(\Gamma_e)$. It is assumed that k_0 is not an interior Dirichlet eigenvalue. In terms of the far-field operator $A_\infty : L_2(\Gamma_i) \to L_2(\hat{\Omega}_n)$, we have, from Eq. (9.4)

$$(A_\infty \phi)(\hat{x}) = u_\infty(\hat{x}) = \eta \int_{\Gamma_i} \phi(y) \mathrm{e}^{-ik_0 \hat{x} \cdot y} \mathrm{d}\Gamma_i(y), \tag{9.6}$$

where η is a complex constant that depends on the dimension of the space. For example, in two space dimension: $\eta = \{(1+i)/4\sqrt{(\pi k_0)}\}$. The densities ϕ and ψ are to be determined by solving Eq. (9.6) with $u_\infty(\hat{x})$ as the scattering data. Being of the first kind, Eq. (9.6) must be regularized. The corresponding regularized solution is obtained by solving the following

optimization problem: *find* $\tilde{\phi} \in L_2(\Gamma_i), \tilde{\psi} \in L_2(\Gamma_e), \tilde{f} \in U$ *such that the functional*

$$\begin{aligned}
\chi(\tilde{\phi}, \tilde{\psi}, \tilde{f}, u_\infty, \alpha, \beta, \gamma) &= M(u_\infty, \alpha) \\
&= \inf\{\chi(\phi, \psi, f, u_\infty, \alpha, \beta, \gamma)\}, \\
&\forall \quad \phi \in L_2(\Gamma_i), \psi \in L_2(\Gamma_e), f \in U,
\end{aligned}$$

where χ *is defined as*

$$\begin{aligned}
\chi(\phi, \psi, f, u_\infty, \alpha, \beta, \gamma) &= \|A_\infty \phi - u_\infty\|^2_{L_2(\hat{\Omega}_n)} + \alpha\big[\|\phi\|^2_{L_2(\Gamma_i)} + \|\psi\|^2_{L_2(\Gamma_e)}\big] \\
&\quad + \beta\|(S_i\phi - S_e\psi + u^{\text{inc}}) \circ f\|^2_{L_2(\hat{\Omega}_n)} \\
&\quad + \gamma\|\partial_n(S_i\phi - \{S_e\psi + u^{\text{inc}}\}) \circ f\|^2_{L_2(\hat{\Omega}_n)}.
\end{aligned}$$

α, β, γ are positive constants and $\circ$ indicates a composition operation meaning that $(g \circ h)$ is the function g evaluated on the boundary represented by the parameterization h.

Note that the interior field u_i given by Eq. (9.5) has been explicitly introduced in the functional *via* $S_e\psi$. This is necessary since Eq. (9.6) does not contain the interior field directly although the scattering amplitude depends upon it indirectly. The term involving α is the Tikhonov regularization term, whereas those involving β and γ control the deviations from the boundary conditions which are

$$S_i\phi - S_e\psi + u^{\text{inc}} = 0$$

$$\partial_n(\mathcal{F}_-S_i\phi - \mathcal{F}_+S_e\psi + u^{\text{inc}}\}) = 0,$$

where $\mathcal{F}_\pm$ were defined in Section 7.8.

A parameterization f is considered to be *admissible* if there exist densities $\{\phi, \psi\}$ such that the triple $\{\phi, \psi, f\} \in L_2(\Gamma_i) \times L_2(\Gamma_e) \times U$ is a minimizing triple, that is, $\chi(\phi, \psi, f, u_\infty, \alpha) = M(u_\infty, \alpha)$. Define by $Z(u_\infty)$ the set of all admissible parameterizations for the scattering amplitude u_∞. Then the main results of the method are:

i. The continuous dependence of the scattering amplitude on the admissible solution. Let $u_\infty^{(m)} \to u_\infty \in L_2(\hat{\Omega}_n), m \to \infty$ and $f^{(m)} \in Z(u_\infty^m)$. Then there exists a convergent subsequence of $f^{(m)}$ and every limit point of the sequence $\{f^{(m)}\}$ lies in $Z(u_\infty)$.

ii. $\text{limit}_{\alpha \to 0} M(u_\infty, \alpha) \to 0$, and

iii. Let $\{\alpha_n\}$ be a null sequence (that is, a sequence that tends to zero) of regularization parameters corresponding to which there is a sequence $\{\phi_n, \psi_n, f_n\}$ of solutions of the optimization problem. Then there exists a convergent subsequence of $\{f_n\}$ and every limit point represents a parameterization of some boundary Γ^* such that the boundary conditions are satisfied with the exact scattered field and the regularized interior field u_i^*.

Numerical reconstructions using the method of potentials on *a priori* surfaces are reported by Zinn (see citations above) for both transmission and Dirichlet obstacles. Moreover, Zinn also reported reconstructions using groupings of incident waves for multidirectional incidence. However, because of the lack of uniqueness of both the scattering (under practical circumstances) and the optimization problem, it is not expected that either u_i^* coincides with the true field in the interior or the parameterization f^* is that of the true boundary. These conclusions are typical of the optimizational solutions of inverse obstacle scattering problems.

9.3 The Method of Superposition of Incident Fields

The method was developed in a series of papers by Colton and Monk and are discussed in the 1992 text by Colton and Kress (see footnoteon p. 281). A numerical application of the Colton–Monk method is given by Misici and Zirilli (1994). Let us demonstrate the basic idea by considering the Dirichlet boundary condition. The method uses a special solution of the Helmholtz equation, the so-called *Herglotz wavefunctions* $v(y)$ defined as

$$v(y) = \int_{\hat{\Omega}_3} g(\hat{x}) e^{ik\hat{x}\cdot y} d\hat{x}, \tag{9.7}$$

where $g(\hat{x}) \in L_2(\hat{\Omega}_3)$ is called the *Herglotz kernel*. $v(y)$ is an entire function in R^3. Now the scattering amplitude for the Dirichlet problem is given by

$$u_\infty(\hat{x}) = -\frac{1}{4\pi} \int_\Gamma (\partial_n u)(y) e^{-ik\hat{x}\cdot y} d\Gamma(y). \tag{9.8}$$

Multiplying Eq. (9.8) by $g^*(\hat{x})$ and integrating over the unit sphere $\hat{\Omega}_3$ yields

$$\int_{\hat{\Omega}_3} u_\infty(\hat{x}, k) g^*(\hat{x}) d\hat{x} = -\frac{1}{4\pi} \int_\Gamma (\partial_n u)(y) v(y) d\Gamma(y). \tag{9.9}$$

Suppose that $v(y)$ is a solution of the interior problem

$$(\Delta + k^2) v(y) = 0, \ y \in \Omega, \tag{9.10}$$

$$v(y) = -\frac{e^{ik|y|}}{|y|}, \ y \in \Gamma. \tag{9.11}$$

It is assumed that $k \notin \sigma_D$, the spectrum of the Dirichlet Laplacian. Inserting Eq. (9.11) into Eq. (9.9) gives

$$\begin{aligned}
\int_{\hat{\Omega}_3} u_\infty(\hat{x}, k) g^*(\hat{x}) d\hat{x} &= \frac{1}{4\pi} \int_\Gamma (\partial_n u)(y) \frac{e^{ik|y|}}{|y|} d\Gamma(y) \\
&= \frac{1}{4\pi} \int_\Gamma \left[(\partial_n) u(y) \frac{e^{ik|y|}}{|y|} - u(y) \partial_n \frac{e^{ik|y|}}{|y|} \right] d\Gamma(y),
\end{aligned} \tag{9.12}$$

where the boundary condition (9.11) was used along with the Dirichlet condition on the boundary. Note that Eq. (9.12) is the Helmholtz representation of the scattered field at the origin of a star-like scatterer and equals the value of the incident wavefunction there, which is unity. Therefore,

$$\int_{\hat{\Omega}_3} u_\infty(\hat{x}, k) g^*(\hat{x}) d\hat{x} = 1. \tag{9.13}$$

The function $g^*(\hat{x})$ depends upon the boundary conditions used. We really should have written $g_D^*(\hat{x})$ instead of $g^*(\hat{x})$ in order to emphasize that the Herglotz kernel pertains to the Dirichlet condition.

The procedure for solving the inverse problem of reconstructing a Dirichlet boundary from the scattering amplitudes runs thus.

  i. Use Eq. (9.13) to obtain the Herglotz kernel from the data which consist of the scattering amplitudes.

ii. Having obtained the Herglotz kernel, determine the Herglotz wavefunction $v(y)$ from Eq. (9.7), and finally,

iii. Use this knowledge of the Herglotz wavefunction $v(y)$ to determine the boundary Γ using Eqs. (9.10) and (9.11).

The procedure remains the same for other boundary conditions.

The final step iii which constitutes the solution of the inverse problem is to be accomplished by transforming the problem into one of constrained nonlinear optimization. The procedure is basically the same as that described in the previous section. Let U_1 and U_2 be two compact sets in $C(\Gamma)$ to which the functions g and the parameterization f are assumed to belong, respectively. There are two minimization problems, one over U_1 for determining g and the other over U_2 for the boundary. These are:

optimization problem OP 1: *Find $g(\hat{x}) \in U_1$ such that*

$$\min_{g \in U_1} \left| \int_{\hat{\Omega}_3} u_\infty(\hat{x}, k) g^*(\hat{x}) d\hat{x} - 1 \right|^2 = M_1, \qquad (9.14)$$

and

optimization problem OP 2: *With g from OP 1, find $f(\hat{x})$ such that* $\min_{f \in C(\hat{\Omega}_3)}$

$$\alpha ||v(f(\hat{x})\hat{x})||^2_{L_2(\Gamma)} + \beta \left| \int_{\hat{\Omega}_3} v(f(\hat{x})\hat{x})|f(\hat{x})| + e^{-ik|f(\hat{x})|}v \right|^2 d\hat{x} = M_2. \qquad (9.15)$$

The existence of $M_{1,2}$ follows from the compactness of $U_{1,2}$ and the continuity of the integrals (9.14) and (9.15).

Any solution to the simultaneous minimization problems OP 1 and OP 2 yielding a boundary $f(\hat{x}) \in U_2$ is called *admissible*. On the other hand, any parameterization $\tilde{f} \in U_2$ is called *optimal* if

$$\inf_{g \in L_2(\hat{\Omega}_3)} \chi(g, \tilde{f}, \alpha) = \inf_{g \in L_2(\hat{\Omega}_3), f \in U_2} \chi(g, f, \alpha) = m(\alpha),$$

where χ is the sum of the functionals of Eqs. (9.14) and (9.15).

The basic results proved in Colton and Kress (1992) for the method of superposition of incident fields are the following.

Existence *The set of admissible solutions is not empty. That is, for each $\alpha > 0$, there exists a solution $f(\hat{x})$ of the optimization problem OP 1 plus OP 2 such that*

$$\inf_{g \in L_2(\hat{\Omega}_3)} \chi(g, \tilde{f}, \alpha) = \inf_{g \in L_2(\hat{\Omega}_3), f \in U_2} \chi(g, f, \alpha) = m(\alpha).$$

Convergence *i. The objective function converges:* $\lim_{\alpha \to 0} m(\alpha) = 0$.

ii. Let $\{\alpha_n\}$ be a null sequence and $\{f_n\}$ be the corresponding sequences of optimal surfaces for the regularization parameter α_n. Then there exists a convergent subsequence of f_n and every limit point f^* of which represents a surface on which the boundary condition (9.11) is satisfied.

Again as in the case of the method of potential of Section 9.2, it is only possible to have convergent subsequences which are not guaranteed to coincide with the exact boundary and the optimization problems do not have unique solutions. Colton and Monk have numerically reconstructed various Dirichlet and transmission boundaries.

We would like to point out that results such as those in the above two sections are typical of optimizational methods and should remain in force for the boundary variation method (to be described below) also. These conclusions, therefore, will not be repeated.

9.4   The Method of Wavefunction Expansion

The method was used by Angell *et al.*(1989) and was developed later by Jones and Mao (1988, 1989). In the optimizational methods, it is difficult to associate the scattering amplitude to the scattering surface without first solving the forward problem. The idea behind the method of wavefunction expansion is to obtain an approximate solution of the inverse scattering problem of a boundary without having to solve a succession of direct problems[2] and can be described as follows. Let Γ be the surface to be recovered from the scattering amplitude data. Now outside a sphere S circumscribing Γ, the scattered field can be represented in terms of the outgoing wavefunctions *via* the expansion

$$u^{\mathrm{sc}}(x) = \sum_{\ell=0}^{\infty} \sum_{m=-\ell}^{\ell} \beta_{\ell m}^0 h_{\ell}^{(1)}(k_0|x|) Y_{\ell m}(\hat{\theta}). \tag{9.16}$$

The following expression for the scattering amplitude $u_\infty(\hat{x})$ follows from Eq. (9.16), namely

$$u_\infty(\hat{x}) = \frac{1}{k_0} \sum_{\ell=0}^{\infty} \sum_{m=-\ell}^{\ell} \beta_{\ell m}^0 (-i)^{(\ell+1)} Y_{\ell m}(\hat{\theta}). \tag{9.17}$$

In order to be specific, let us consider the Neumann boundary condition. Using the Helmholtz representation in the exterior (Theorem 6.2) and the spherical wave expansion of the free-space Green's function (see Section 7.9), the coefficients $\beta_{\ell m}^0$ are found to be given by

$$\beta_{\ell m}^0 = ik_0 \frac{(2\ell+1)(\ell-|m|)!}{4\pi(\ell+|m|)!} \int_{\Gamma} u(y)\partial_n(y)[j_\ell(k|y|)]Y_{\ell m}^*(\hat{\theta}) d\Gamma(y). \tag{9.18}$$

---

[2]Note that this is also the case for the method of superposition of incident fields of the previous section.

The complex constants $\beta_{\ell m}^0$, therefore, depend on the boundary.

It can be seen from Eq. (9.18) that the calculation of $\beta_{\ell m}^0$ requires that the *total* field u on the surface be known. In general, this field can be evaluated by solving the boundary integral equation

$$u = u^{\text{inc}} + 2 \int_\Gamma u(y)\partial_n(y)G^0(x, y; k_0)\mathrm{d}\Gamma(y). \tag{9.19}$$

However, it is known (Vekua, 1967; see Millar, 1973, for a detailed discussion) that the outgoing wavefunctions $h_\ell^{(1)}(k|y|)Y_{\ell m}(\hat{\theta})$ form a complete set in terms of which any function which is square-integrable on the boundary can be expanded. One can, therefore, express the scattered field u^{sc} on Γ as

$$u^{\text{sc}}(y) = \sum_{\ell=0}^\infty \sum_{m=-\ell}^\ell \gamma_{\ell m} h_\ell^{(1)}(k_0|y|)Y_{\ell m}(\hat{\theta}), \, x \in \Gamma, \tag{9.20}$$

instead of using the boundary integral equation (9.19) directly.

Now it is also known that the representation (9.16) cannot be continued all the way from S down to the boundary Γ except under very restricted conditions. The most important of these is the requirement that the scatterer must be canonical in shape, i.e., primarily circles and spheres. The assumption that (9.16) is valid in all $R^3 \setminus \Omega$ was made by Rayleigh (1907) and has come to be known as the *Rayleigh hypothesis*. A brief summary of these issues are given in Appendix A.9.3. The coefficients $\gamma_{\ell m}$ in Eq. (9.20) are thus generally not the same as $\beta_{\ell m}^0$ of Eq. (9.16) and cannot, therefore, be calculated using Eq. (9.18). These are to be evaluated by minimizing the L_2-norm of the difference between the normal derivative of Eq. (9.20) (because of the Neumann boundary) and the given boundary data. For the Neumann condition, therefore, $\gamma_{\ell m}$ must be understood in the sense of the functional

$$J_f = \|\sum_{\ell=0}^\infty \sum_{m=-\ell}^\ell \partial_n(y)[\gamma_{\ell m} h_\ell^{(1)}(|y|)Y_{\ell m}(\hat{\theta}) + u^{\text{inc}}]\|_{L_2(\Gamma)}^2 \tag{9.21}$$

being minimum. The subscript f on J emphasizes the dependence of J on the parameterization $r(y) = \hat{y}f(\hat{y})$.

Let us denote the least-square solution of Eq. (9.21) by $\beta_{\ell m}$ without the superscript 0 in order to emphasize that $\beta_{\ell m}$ is different from $\beta_{\ell m}^0$. It is, therefore, necessary to consider a second functional J_β to describe the minimization of the discrepancy between $\beta_{\ell m}$ and $\beta_{\ell m}^0$. The total objective function J_T, the minimizer of which would constitute the inverse solution, is the sum

$$\begin{aligned}
J_T &= J_\beta + J_f \\
&= \frac{1}{k_0^2}\eta\left|\sum_{\ell=0}^\infty \sum_{m=-\ell}^\ell (-\mathrm{i})^{(\ell+1)}(\beta_{\ell m} - \beta_{\ell m}^0)Y_{\ell m}(\hat{\theta})\right|_{L_2(\Gamma_0)}^2 \\
&\quad + \left|\sum_{\ell'=0}^\infty \sum_{m'=-\ell'}^{\ell'} \gamma_{\ell'm'}\partial_n(y)\left[h_\ell^{(1)}(ky)Y_{\ell m}(\hat{\theta}) + u^{\text{inc}}\right]\right|_{\Gamma'}^2,
\end{aligned}$$

where η is a weighting factor and Γ_0 is the surface of any ball of radius $R_0 \geq R$, R being the radius of the circumscribing sphere S and Γ' is the predicted boundary. Upon using the orthogonality relation of the Legendre functions, namely

$$\int_0^{2\pi} P_l^m P_{l'}^m \sin\theta d\theta = \frac{2(l+m)!}{(2l+1)(l-m)!}\delta_{ll'}, l \geq m,$$

the functional J_β can be reduced to the form

$$J_\beta = \frac{\eta}{k_0^2} \sum_{\ell=0}^{\infty} \sum_{m=-\ell}^{\ell} \frac{4\pi}{(2l+1)}|\beta_{\ell m} - \beta_{\ell m}^{02}| + \text{constant}.$$

The constant term can be disregarded.

A Fourier parameterization of the surface is introduced at this point, namely

$$f(\hat{y}) = \sum_{p=0}^{\infty} \sum_{q=-p}^{p} f_{pq}Y_{pq}.$$

The functional J is then $J_T = J_T(f_{pq}, \beta_{\ell m}, \gamma_{\ell m})$. The approximate solution of the problem is given by the minimizer of J_T. It can be appreciated that numerically, the problem of recovering an obstacle of an arbitrary shape using the method of wavefunction expansion is at best very involved, especially if the dimension of the problem exceeds two. The Jacobian matrix is of large dimension. Jones and Mao have reported some numerical results for simple axisymmetric-shaped bodies.

Methods using wave functions expansion have been applied by Gilbert, Xu and coworkers (1991, 1993, 1997, 1999) for scattering and inverse scattering of obstacles in a waveguide with a view to applications in shallow-water oceanic studies. For a more recent study of this group for obstacle inversion in a waveguide based on the Rayleigh expansion and approximation of an obstacle boundary as intersections of canonical shapes, we refer the reader to a recent paper by Buchanan, et al., (2000).

Next we present the method of boundary variation.

9.5   The Method of Boundary Variation and Shape Differentiation

The method of boundary variation starts out by assuming that the shape of the scatterer can be obtained by superimposing a deformation of an arbitrary shape on an otherwise known geometry. The scattered field is first calculated in a perturbative way by assuming that the deformation is sufficiently small so that a power series representation of the field in terms of the magnitude of the perturbation can be written. The field thus obtained is then extrapolated to larger perturbations. This gives rise to two essential component parts of the method: the differentiation of the scattered field with respect to the boundary of the scatterer-the so-called *shape derivative* (Simon, 1980; Pirronneau, 1984; Haug, *et al.*, 1986; Delfour, 1990; Sokolowski and Zolesio, 1992) and an extrapolative rational function approximation such as the *Padé approximation* (Baker and Graves-Morris, 1996; Petrushev and Popov, 1987) which is used here. The manner in which the two combine is described below. A brief discussion on shape differentiation is presented in Appendix A.9.1 while a summary review of the Padé approximation appears in Appendix A.9.2 at the end of the chapter.

Essentially, one considers a vector field $\mathbf{V} : R^3 \rightarrow R^3$ and a linear transformation $\mathbf{T}_t$ such that a domain $\Omega_0 \subset R^3$ is deformed to another domain $\Omega_t \subset R^3$.

$$\Omega_t = \mathbf{T}_t \Omega_0 = \{x | x = (I + t\mathbf{V})(X), \ \forall X \in \Omega_0\}.$$

$t \in R^1$ is small. Moreover, $\Omega_{t=0} = \Omega_0$. This is schematically illustrated in Figure 9.1.

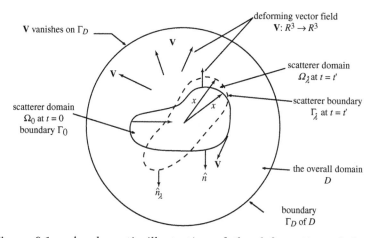

Figure 9.1 – A schematic illustration of the deformation of the scatterer. The overall domain D contains $\bar{\Omega}_\lambda = \Omega_\lambda \cup \Gamma_\lambda$, $\forall \lambda \in [0, \lambda_0) \in R^1$. The arrows indicate the vector field $\mathbf{V}$ which deforms the obstacle. $\mathbf{V}$ is assumed to vanish on Γ_D. Moreover, $\Omega_{\lambda=0} \equiv \Omega_0$. All domains are assumed to be simply connected. $\hat{n}$ is the outward unitary surface normal.

Consider an exterior scattering problem on both Ω_0 and Ω_t for which the solutions are u_0 and u_t, respectively. The shape derivative $u^{[1]}$ of the scattered field is defined by the *Gâteaux differential* (Nashed, 1971; Tapia, 1971)

$$u^{[1]} = \underset{t \to 0}{\text{limit}} \ \frac{u_t(X + t\mathbf{V}(X)) - u_0(X)}{t}.$$

The partial derivative $u^{(1)}$ is given by

$$u^{(1)} = \underset{t \to 0}{\text{limit}} \ \frac{u_t(X) - u_0(X)}{t}.$$

The two derivatives are connected by the formula

$$u^{[1]} = u^{(1)} + \mathbf{V} \cdot \nabla u = (\partial_t + \mathbf{V} \cdot \nabla)u = D_t u.$$

The higher-order derivatives $u^{[m]}$ and $u^{(m)}, m > 1$, can be similarly defined by repeated applications of the operator D_t and ∂_t provided that the derivatives exist. It is the partial derivatives $u^{(m)}, m \geq 1$, of the scattered field which are of interested to us. But their determination involves solving the scattering problem for $u^{[m]}$. This will be elaborated below. However, the following remarks are necessary for the future development of the arguments.

Consider the scattering problem (9.1) on a deformed boundary Γ_t, the corresponding reference surface being Γ_0. Let u_t be the scattering solution for Γ_t and u_0 for Γ_0. u_0 is assumed to be known either analytically or from numerical calculations. Now given an arbitrary deformation described by $\mathbf{T}_t$, the scattered field exterior to the overall domain D can be written as

$$u_t(x) = \sum_{\vec{n}} u_{t\vec{n}}(\mathbf{V}) \Phi_{\vec{n}}, \tag{9.22}$$

where the subscript $\vec{n}$ represents the number of indices which equals the dimension of the physical space less one. $\Phi_{\vec{n}} = H_n^1(k|x|)e^{in\theta}$ in R^2 and $\Phi_{\vec{n}} = h_n^1(k|x|)Y_{nm}(\hat{\theta})$ in R^3 are the outgoing wavefunctions. As noted in Section 9.4, for an arbitrary deformation of the surface, the coefficients $u_{t\vec{n}}$ cannot be determined by extending Eq. (9.22) to the boundary and then applying the boundary conditions. In other words, the Rayleigh hypothesis does not apply. However, earlier research (Petit and Cadilhac, 1966; Millar, 1973) on diffraction gratings concluded that if the ratio h/d (h is the height and d is the constant of the grating) is less than 0.142, then the Rayleigh series could be extended to the boundary itself, but not otherwise. Similar conclusions were drawn by Van den Berg and Fokkema (1979) for surfaces with $\cos n\theta$ parameterization. Again, see Appendix A.9.3.

Recently, these results were generalized in a fundamental way by Bruno and Reitisch (1992)[3] in which the authors established the analyticity properties of the scattering solution. Their basic result is as follows. Let $\mathbf{V}$ be an analytic vector field. In actual computations, $\mathbf{V}$ is usually the vector $\hat{n}f(\hat{\theta})$, where $f(\hat{\theta})$ is an analytic function of angular coordinates $\hat{\theta} \in \hat{\Omega}_j, j = 2, 3$, and $\hat{n}$ is the unit normal to Γ, a reference surface. The results of Bruno and Reitisch can be stated as: *given any t, say, $t_0 \in R^1$, there exists a number τ such that the scattering solution $u(x,t)$ is analytic in both x and t for $|t - t_0| < \tau$ and $|x - X| < \tau, \forall X \in \Gamma_0$. Moreover, away from the scatterer (more precisely, outside the overall domain D (Figure 9.1) in the context of the present discussion) the solution is analytic for any t.* Mathematically, it says that the transmission equation, for example, can be inverted in the space of analytic functions under the conditions stated. It further implies that the solution can be extended analytically in the complex t-plane around the real axis, the width of the extension varying along the axis.

---

[3]A related paper is by Chen and Friedman, 1991, for Maxwell's equation.

The above result is fundamental for several reasons. First, it generalizes the earlier results on periodic diffraction gratings to gratings of arbitrary shapes. Second, for small values of the deformation parameter t, the Rayleigh type expansion has been used in the past (Yeh, 1964; Erma, 1968). However, the validity of these expansions has always been in question. The above analyticity property of the solution in both x and t allows us to extend the solution to the boundary and apply the boundary conditions for evaluating the Rayleigh series. Third, and perhaps the most important is the fact that such an extension to the boundary leads to straightforward numerical schemes for solving scattering problems.

The last point is worth an elaboration since it has a direct bearing on implementations in practice. The analyticity property by which the extension to the boundary is possible holds only for small deviations τ, albeit for a fixed arbitrary value t_0. In other words, the width of the strip of analyticity around the real axis is small at any value $t = t_0$ on the real axis. However, the boundary of the obstacle may be arbitrarily deformed. This, therefore, necessitates that the width of the strip of analyticity be extended by using the available existing techniques for this purpose. The method of the Padé approximation is used in the present case. *The procedure for solving the forward problem, therefore, consists of two steps: i. find the coefficients $u_{t\bar{n}}$ in Eq. (9.22) assuming small values of t; and ii. Extrapolate the result to larger t by using the technique of the Padé approximation.* The details are given below.

First consider that Γ_t is only a slight perturbation of Γ_0. From the analyticity result, the scattered field u_t has a Taylor series around $t = 0$, namely

$$u_t(x; t, f(\hat{\theta})) = \sum_{m=0}^{\infty} \frac{1}{m!} u^{(m)}(x; f(\hat{\theta}))t^m. \tag{9.23}$$

The coefficients $u^{(m)}, \forall m$, of the Taylor series (9.23) (which are the partial derivatives defined earlier) are determined by first solving the system

$$(\Delta + k_0^2)u^{[m]}(x; f(\hat{\theta})) = 0, \ x \in \Omega_e$$

$$\mathcal{B}u^{[m]}(x; f(\hat{\theta})) = \tilde{g}^{[m]}(f(\hat{\theta}), u^{(p)}(x; \mathbf{V})), \ p < m, \tag{9.24}$$

for $u^{[m]}$ and then using the relation between $u^{[m]}$ and $u^{(m)}$ as appears on p. 290. Moreover, in Eq. (9.24) $\mathcal{B}$ is the same boundary operator as in the original problem (9.1). Moreover, $u^{(m)}(x; f(\hat{\theta}))$ satisfies the radiation condition at infinity because the perturbation of the boundary does not affect the decay of Green's function. The general form of the boundary data $\tilde{g}^{(m)}$ is

$$g^{(m)} = g^{[m]} - \sum_{q=0}^{m} \frac{m!}{q!(m-q)!}(\mathbf{V} \cdot \nabla)^{m-q}g^{(q)}.$$

From now onward it will be assumed that the reference domain Γ_0 is canonical so that $u^{(m)}$ can be expanded in terms of the outgoing wavefunctions. Assume also that $f(\hat{\theta})$ has the Fourier series representation

$$f(\theta) = \sum_{\ell=-\infty}^{\infty} f_\ell e^{i\ell\theta} \tag{9.25}$$

for $n = 2$ and

$$f(\hat{\theta}) = \sum_{\ell=0}^{\infty} \sum_{\ell'=-|\ell|}^{|\ell|} f_{\ell\ell'} Y_{\ell\ell'}(\hat{\theta}) \tag{9.26}$$

when $n = 3$. Moreover, any power of f can also be expressed in Fourier series. From the outgoing wavefunction expansion Eq. (9.22) and the Fourier series representations (9.25)–(9.26) of f, the coefficients $u^{(m)}$ of the Taylor series (9.23) can be obtained from the boundary data in Eq. (9.24). The coefficients $u_{t\vec{n}}$ in (9.22) are found to be

$$u_{t\vec{n}} = \sum_{m=0}^{\infty} \frac{1}{m!} u_{\vec{n}}^{(m)}(\mathbf{V}) \mathbf{t^m}, \tag{9.27}$$

a power series expansion. It is this series (9.27) that will be extrapolated to large t by the Padé approximants. It must be emphasized that Eq. (9.22) applies only outside the smallest ball (a circle in two and a sphere in three dimensions) enclosing the obstacle.

Remarks

i. Equation (9.24) shows that a derivative of any order can be obtained essentially by solving the forward problem for the reference boundary albeit with different boundary data. The Jacobian matrix of the scattered field (with respect to the boundary) can, therefore, be obtained by solving a series of Helmholtz problems in the same exterior domain which is frequently canonical. This can result in a substantial simplification in computations.

ii. The derivatives are obtained recursively. The boundary data $g^{[m]}$ for the m-th order derivative $u^{[m]}$ contain only derivatives of orders strictly less than m.

iii. It is not necessary to incorporate the boundary conditions in the objective function to be minimized when using Newton-type methods, and finally,

iv. There is no nonuniqueness in the solution owing to the interior eigenvalues.

The mathematical expressions for the coefficients $u^{(m)}$ are listed below for various boundary conditions. We list only the two-dimensional cases. Results in three dimensions will appear in the future. The Fréchet differentiability of $u^{(1)}$ was established rigorously for the impedance and transmission problems by Hettlich (1995).

9.5.1   Dirichlet and Neumann Problem In Two-dimension

$$f(\theta) = \sum_{\ell=-L}^{L} f_\ell e^{i\ell\theta}.$$

$$f^p(\theta) = \sum_{\ell=-pL}^{pL} f_{p,\ell} e^{i\ell\theta}.$$

$$\xi = k_0 r_0.$$

$$u^{sc}(x) = \sum_{n=-\infty}^{\infty} \left[\sum_{m=0}^{\infty} u_m^{(n)} t^m \right] (-i)^n H_n^{(1)}(k_0|x|) e^{in\theta}.$$

For the Dirichlet problem, the coefficients $u_m^{(n)}$ are:

$$
\begin{aligned}
u_n^{(m)} &= \frac{1}{H_n^{(1)}(\xi)} \left[k_0^m \sum_{p=n-mL}^{h+mL} f_{m,n-p}(-i)^{p-n} J_P'(\xi) \right.\\
&\left. + \sum_{\ell=0}^{m-1} k_0^{m-\ell} \sum_{p=n-(m-\ell)L}^{n+(m-\ell)L} k_0^{m-\ell} f_{m-\ell,n-p}(-i)^{p-n} u_\ell^{(p)} H_p^{(m-\ell)}(\xi) \right].
\end{aligned}
$$

For the Dirichlet problem, the expression for the first derivative $u^{(1)}$ was also derived by Kirsch (1993). In addition, Kirsch gave a rigorous analysis of the solvability of the boundary value problem for $u^{(1)}$ using variational formulation.

The Neumann coefficients $u_n^{(m)}$ are as follows.

$$u_n^{(m)} = -\frac{k_0^m}{H_n^{(1)\prime}}(\xi) \sum_{i=1}^{4} T_i,$$

where

$$T_1 = \left(\frac{1}{m!}\right) \sum_{\ell=-Lm}^{Lm} (-i)^{-\ell} f_{m,\ell} J_{n-\ell}^{(m+1)}(\xi),$$

$$
\begin{aligned}
T_2 &= \left(\frac{1}{m!}\right) \sum_{j=0}^{m-1} \sum_{\ell=-Lm}^{Lm} (-i)^{-\ell}(-1)^j \\
&\quad \cdot \{(m-j-1)!\}^{-1}(1+j)\ell(n-\ell)\xi^{-(2+j)} f_{m,\ell} J_{n-\ell}^{(m-j-1)}(\xi),
\end{aligned}
$$

$$
\begin{aligned}
T_3 &= \sum_{j=0}^{m-1} \sum_{\ell=-L(m-j)}^{L(m-j)} (-i)^{-\ell} \\
&\quad \cdot \{(m-j)!\}^{-1} k_0^{-j} f_{m-j,\ell} u_j^{(n-\ell)} (H_{n-\ell}^{(1)})^{(m-j+1)}(\xi),
\end{aligned}
$$

$$T_4 = \sum_{j=0}^{m-1} \sum_{k=0}^{m-j-1} \sum_{\ell=-(m-j)L}^{(m-j)L} (-i)^{-\ell}(-1)^k(1+k)\ell(n-\ell)k_0^{-j}\xi^{-(2+k)}$$

$$\times [(m-j)\{(m-j-k-1)!\}^{-1}]f_{m-j,\ell}u_j^{(n-\ell)}(H_{n-\ell}^{(1)})^{(m-j-k-1)}(\xi).$$

9.5.2 Transmission Problem in Two-dimensions

$$\xi_\pm = k_\pm r_0.$$

$$\mathcal{F} = \frac{\mathcal{F}_-}{\mathcal{F}_+},$$

where $\mathcal{F}_\pm$ were defined in Chapter 7. θ_0 is the angle of incidence.

$$u_+(x) = \sum_{n=-\infty}^{\infty} \left[\sum_{m=0}^{\infty} u_{+n}^{(m)}t^m\right](-i)^n H_n^{(1)}(k_+|x|)e^{in\theta_0}.$$

$$u_-(x) = \sum_{n=-\infty}^{\infty} \left[\sum_{m=0}^{\infty} u_{-n}^{(m)}t^m\right](-i)^n J_n^{(1)}(k_-|x|)e^{in\theta}.$$

The coefficients $u_{\pm n}^{(m)}$ are obtained by solving the following two algebraic equations:

$$u_{+n}^{(m)}H_n^{(1)}(\xi_+) - u_{-n}^{(m)}J_n^{(1)}(\xi_-) = -\frac{k_+^m}{m!}\sum_{q=n-mL}^{n+mL}(-i)^{q-n}f_{m,n-q}J_q^{(m)}(\xi_+)e^{i(n-q)\theta_0}$$

$$-\sum_{j=0}^{m-1}\sum q = n - (m-j)L^{n+(m-j)L}(-i)^{q-n}\{(m-j)!\}^{-1}f_{m-j,n-q}$$

$$[(u_+)_q^{(j)}k_+^{m-j}H_q^{(m-j)}(\xi_+) - (u_-)_q^{(j)}k_-^{m-j}Jq^{(m-j)}(\xi_-)],$$

and

$$u_{+n}^{(m)}\ k_+H_n^{(1)}(\xi_+) - u_{-n}^{(m)}\mathcal{F}k_-J_n^{(1)}(\xi_-)$$

$$= -\frac{k_+^{m+1}}{m!}\sum_{\ell=-mL}^{mL}(-i)^{-\ell}f_{m,\ell}e^{i(n-q)\theta_0}$$

$$\left\{J_{n-\ell}^{(m+1)}(\xi_+) + \ell(n-\ell)\frac{d^{(m-1)}}{dr^{(m-1)}}\frac{J_{n-\ell}(\xi_+)}{r^2}\right\}$$

$$-\sum_{j=0}^{m-1}\sum\ell = -(m-j)L^{(m-j)L}(-i)^{-\ell}f_{m,\ell}\left\{k_+^{m-j+1}u_{+,n-\ell}^{(j)}\right.$$

$$\left(H_{n-\ell}^{(m-j+1)}(\xi_+) + \ell(n-\ell)\frac{d^{(m-j-1)}}{dr^{(m-j-1)}}\frac{H_{n-\ell}(\xi_+)}{r^2}\right)$$

$$\left.\mathcal{F}k_-^{m-j+1}u_{-,n-\ell}^{(j)}\left(J_{n-\ell}^{(m-j+1)}(\xi_-) - \ell(n-\ell)\frac{d^{(m-j-1)}}{dr^{(m-j-1)}}\frac{J_{n-\ell}(\xi_{+-})}{r^2}\right)\right\}.$$

The final step involves extrapolating the coefficients to larger values of t by using the technique of the Padé approximation (Appendix A.9.2). The objective function to be minimized is

$$\chi = ||F(\vec{f}) - u_\infty||_2^2 + \alpha|\vec{f}|^2.$$

The analyticity of the representation is equivalent to the continuity of the scattering data with the variation in the boundary. Furthermore, since the Fourier components form a vector in a Euclidean space of finite dimension, the parameterization is restricted to a compact set. The minimization problem, therefore, has a solution. Again, the conclusions of the optimizational inversion discussed in the earlier Sections 9.2 and 9.3, namely, that only convergent subsequences are possible.

Finally, we present some numerical results of obstacle reconstructions using this method. Figure 9.2 shows the reconstruction of a sound-hard cloverleaf obstacle parameterized by $f(\theta) = \cos 4\theta$. The conditions under which the reconstruction was performed are described in the figure.

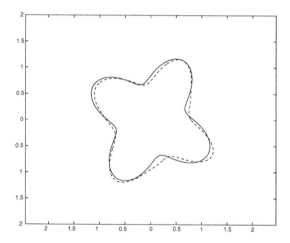

Figure 9.2. The reconstruction of a cloverleaf from the far-field data (8% random noise) collected over a 45° wedge. The number of receivers is 23 and a total of 17 unknown Fourier coefficients were recovered. $f(\theta) = \cos 4\theta$.

Figure 9.3 shows the reconstruction of a penetrable asymmetric-shaped obstacle. The data collection was what has been termed *duostatic* in the figure. A single data point consisted of a narrow aperture (roughly 2°) quasi-monostatic and a bistatic measurement. For more details, see Warner *et al.*, (2000).

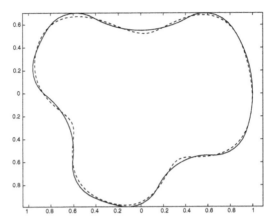

Figure 9.3. Reconstruction of an asymmetric figure from
duostatic data. True $k_{\text{ratio}} = 2.5$ and true $\rho_{\text{ratio}} = 0.333$.
Recovered $k_{\text{ratio}} = 2.634$ and recovered $\rho_{\text{ratio}} = 0.338$. The
highest order Padé approximate used was [6/6]. Frequen-
cies used were $k_+ = 1.0$ through 5.0.

This essentially concludes our discussion of the nonlinear optimizational solutions of inverse scattering problems. In general, the methods have two major
drawbacks: the need to solve the direct problem repeatedly and the occurrence
of the local minima. On the other hand, the optimizational techniques have
the advantage that they can be employed to solve inverse scattering problems
under rather general conditions. For example, no restrictions apply as to what
characteristics of the scatterers can be recovered from the scattering data or
whether the ambient medium is absorbing or not. This notwithstanding, it is
advantageous to design methods where it is not necessary to solve the forward
problem repeatedly (which is computationally intensive) and in which the problem of local minima does not arise. It only stands to reason that we conclude
the final chapter with an introduction to the important developments in this
direction.

9.6 Some Nonoptimizational Methods in Inverse Scattering Theory

In perhaps the most noteworthy of the nonoptimizational methods, the scattering amplitudes (over all directions of incidence and scattering) are combined in
a very special way such that a quantity similar to the characteristic function of
the scatterer domain is recovered. There are two trends in this direction. One
(Colton and Kirsch, 1996) uses the properties of the Herglotz wavefunctions and
the other (Mast *et al.*, 1997; Norris, 1998) is based on the eigensystem of the
far-field scattering operator. Another different and interesting formulation is by
Chen and Rokhlin (1997) who use Riccati-type equations. In this section, we
discuss the first two beginning with the Herglotz wavefunction method.

9.6.1 The Method of Colton and Kirsch

An up-to-date and comprehensive review of the method was published recently by Colton *et al.*, (2000). The review also contains carefully presented basic fundamentals of inverse scattering theory (obstacles and medium) and discusses how to solve the inverse problems with synthetic as well as real data. Let $u_\infty(\hat{x}_i, \hat{x}_s)$ be the scattering amplitude for an incident plane wave $e^{ik \cdot x}$, $\hat{x}_i$, $\hat{x}_s$ denoting unit vectors along the direction of incidence and scattering, respectively. Consider a function $g(\hat{\theta}, z) \in L_2(\hat{\Omega}_3)$, z being a fixed point in the scatterer domain $\Omega \in R^3$. Let $F_\infty : L_2(\hat{\Omega}_3) \to L_2(\hat{\Omega}_3)$ be the far-field scattering operator and consider the equation.

$$(F_\infty g)(\hat{x}, z) = \int_{\hat{\Omega}_3} u_\infty(\hat{x}, \hat{\theta}) g(\hat{\theta}, z) d\hat{\theta}. \tag{9.28}$$

It is then shown that there exists a function $g(\hat{\theta}, z) \in L_2(\hat{\Omega}_3)$, $z \in \Omega$, such that

$$||(F_\infty g)(\hat{x}, z) - G_\infty^0(\hat{x}, z)|| < \epsilon, \; \epsilon > 0, \; \forall \; z \in \Omega. \tag{9.29}$$

In Eq. (9.29), $G_\infty^0(\hat{x}, z)$ is the far-field value of Green's function between the field point along $\hat{x}$ and the source at z. It is observed that

$$||g(\hat{\theta}, z)||_{L_2(\hat{\Omega}_3)}, \; ||v_g||_{L_2(\hat{\Omega}_3)} \to \infty \text{ as } z \to \Gamma,$$

Γ being the boundary of the scatterer and the Herglotz wave function v_g is, as earlier,

$$v_g(z) = \int_{\hat{\Omega}_3} e^{ik\hat{k} \cdot z} g(\hat{k}, z) d\hat{k}. \tag{9.30}$$

If we use the asymptotic ($|x| \to \infty$) form of the free-space Green's function as the L.H.S. of Eq. (9.28), we obtain the *far-field equation*

$$\frac{1}{4\pi} e^{-ik\hat{x}_i \cdot z} = \int_{\hat{\Omega}_3} u_\infty(\hat{x}, \hat{\theta}) g(\hat{\theta}, z) d\hat{\theta}. \tag{9.31}$$

Now by Rellich's theorem (Chapter 6), the scattered field u^{sc} is uniquely determined by the scattering amplitude. Then from Eq. (9.31), it follows that

$$G^0(x, z) = \int_{\hat{\Omega}_3} u^{sc}(x, \hat{\theta}) g(\hat{\theta}, z) d\hat{\theta}, \; x \in R^3 \setminus \overline{\Omega}. \tag{9.32}$$

Since the Helmholtz equation is linear (in the incident field) and u^{sc} is the scattered field due to an incident plane wave, the integral in Eq. (9.32) represents a scattered field corresponding to an *incident wavefunction* v_g given by Eq. (9.30). If we consider a Dirichlet problem for the sake of illustration, then from the Dirichlet boundary condition, we have

$$v_g(x) = -G^0(x, z), \; x \in \Gamma \tag{9.33}$$

and $z \in \Omega$ is a fixed point. Equation (9.33) clearly demonstrates that as $z \to x$ on the boundary, v_g becomes unbounded and so does $g(\cdot, z)$. Therefore,

$$\lim_{z \to \Gamma} \|g(\cdot, z)\|_{L_2(\Omega)} \to \infty. \tag{9.34}$$

Equation (9.34) then furnishes an inversion scheme. First, solve the far-field equation (9.31) for $g(\cdot, z)$. Enclose the scattering region Ω in some larger domain $\tilde{\Omega}$ and determine $g(\cdot, z)$ pointwise in $\tilde{\Omega}$. The points on which $g(\cdot, z)$ becomes unbounded then delineate the boundary of the scattering region. Essentially, this constitutes the method of Colton and Kirsch.

There are a few remarks to be made. First, the far-field equation (9.31) is an integral equation of the first kind and is, therefore, ill-posed. The equation must be regularized. More importantly, it is not clear why the singularity should not occur in the exterior of Ω also. In other words, the behavior of g in the exterior $R^3 \setminus \Omega$ is not clear. As an answer to the question, Kirsch (1998, 1999) furnished some characterization of the spectral data. Specifically, Kirsch showed that the function g is to be calculated according to the formula $(F_\infty F_\infty^*)^{\frac{1}{4}} g = \Psi_\infty$ instead of the straightforward far-field equation (9.31). Finally, we note that if M is the number of directions of the incident fields distributed around the scatterer, N the number of receivers and P the number of searching coordinates $\tilde{x}$ in $\tilde{\Omega}$, then the size of the inversion is MNP in this method. A large amount of data is, therefore, necessary. In other words, as it stands now, the method essentially requires the knowledge of the full scattering matrix (Norris, 1998). On a minor note, it is necessary to know the location of the scatterer at least roughly in order to specify the larger domain $\tilde{\Omega}$ over which to sample $g(\cdot, z)$.

It was mentioned that the far-field equation (9.31) is ill-posed and must be regularized. Tikhonov's method (Chapter 3) can be applied using the Euler–Lagrange equations

$$(F_\infty F_\infty^* + \beta I)\tilde{g} = F_\infty^* \tilde{g}$$

if F_∞ is used and

$$(\hat{F}_\infty \hat{F}_\infty^* + \beta I)\tilde{g} = \hat{F}_\infty^* \tilde{g}$$

if the operator $(F_\infty F_\infty^*)^{\frac{1}{4}}$ is used. In the above, $\hat{F}_\infty = (F_\infty F_\infty^*)^{\frac{1}{4}}$ and $\tilde{g}$ is the numerical approximation of g. In either case, the regularized solution is given by

$$\|\tilde{g}\|^2 = \sum_{j=1}^{N} \left(\frac{\lambda_j}{\lambda_j + \beta} \right) |Z^* \Psi_\infty|^2,$$

where λ_j are the singular values and Z^* the singular function found by the SVD of the matrix for F_∞ or $\hat{F}_\infty$. If Morozov's discrepancy principle is to be applied for determining the regularization parameter β, then the equation to be used is

$$0 = \sum_{j=1}^{N} \frac{\beta^2 - \epsilon^2 \lambda_j^2}{(\lambda_j + \beta)^2} |Z^* \Psi_\infty|^2,$$

where ϵ is the noise level.

The method has some striking advantages: (1) it is independent of the geometry of the scatterer; (2) no boundary condition needs to be specified; and (3) no repeated solution of the forward problem is necessary. However, it is somewhat less versatile (at least at the moment) than the optimizational methods in that since no boundary condition is specified, it is only possible to reconstruct the shape of the obstacle with this technique. Also, as already pointed out, a large amount of data is necessary and the geometry of data collection (the directions of incidence all around the object) may not be the most desirable for many important applications such as underwater acoustics. It is also clear that no absorption in the surrounding medium is permissible. Finally, the numerical results show that the function g may remain high not only on the boundary as the theory predicts, but in the interior as well which may result in the reconstructed boundary being not sharply defined. For more rigorous details, the reader must consult the references cited above. We would also like to mention that Xu (1999) has applied essentially the same procedure to the reconstruction of a 2D obstacle in a two-dimensional waveguide. The scattering data were simulated on a line above the scatterer. The point sources were also positioned on the same line and the obstacle was supposed to be in a square box.

9.6.2 The Method of Eigensystem of the Far-field Operator

There exist closely related studies which exploit the eigensystem of the scattering operator. We mention the papers by Mast *et al.*(1997) and Norris (1998). The first mentioned was applied to the inverse medium problem, whereas the second pertained to the problem of inverse obstacle scattering. Note that the method of Colton and Kirsch described above can also be applied to the inverse medium case. Again consider the *incident-wave function*

$$E(x) = \int g(\theta)e^{ik\theta \cdot x}d\theta, \qquad (9.35)$$

where $g(\theta)$ is the *incident-wave distribution* and the integral is over $[-\pi, \pi]$ in two dimensions. Essentially, $E(x)$ is Herglotz. The corresponding scattered field is given by

$$u_e^s(x) = \int g(\theta)u^s(x, \theta)d\theta, \qquad (9.36)$$

where $u^s(x, \theta)$ is the scattered field for a plane wave incident along the direction θ. The far-field scattering operator F is then given by

$$(Fu_\infty)(\hat{x}) = \int g(\theta)u_\infty(\hat{x}, \theta)d\theta, \qquad (9.37)$$

where

$$u_\infty(\hat{x}, \theta) = \int e^{-ik\hat{x} \cdot x}q(x)u^s(x, \theta)dx \qquad (9.38)$$

is the scattering amplitude corresponding to $u^s(x, \theta)$. q is the potential or inhomogeneity to be recovered. The far-field operator F is related to the standard scattering matrix S by

$$\frac{i}{4\pi}F = I - S.$$

It is well-known (Newton, 1982) that F is compact, has a countable number of discrete eigenvalues with zero as the limit point, and has eigenfunctions $\{f_i\}$ which form an orthonormal basis over $L_2[-\pi, \pi]$. Moreover, F is unitary (the inhomogeneity $q(x)$ in the Lippmann-Schwinger equation is assumed to be real) and the eigenvalues are distributed on a circle in the complex plane centered at $-4\pi i$.

Using Eqs. (9.37) and (9.38), the incident-wave function of Eq. (9.35) and the scattered field $u_e^s(x)$ of Eq. (9.36) can be related to the potential $q(x)$ as

$$E_i(x) = \frac{2\pi}{\lambda_i} \int J_0(k|x - y|)u_{E_i}^s(y)q(y)dy \tag{9.39}$$

and

$$u_{E_i}^s(x) = \frac{1}{\lambda_i} \int < u^s(x,\theta), e^{ik\theta \cdot y} > u_{E_i}^s(y)q(y)dy, \tag{9.40}$$

where $<,>$ is the inner product on $L_2[-\pi, \pi]$. From the eigensystem of the scattering operator F and upon using Eqs. (9.39) and (9.40) we obtain

$$< Ff_i, f_j >= \delta_{ij}\lambda_i = \int u_{E_i}^s(x)E_i^*(x)q(x)dx, \ i, j = 1, 2, \cdots. \tag{9.41}$$

The inversion procedure of Mast *et al.*is to solve the variational problem
minimize $||q||_W^2$ *subject to the constraint given by Eq. (9.41) above:*

$$||q||_W^2 = \int |q(x)|^2 W(x)dx,$$

the weight function being $W(x) = (1 + x^2)^\delta$, $\delta > 0$. It is also assumed that the potential falls off rather rapidly with distance such that $|q(x)| \leq C(1+|x|)^{-(1+\delta)}$. The final result is the following expression for the sought-for potential q, namely

$$q(x) = \frac{1}{W(x)} \sum_i \sum_j Q_{ij}F_i(x)\overline{F}_j^*(x),$$

where $\overline{F}^*(x)$ means

$$\overline{F}^*(x) = \int g^*(\theta)u^s(x, \theta + \pi)d\theta.$$

For further implementational and numerical details, the paper cited above should be consulted.

The work of Norris (1998) is basically the method of Colton and Kirsch (1996) discussed above, but from the point of view of the eigensystem of the far-field scattering operator as in Mast *et al.* In a way, it is a combination of the two methods. Norris finds an indicator function which is large in the exterior, but bounded inside the scatterer. The problem of why the indicator function cannot be unbounded inside the obstacle also arises for Norris. However, for a circle, he shows the boundedness inside, but no theoretical conclusion is drawn for obstacles of other geometries. However, the author presents some interesting physical arguments.

9.7   Appendix A.9.1

The Shape Derivative of the Scattered Field

Let $V : R^3 \rightarrow R^3$ be a vector field with compact support in an overall domain D, the normal derivative of V vanishing on the boundary of D (see Figure 9.1). Let $T_t : R^3 \rightarrow R^3$ be a translation operator defined by

$$x = T_t X = (I + tV)(X), \ t > 0$$

moving a point $X \in R^3$ to a new point $x \in R^3$, the parameter $t > 0$ being small. Similarly, if Ω_t is the image of a domain Ω_0 under T_t, then

$$\Omega_t = T_t(\Omega_0) = \{T_t X, \forall X \in \Omega_0\}$$

and $\Omega_0 = T_0(\Omega_0)$. The transformation T_t is called the *perturbation of the identity* and is *diffeomorphic of order* C^k if V has C^k-regularity. The field V must be local, that is, must depend upon the position X in order for the reference domain Ω_0 to be deformed.[4]

Consider a boundary value problem in Ω_t

$$(A\psi_t)(x) = f(x), \ x \in \Omega_t$$

$$\psi_t(x) = g(x), \ x \in \Gamma_t.$$

Let $\psi_0(X)$ be the solution of the same problem, but posed in Ω_0 and Γ_0. Define:

$$\begin{aligned}
\psi^{[1]} &= \lim_{t \to 0} \frac{\psi_t(\Omega_t) - \psi(\Omega)}{t} \\
&= \lim_{t \to 0} \frac{\psi_t(X + tV(X)) - \psi(X)}{t} \\
&= \frac{d}{dt}\psi(x \circ T_t)|_{t=0}, \quad (1)
\end{aligned}$$

where, as before, $\circ$ denotes the composition $x \circ T_t = T_t X$ and the subscript t on ψ emphasizes its explicit dependence on t. The limit $\psi^{[1]}$ in Eq. (1), if it

---

[4]The deformation x of X can also be obtained in a more general way (Delfour, 1990) by solving the so-called *autonomous system* given by the initial value problem

$$\frac{dx(t)}{dt} = V(x(t)), t > 0; x(0) = X.$$

exists, is referred to as the *shape derivative* of ψ with respect to the domain Ω_0. It is essentially a *Gateaux* derivative (Nashed, 1971; Tapia, 1971) and is actually a differential. It can also be viewed as a directional differential along **V**. However, the functions ψ_t depend explicitly on t and are to be considered different functions at different values of t.[5] The higher-order derivatives

$$\psi^{[m]} = \frac{\mathrm{d}^m}{\mathrm{d}t^m}\psi(x \circ \mathbf{T}_t)|_{t=0}$$

can be similarly defined provided again that the derivatives exist.

Let us consider two examples of shape derivatives.

Example 1

Consider the shape differentiation of the domain functional

$$\ell(\Omega_t) = \int_{\Omega_t} \phi_t(x)\mathrm{d}x \tag{2}$$

of function ϕ. Equation (2) can also be written as

$$\ell(\Omega_t) = \int_{\Omega_0} (\phi_t \circ T_t)(X) DJ_T \mathrm{d}X, \tag{3}$$

where

$$DJ_T = \det|\mathbf{J}_T| = \det|I + t\mathbf{J}_\mathbf{V}|,$$

where **J** is the Jacobian matrix. By direct calculation,

$$\frac{\mathrm{d}}{\mathrm{d}t}DJ_T\Big|_{t=0} = \nabla \cdot \mathbf{V}(X). \tag{4}$$

Differentiating Eq. (3) in t, setting $t = 0$ and using Eq. (4) gives

$$
\begin{aligned}
\ell^{[1]} &= \frac{\mathrm{d}}{\mathrm{d}t}\ell(\Omega_t)\Big|_{t=0} \\
&= \int_{\Omega_0} \left[\frac{\mathrm{d}}{\mathrm{d}t}(\phi_t \circ T_t) + \phi\left(\frac{\mathrm{d}}{\mathrm{d}t}DJ_T\right)\right]\Big|_{t=0}\mathrm{d}X \\
&= \int_{\Omega_0} [\phi^{[1]} + (\phi\nabla \cdot \mathbf{V}](X)\mathrm{d}X.
\end{aligned}
\tag{5}
$$

Since ϕ_t depends explicitly on t, we have

$$\phi^{[1]}(X) = \frac{\partial}{\partial t}\phi(X) + [\mathbf{V} \cdot \nabla\phi](X) = \phi^{(1)}(X) + [\mathbf{V} \cdot \nabla\phi](X). \tag{6}$$

The quantity $\phi^{(1)}(X)$ is the partial derivative of ϕ at X and is defined by

$$\phi^{(1)}(X) = \underset{t\to 0}{\text{limit}}\,\frac{\phi_t(x) - \phi(X)}{t} = \underset{t\to 0}{\text{limit}}\,\frac{\phi_t(T_tX) - \phi(X)}{t}. \tag{7}$$

---

[5] The derivative is akin to the Lie derivative of differential geometry (Boothby, 1970). It is also the so-called Lagrangian, total, substantial or Stokes' derivative of the continuum mechanics of deformable media (Chapter 6).

Inserting Eqs. (6) and (7) into Eq. (5) gives the shape derivative $\ell^{[1]}$ as

$$\ell^{[1]} = \int_{\Omega_0} [\phi^{(1)} + \nabla \cdot \phi \mathbf{V}](X) \mathrm{d}X.$$

Remarks

i. From Eq. (6)

$$\phi^{(1)}(X) = \phi^{[1]}(X) - (\mathbf{V} \cdot \nabla \phi)(X).$$

The above equation will be considered to be the definition of the partial derivative $\phi^{(1)}(X)$.

ii. Care is to be exercised in defining the partial derivative. Because the undeformed and deformed quantities are both considered at the same point in the unperturbed domain which is in motion, such a comparison may be meaningless. It is, therefore, necessary that the deformed function ϕ_t can be extended in a neighborhood of Γ_t. The existence of an extension operator and the independence of the derivative of the extension are proven in Sokolowskii and Zolesio (1992). Moreover, there exists the so-called *structure theorem* (Delfour, 1990) in shape calculus according to which the existence of the partial derivative $\phi^{(1)}$ is guaranteed once the shape derivative $\phi^{[1]}$ is shown to exist.

Example 2

In this example we calculate $\hat{n}^{[1]}(X)$, the shape derivative of the normal vector on the boundary. This quantity appeared in the coefficient calculations of Section 9.5.2. Consider the unit normal $\hat{n}(x)$ to Γ_0 at a point X on it. The oriented area $\mathbf{d\Gamma_0}$ around X is given by

$$\mathbf{d\Gamma_0} = \hat{n} d\Gamma_0 = \mathbf{du} \times \mathbf{dv}.$$

$\mathbf{du}$ and $\mathbf{dv}$ are two tangent vectors at $x \in \Gamma$. The corresponding transformed surface element is

$$\mathbf{d\Gamma_t} = T_t \mathbf{d\Gamma_0} = \hat{n}_t d\Gamma_t = \mathbf{du_t} \times \mathbf{dv_t}. \tag{8}$$

Consider the quantity $\mathbf{J}_T^* \mathbf{d\Gamma_t}, \mathbf{J}_T^*$ being the transpose of $\mathbf{J}_T$. Using Einstein's summation convention throughout, we obtain

$$\begin{aligned}
\mathbf{J}_T^* \mathbf{d\Gamma_t} &= \mathbf{J}_T^* \hat{n}_t d\Gamma_t \\
&= \hat{e}_i (\mathbf{J}_T^*)_{ij} (\mathbf{d\Gamma_t})_j \\
&= \hat{e}_i (\mathbf{J}_T^*)_{ij} (\mathbf{du_t} \times \mathbf{dv_t})_j \\
&= \hat{e}_i \epsilon_{jkl} (\mathbf{J}_T)_{ji} (\mathbf{J}_T)_{km} (\mathbf{J}_T)_{ln} (\mathbf{du_t})_m (\mathbf{dv_t})_n \\
&= \hat{e}_i |\mathbf{J}_T| \epsilon_{imn} (\mathbf{du})_m (\mathbf{dv})_n \\
&= |\mathbf{J}_T| \hat{e}_i (\mathbf{du} \times \mathbf{dv})_i \\
&= |\mathbf{J}_t| (\mathbf{du} \times \mathbf{dv}) = |\mathbf{J}_T| \hat{n} d\Gamma. \tag{9}
\end{aligned}$$

ϵ_{jkl} is the *Levi-Civita symbol* and use was made of the well-known determinental relation (see, for example, Dettman, 1969)

$$\epsilon_{i_1 i_2 \cdots i_n} a_{i_1 j_1} a_{i_2 j_2} \cdots a_{i_n j_n} = |\mathbf{A}| \epsilon_{j_1 j_2 \cdots j_n}.$$

From Eqs.(8) and (9)

$$\hat{n}_t d\Gamma_t = |\mathbf{J}_T| \mathbf{J}_T^{-*} \hat{n} d\Gamma,$$

where $\mathbf{J}_T^{-*} = (\mathbf{J})_T^*{-1} = (\mathbf{J}_T^{-1})^*$. Using the fact that $d\Gamma_t = |\mathbf{J}_T| d\Gamma$, it follows at once that

$$\hat{n}_t = \frac{\mathbf{J}_T^{-*} \hat{n}(x)}{||\mathbf{J}_T^{-*} \hat{n}(x)||}. \tag{10}$$

Therefore, upon differentiating Eq. (10) in t, we obtain

$$\hat{n}^{[1]} = \left. \frac{\alpha - \beta}{||\mathbf{J_T}^{-*} \hat{n}||^2} \right|_{t=0}, \tag{11}$$

where

$$\alpha = (\mathbf{J}_T^{-*})^{[1]} \hat{n} ||\mathbf{J_T}^{-*} \hat{n}|| \tag{12}$$

and

$$\beta = \mathbf{J}_T^{-*} \hat{n} ||\mathbf{J_T}^{-*} \hat{n}||^{[1]}. \tag{13}$$

Differentiating the identity relation: $\mathbf{J}_T \mathbf{J}_T^{-1} = I$ gives

$$\left(\mathbf{J}_T \mathbf{J}_T^{-1}\right)^{[1]} = (\mathbf{J}_T)^{[1]} (\mathbf{J}_T^{-1})_{t=0} + (\mathbf{J}_T^{-1})^{[1]} (\mathbf{J}_T)_{t=0} = 0$$

from which

$$(\mathbf{J}_T^{-1})^{[1]} = -\mathbf{J}_T^{[1]} = -\mathbf{J_V}(X).$$

Therefore,

$$\left((\mathbf{J}_T^{-1})^*\right) = \left(\mathbf{J}_T^{-*}\right)^{[1]} = -\mathbf{J_V}^*(X). \tag{14}$$

Also

$$\left[||\mathbf{J}_T^{-*} \hat{n}||\right]^{[1]} = \left. \frac{d}{dt} < \mathbf{J}_T^{-*} \hat{n}, \mathbf{J}_T^{-*} \hat{n} > \right|_{t=0} = - < \mathbf{J_V} \hat{n}, n > . \tag{15}$$

Substituting Eqs. (14) and (15) into Eqs. (12) and (13), respectively, and noting that as $t \to 0, \mathbf{J}_T \to I$, the quantities α and β are obtained. Using α and β thus determined into Eq. (11) and simplifying, we finally obtain the shape derivative of the normal vector, namely

$$\hat{n}^{[1]} = < \hat{n}, \mathbf{J_V}^* \hat{n} > \hat{n} - J_V^* \hat{n}. \tag{16}$$

Next consider the case when the velocity field is along the normal direction, that is, $\mathbf{V} = |\mathbf{V}| \hat{n}$. If the shape derivative is linear and continuous in the velocity field, then it depends only upon the normal component of $\mathbf{V}$. Intuitively, this reflects the fact that a tangential field on the boundary does not result in a distortion of the domain. In this case

$$\mathbf{J_V}^*\hat{n} = \nabla|\mathbf{V}|$$

and Eq. (16) simplifies to

$$\hat{n}^{[1]} = \hat{n} < \hat{n}, \nabla|\mathbf{V}| > -\nabla|\mathbf{V}| = \mathrm{Grad}|V\mathbf{V}. \tag{17}$$

$\mathrm{Grad}|\mathbf{V}|$, which is the surface gradient of $|\mathbf{V}|$, will be denoted by ∇_Γ.

Before leaving this section, we shape differentiate the normal derivative $\partial_n\phi = \hat{n} \cdot \nabla\phi$ of a function defined in Ω. The result is

$$
\begin{aligned}
[\hat{n} \cdot \nabla\phi]^{[1]} &= \hat{n}^{[1]} \cdot \nabla\phi^{[0]} + \hat{n}^{[0]} \cdot [\nabla\phi]^{[1]} \\
&= -\nabla_\Gamma \cdot [\nabla\phi]^{[0]} + |\mathbf{V}|\hat{n} \cdot [[\nabla\phi]^{(1)} + \hat{n} \cdot [\nabla\nabla\phi]^{[0]}] \\
&= \hat{n} \cdot \nabla\phi^{(1)} - \nabla_\Gamma|\mathbf{V}| \cdot \nabla\phi + |\mathbf{V}|\hat{n} \cdot \nabla\nabla\phi \\
&= \hat{n} \cdot \nabla\phi^{(1)} - \nabla_\Gamma|\mathbf{V}| \cdot \nabla\phi + |\mathbf{V}|\partial_n^2\phi. \tag{18}
\end{aligned}
$$

Use was made of Eqs. (17) and (6) in obtaining the second step of the above derivation. Equation (18) was used in deriving the expressions for $u^{(m)}$ in Section 9.5. Note that $(\nabla\phi)^{(1)} = \nabla\phi^{(1)}$ because the variables X and t are independent of each other.

9.8 Appendix A.9.2

The Padé Approximation

The overall purpose of the Padé approximation in our case is to extract the values of the coefficients $u_{t\bar{n}}$ from its Taylor series (9.22) beyond the radius of convergence of the series. In brief, the approximation can be described as follows. Let

$$f(z) = \sum_{n=0}^{\infty} f_n z^n, \tag{19}$$

be a formal power series and L, M two nonnegative integers. We say that a Padé approximant $\pi_{LM} = [L/M] = P_L \backslash Q_M$ of order LM exists if there are two polynomials $P_L \in P(L)$ and $Q_M \in P(M)$ such that
i.

$$Q_M(z)f(z) - P_L(z) \sim O(z^{L+M+1}). \tag{20}$$

and ii. $Q_M(0) = 1$.
Let

$$P_L(z) = \sum_{j=0}^{L} a_j z^j, \text{ and } Q_M(z) = \sum_{j=0}^{M} b_j z^j, \; b_0 = 1. \tag{21}$$

We substitute the polynomials P_L and Q_M given in Eq. (3) into the expression (2) and obtain

$$\left(\sum_{j=0}^{M} b_j z^j\right)\left(\sum_{n=0}^{\infty} f_n z^n\right) - \left(\sum_{j=0}^{L} a_j z^j\right) = O(z^{L+M+1}).$$

If we set the coefficients of the powers $z^k, k = 0, \cdots, L + M$, on the L.H.S. of the above identity to be zero, then we obtain

$$a_i = f_i b_0 \sum_{j=1}^{i} f_{j-1} b_j, \tag{22}$$

$$a_L = f_L b_0 + \sum_{j=1}^{\min(L,M)} f_{j-1} b_j, \tag{23}$$

and

$$\mathbf{b} = b_0 \mathbf{M}^{-1} \mathbf{f}, \tag{24}$$

where $\mathbf{b} = (b_1, b_2, \cdots, b_M)^T, \mathbf{d} = (f_{L+1}, f_{L+2}, \cdots, f_{L+M})^T$, and the matrix $\mathbf{M}$ is given by

$$\begin{pmatrix} f_{LM+1} f_{LM+2} \cdots f_L \\ f_{LM+2} f_{LM+3} \cdots f_{L+1} \\ \cdots \cdots \\ \cdots \cdots \\ f_L f_{L+1} \cdots f_{L+M-1} \end{pmatrix}$$

It is assumed that $f_j = 0$ if $j < 0$. $b_0 = 1$ and the system (6) always has a solution. Solving the linear system (6) for $\mathbf{b}$ and substituting in Eqs. (4) and (5) yields the coefficient vector $\mathbf{a}$ for the numerator matrix P_L. Conditions i and ii above are equivalent to the condition that the linear system (6) has a solution with $b_0 = 1$. Sometimes the approximation is formulated as
    find $P_L \in P(L)$ and $Q_M \in P(M)$ such that

$$f(z) - \frac{P_L}{Q_M} = O(z^{L+M+1}).$$

This is not necessarily equivalent to the definition (2) above. An example is furnished by the function $f(z) = 1 + z^2$.

As we just mentioned, a solution of system (6) always exists. As to the convergence, there are two types (Petrushev and Popov, 1987): direct and converse. In the direct type, the following question is asked: *given the knowledge of power series (1) (for example, the number of poles in some domain), what can be said about the corresponding Padé approximants?* The converse theorem runs thus: *given that something is known about the Padé approximation of a function f, e.g., the number and locations of the poles, what can be said about the function f?* We mention only the classical direct theorem of Montessus de Ballore (Baker and Graves-Morris, 1996; Petrushev and Popov, 1987). The theorem says that if the function $f(z)$ is meromorphic in a disc D_m of radius R_m centered at $z = 0$ having k-distinct poles at $z_1, z_2, \cdots, z_k$ with multiplicities $p_1, p_2, \cdots, p_k$ such that $\sum_{i=1}^{k} p_i = m$, then the sequence of $[L/M]$ Padé approximants, M fixed, converges uniformly to f as $L \to \infty$ on each compact subset of $D_m \setminus z_1, z_2, \cdots, z_k$. Moreover, the poles of the denominator Q_M tend to the poles of the function f.

Now the Padé approximants are rational functions. But are they the best rational function approximations (see Chapter 4 for the definition of the best approximation)? It is known that finding the best rational function approximation is difficult. However, it can be shown that the Padé approximants are the best local rational function approximations. Let $f(z)$ be analytic at the origin and consider a disc $|z| \le \epsilon$. Let $R^{[L/M]}(\epsilon, z)$ be the best rational function approximation of type $[L/M](z)$ on $|z| \le \epsilon$. Then under the definitions i and ii of the Padé approximation, it can be shown (Baker and Graves-Morris, 1996) that as $\epsilon \to 0$, $R^{[L/M]}(\epsilon, z) \to [L/M]_f(z)$ uniformly on any compact set containing no pole of $[L/M]_f(z)$. The subscript f was added to $[L/M](z)$ in order to emphasize that it is the approximation of the function $f(z)$.

9.9 Appendix A.9.3

Some Remarks Concerning the Rayleigh Hypothesis

We mentioned in the main text that the assumption, namely, that the field representation (9.16) is valid in all $R^3 \setminus \Omega$, that is, everywhere in R^3 including the boundary of the obstacle, was introduced by Rayleigh and is called the Rayleigh hypothesis. If this conjecture is considered to be true, then it should be possible to write Eq. (9.16), for example, even in the grooves of a diffraction grating. This, however, turns out not to be the case in general. As a matter of fact, the validity of Rayleigh's hypothesis has been called into question (Deriugin, 1952; Lysanov, 1958; Lippmann, 1953; Fortuin, 1970; Barantsev, 1971).

In general, the hypothesis is neither true nor false, and its validity depends upon the boundary in question. But what is known with certainty is that Rayleigh's hypothesis cannot apply to boundaries which are nonsmooth. There is a popular notion that the inapplicability of the hypothesis for bodies of arbitrary shapes is due to the simultaneous existence of both incoming and outgoing waves. However, as observed by Millar (1973), this cannot be the reason simply because of the fact that sinusoidal diffraction gratings do satisfy Rayleigh's hypothesis provided that $kb < 0.142$ as was pointed out in the text. The primary consideration in evaluating the applicability of the hypothesis lies in the analytic continuation of the wave field inside the smallest sphere circumscribing the obstacle including the interior. The analysis (Millar, 1971, 1973; van den Berg and Fokkema, 1979) reveals that the hypothesis is actually applicable all the way to the convex hull of the field singularities that lie interior to the boundary, if it exists. By a singularity of the field, we mean the following. The Helmholtz solution in the exterior is a real analytic function in its arguments and, therefore, can be continued analytically. It is the singularity in this continuation that is meant here. Put alternatively, the diameter of the hull is the diameter of the smallest sphere inside the scatterer on which the series (9.16) ceases to converge.

As mentioned previously, the problem of determining the domain of validity of Rayleigh's hypothesis was considered by several authors. We mention the work of Millar (1970, 1971), vanden Berg and Fokkema (1979) and Kyurkchan et al. (1996) in this connection. The last mentioned is an excellent review of the subject. Millar (1970) considered the problem of harmonic functions first, and, subsequently (1971), generalized the method to the Helmholtz equation. He considers a point P_0 outside the circumscribing ball (sphere or circle) and a function Φ harmonic outside another ball of radius ρ with P_0 as the center. The function Φ is expressed in Fourier series in the angular variable(s), the coefficients varying inversely with the radial distance. The series converges for some $\rho > \rho_{\min}$ which then gives at least one of the singularities of Φ. The point P_0 is then allowed to move on a ball around the obstacle. The resulting singularities obtained in this way describes an envelope which lies inside the scatterer and encloses the singularities of the analytic continuation of the wavefield.

In another method, the angular variable θ is considered to be complex, namely, $\theta = \theta_1 + i\theta_2$, and the Helmholtz representation is analyzed in the complex θ-plane. The scattering solution is found to be analytically continuable in the entire θ-plane. Moreover, as $|\theta_2| \to \infty$, i.e., near the infinite point of the plane, the scattering amplitude asymptotically coincides with an entire function of exponential type of degree σ. In other words,

$$\lim_{R \to \infty} \frac{ln\{\max_{|z|=R} |f(z_\mp)|\}}{R} = \sigma_\mp \leq \frac{kr_{\max}}{2},$$

where $z_\pm = e^{|\theta_1| \pm i\theta_2}$, $R = \exp(|\theta_2|)$, and $r_{\max}$ is the maximum distance of the points of the boundary from the origin. Alternatively, $r_{\max}$ is the minimum diameter of the circumscribing ball. It is the $\max(\sigma_+, \sigma_-)$ that determines the radius of the envelope enclosing the field singularities. Note that by the bound given above, $r_{\max}$ is inside the scatterer boundary. This procedure is explained in details by Kyurkchan et al. (1996).

The above results are closely linked to the following well-known result which we state in the form of a theorem below.

Theorem 9.1 Let $f(\hat\theta) \in L_2(\hat\Omega_3)$ be analytically continuable in the entire complex plane $\theta = \theta_1 + i\theta_2$. Let B_r be a ball of radius r such that it is the smallest ball containing the obstacle. Define

$$\zeta = e^{|\theta_2| \pm i\theta_1}.$$

If asymptotically, i.e., $|\theta_2| \to \infty$,

$$f(\hat\theta) = \tilde{f}(\zeta, \phi)\left[1 + O\left(e^{-|\theta_2|}\right)\right],$$

where $\tilde{f}(\zeta, \phi)$ is an entire function of degree not exceeding $kr/2$, then $f(\hat{\theta})$ is the scattering amplitude of a wavefield.

A similar theorem obtains in two dimensions. The theorem is important in inverse problems for two accounts: in establishing uniqueness and in recovering the scatterer boundary from scattering amplitudes. For an exhaustive account, the reader is referred to Colton and Kress (1983).

A crucial consideration in discussing Rayleigh's hypothesis is Vekua's theorem which was mentioned in Section 9.4 of the main text. We know that if Rayleigh's hypothesis applies, then the coefficients in the Rayleigh series (α_ℓ in Eq. (6.40)) is given by the Helmholtz representation

$$\alpha_\ell = \frac{i}{8} \int_\Gamma \{u^{sc}\partial_n - \partial_n u^{sc}\} H_\ell^{(1)}(kr) e^{-i\ell\theta} ds,$$

where s is the arc-length along the boundary. We have considered a general case and written the above expression in two space dimensions simply for the sake of illustration. The above formula for α_ℓ is valid all the way down to the boundary Γ itself. However, in the case that the hypothesis is not applicable (most likely to be if the scatterer has a noncanonical shape), α_ℓ cannot be determined by the above Helmholtz representation. In such situations, we apply Vekua's theorem regarding the completeness of the metaharmonic functions over any Lyapunov boundary on which the field is square integrable and still express the solution in terms of the outgoing wavefunctions $H_\ell^{(1)}(kr) e^{-i\ell\theta}$. In other words, we write the solution as

$$u(x) = \sum_{\ell=-\infty}^{\infty} u_\ell H_\ell^{(1)}(kr) e^{-i\ell\theta}.$$

However, the coefficients $\{u_\ell\}$ will neither be given by the Helmholtz formula as for α_ℓ nor would they equal $\{\alpha_\ell\}$. The constants $\{u_\ell\}$ are to be determined in the L_2-sense *via* the minimization of the norm functional

$$||u_d - \sum_{\ell=-L}^{L} u_\ell(L) H_\ell^{(1)}(kr) e^{-i\ell\theta}||_{L_2(\Gamma)},$$

where u_d is the boundary data. Note that we have written $u_\ell(L)$ in order to emphasize that these coefficients are functions of L, the number of terms kept in the expansion.

It is not difficult to show (Millar, 1973) that the expansion of u with coefficients $u_\ell(L)$ determined by minimizing the norm

$$||u_d - \sum_{\ell=-L}^{L} u_\ell(L) H_\ell^{(1)}(kr) e^{-i\ell\theta}||_{L_2(\Gamma)}$$

is uniformly convergent in any closed set of the exterior domain. However, the number L would increase as one approaches the actual boundary Γ because of the presence of the Hankel function. This essentially reflects the instability in backpropagating a forward propagated wave, the fundamental source of ill-posedness in holographic problems, as was demonstrated in Example 5 of Chapter 2.

Chapter 10

Bibliography

M. J. Ablowitz and P.A. Clarkson (1991). *Solitons, Nonlinear Evolution Equations and Inverse Scattering*, Cambridge University Press, Cambridge.

M. Ablowitz and H. Segur (1981). *Solitons and Inverse Scattering Transform*, SIAM Publications, Philadelphia.

F.F. Abraham, J.Q. Broughton, N. Bernstein and E. Kaxiras (1998). Comput. Phys., **12**, pp. 538-546.

M. Abramowitz and I.A. Stegun (1964). *Handbook of Mathematical Functions*, Dover Publications, New York.

S. Agmon (1975). "Spectral properties of Schrödinger operators and scattering theory", Ann. Scuola Norm. Sup. Pisa, **2**, pp. 151-218.

N.I. Akhiezer and I.M. Glazman (1961) *Theory of Linear Operators in Hilbert Spaces*, Frederick Unger Publishing Co., New York.

R.C. Allen, W.R. Bolland, V. Faber and G.M. Wing (1985). "Singular values and condition numbers of Galerkin matrices arising from linear integral equations of the first kind", J. Math. Anal. Appl., **109**, pp. 564-590.

L. Alvarez, F. Guichard, P.L. Lions and J.M. Morel (1993). "Axioms and fundamental equations of image processing", Arch. Rational Mech. Anal., **123**, pp. 199-257.

K.A. Ames and B. Straughan (1997). *Non-standard and Improperly Posed Problems*, Academic Press, San Diego.

T.S. Angell, R.E. Kleinman and G.F. Roach (1987). "An inverse transmission problem for the Helmholtz equation", Inverse Problems, **3**, pp. 149-180.

G. Anger (1990). *Inverse Problems in Differential Equations*, Plenum Press, New York.

Yu.E. Anikonov, B.A. Bubnov and G.N. Erokhin (1997). *Inverse And Ill-posed Source Problems*, Ridderprint bv, Ridderker Utrecht, The Netherlands.

P. Anselone (1971). *Collectively Compact Operator Approximation: Theory and Applications*, Prentice Hall, Engelwood–Cliffs.

F.V. Atkinson (1949). "On Sommerfeld's "radiation" condition", Phil. Mag., **40**, pp. 645–651.

K.E. Atkinson (1989). *An Introduction to Numerical Analysis*, John Wiley, New York.

K.E. Atkinson (1997). *The Numerical Solution of Integrals Equations of the Second Kind*, Cambridge University Press, Cambridge.

I. Babuska and A.K. Aziz (1972). *Survey lectures on the Mathematical Foundations of the Finite Element Method*, Academic Press, New York.

C.T.H. Baker (1977). *The Numerical Treatment of Integral Equations*, Oxford University Press, Oxford.

G.A. Baker and P. Graves–Morris (1996). *Padé Approximants*, Addison–Wesley.

A. Bakushinsky and A. Goncharsky (1994). *Ill -posed problems: Theory And Applications*, Dordrecht, Kluwer Academic Publishers, 1994.

C.A. Balanis (1997). *Antenna Theory: Analysis and Design*, John Wiley, New York.

H.P. Baltes, ed. (1978), *Inverse Source Problems in Optics*, Topics in Current Physics, **20**, Springer–Verlag, Berlin.

N.V. Banichuk (1990). *Introduction to Optimization of Structures*, Springer–Verlag, New York.

A. Banos (1966). *Dipole Radiation in the Presence of a Conducting Half-space*, Pergamon Press, Oxford.

R.H.T. Bates, V.A. Smith and R.D. Murch (1991). "Manageable multidimensional inverse scattering theory", Physics Reports, **201**, pp. 185–277.

A. Bayliss and E. Turkel (1980). "Radiation boundary conditions for wave-like equations", Comm. Pure and Appl. Math., **33**, pp. 707–725.

J.V. Beck, B. Blackwell and C.R. St. Clair, Jr. (1985). *Inverse Heat Conduction Ill-posed Problems*, John Wiley and Sons, New York.

J.A. Bennett and M.E. Botkin (1986). *The Optimum Shape: Automated Structural Design*, Plenum Press, New York.

P. M. vanden Berg and J.T. Fokkema (1979). "The Rayleigh hypothesis in the theory of diffraction by a cylindrical surface", IEEE Trans. Ant. Propag., **AP–17**, pp. 577–583.

M. Bertero (1989). "Linear inverse and ill-posed problems", *Advances in Electronics and Electron Physics*, **75**, pp. 1–120, Academic Press, New York.

A. Björck (1996). *Methods for Least Squares Problems*, SIAM, Philadelphia.

N. Bleistein and J. Cohen (1977). "Nonuniqueness in the inverse source problem in acoustics and electromagnetics", J. Math. Phys., **18**, pp. 194–201.

Boothby (1970). *An Introduction to Differentiable Manifolds and Riemaninan Geometry*, Academic Press, New York.

H. Brakehage and P. Werner (1965). "Über das Dirichletsche Aussenraumproblem für die Helmholtzsche Schwingungsgleichung", Arch. Math., **10**, pp. 325–329.

O. Bruno and F. Reitich (1995). "A new approach to the solution of problems of scattering by bounded obstacles", Proc. SPIE, **2192**, pp. 20–27.

O. Bruno and F. Reitich (1992). "Solution of a boundary value problem for Helmholtz equation via variation of the boundary into the complex plane", Proc. Roy. Soc. Edinburgh, **122A**, pp. 317–340.

J.L. Buchanan, R.P. Gilbert and A. Wirgin (2000). "Identification, by the intersecting canonical domain method, of the size, shape and depth of a soft body of revolution located within an acoustic waveguide", Inverse Problems **16**, pp. 1709–1926.

R.C. Buck (1956). *Advanced Calculus*, McGraw–Hill Book Company, New York.

A.J. Burton and G.F. Millar (1971). "The application of integral equation methods to the numerical solution of some exterior boundary-value problems", Proc. Roy. Soc. London, **A323**, pp. 201–220.

A.P. Calderón (1980). "On an inverse boundary value problem", *Seminar on numerical analysis and Its Applications to Continuum Physics*, Soc. Brasileria de Matematica, pp. 65–73.

J.R. Cannon (1984). *One-Dimensional Heat Equation*, Addison–Wesley, Reading.

C.D. Cantrell (2000). *Modern Mathematical Methods for Physicists and Engineers*, Cambridge University Press, Cambridge.

P. Carrion (1987). *Inverse Problems and Tomography in Acoustics and Seismology*, Penn Publishing, Atlanta.

G. Caviglia, A. Morrow and C. Giacomo (1992). *Inhomogeneous Waves in Solids and Fluids*, World Scientific, River Edge.

K. Chadan and P. C. Sabatier (1989). *Inverse Problems in Quantum Scattering Theory*, Springer–Verlag, New York.

Yu Chen and V. Rokhlin (1997). "On the Riccati equation for the scattering matrices in two dimensions", Inverse Problems, **13**, pp. 1–13.

X. Chen and A. Friedman (1991). "Maxwell's equations in a periodic structure", Trans. Am Math. Soc., **323**, pp. 465–506.

M. Cheney (1984). "Inverse scattering in dimension two", J. Math. Phys., **25**, pp. 94–102.

M. Cheney and J.H. Rose (1986). "Three-dimensional inverse scattering for the classical wave-equation with variable speed", in *Rev. Prog. NDE*, vol. 5, pp. 1–5, eds. D.O. Thompson and D.E. Clementi.

M. Cheney, J.H. Rose and B. De Facio (1988). "Three dimensional inverse scattering for the wave-equation: weak scattering approximation with error estimates," Inverse Problems, **9**, pp 435–447.

R.V. Churchill (1963). *Fourier Series and Boundary Value Problems*, McGraw–Hill Book Company, New York.

R.V. Churchill (1960). *Complex Variables and Applications*, McGraw–Hill Book Company, New York.

J. A. Cochran (1972). *The Analysis of Linear Integral Equations*, McGraw–Hill Book Company, New York.

S. Coen, M. Cheney and A. Weglein (1984). "Velocity and density of a two-dimensional acoustic medium", J. Math. Phys., **25**, pp. 1857–1861.

D. Cohen (1968). "Magnetoencephelography: evidence of magnetic field produced by alpha rhythm currents", Science, **161**, pp. 784–786.

L. Collatz (1966). *Functional Analysis and Numerical Mathematics*, Academic Press, New York.

D. Colton and R. Kress (1983). *Integral Equation Methods In Scattering Theory*, John Wiley, New York.

D. Colton and R. Kress (1992). *Inverse Acoustic and Electromagnetic Scattering Theory*, Springer–Verlag, Berlin.

D. Colton and A. Kirsch (1996). "A simple method for solving inverse scattering problems in the resonance region", Inverse problems, **12**, pp. 383–393.

D. Colton, M. Piana and R. Potthast (1997). "A simple method using Morozov's discrepancy principle for solving inverse scattering problems", Inverse problems, **13**, pp. 1477–1493.

D. Colton, J. Coyle and P. Monk (2000). "Recent developments in inverse scattering theory", SIAM Review, pp. 369–414.

J.B. Conway (1990). *Course in Functional Analysis*, Springer–Verlag, New York.

E.T. Copson (1975). *Partial Differential Equations*, Cambridge University Press, Cambridge.

I.J.D. Craig and S.C. Brown (1986). *Inverse Problems in Astronomy*, Adam Hilger, Bristol.

J. Cullum (1979). "The effective choice of the smoothing norm in regularization", Math. Comut., **33**, pp. 149 –170.

A.R. Davies (1992). "Optimality in regularization", pp. 393–410, in: *Inverse Problems in Scattering and Imaging*, eds. M. Bertero and E.R. Pike, Adam Hilger, Bristol.

R. Dautray and J.L. Lions (1990). *Mathematical Analysis and Numerical Methods*, Springer–Verlag, Berlin.

E.B. Davies (1995). *Spectral Theory and Differential Operators*, Cambridge University Press, Cambridge.

S. Deans (1983). *The Radon Transform and its Applications*, John Wiley and Sons, New York.

L. Debnath and P. Mikushinski (1998). *Introduction to Hilbert Space With Applications*, Academic Press, Boston.

C. De Mol (1992). "A critical survey of regularized inversion methods", in: *Inverse Problems in Scattering and Imaging*, eds. M. Bertero and E.R. Pike, pp. 345–370, Adam Hilger, Bristol.

M. Defrise and C. de Mol (1987). "A note on stopping rules for iterative regularization methods and filtered SVD", in *Inverse Problems: An Interdisciplinary Study* edited by P.C. Sabatier, Academic Press, New York, pp. 261–268.

M. Delfour and G. Sabidussi (eds.) (1992). *Shape Optimization and Free Boundaries*, Kluwer Academic publishers, Dordrecht.

L.N. Deriugin (1952). "Equations for coefficients of wave solutions from aperiodically uneven surface", Dokl. Akad. Nauk. SSSR, **87**, pp. 913–916.

J.W. Dettman (1969). *Mathematical methods in Physics and Engineering*, McGraw–Hill, New York.,

E. Deuflhard and E. Hairer (1983) Ed., *Numerical Treatment of Inverse Problems in Differential and Integral Equations*, Springer–Verlag, Berlin.

A.J. Devaney and E. Wolf (1973). "Radiating and nonradiating classical current distributions and the fields they generate", Phys Rev., **D8**, pp. 1044–1047.

E. DiBenedetto (1995). *Partial Differential Equations*, Birkhauser, Boston.

A. Doicu, Y. Eremin and T. Wriedt (2000). *Acoustic and Electromagnetic Scattering Analysis*, Academic Press, San Diego.

E.N. Domanskii (1987). "On the equivalence of convergence of a regularizing algorithm to the existence of a solution to an ill-posed problem", Russian Math. Surveys, **42**, pp. 123–144.

P.G. Drazin and R.S. Johnson (1989). *Solutions: an introduction*, Cambridge University Press, Cambridge.

V.T. Duc and D.N. Hao (1994). *Differential Operators of Infinite Order with real arguments and their Applications*, World Scientific, Singapore.

L. Elden (1982). "A weighted pseudoinverse, generalized singular values, and constrained least squares problems", BIF, **22**, pp. 487–501.

H.W. Engl and C.W. Groetsch (1987). *Inverse And Ill-posed Problems*, Academic Press, Boston.

H.W. Engl, M. Hanke and A. Neubauer (1996). *Regularization of Ill-posed Problems*, Kluwer, Dordrecht.

B. Engquist and A. Majda (1977). "Absorbing boundary conditions for the numerical simulation of waves", Math. Comput., **31**, pp. 629–651.

B. Engquist and A. Majda (1979). "Radiation boundary conditions for acoustic and elastic wave calculations", Comm Pure and Appl. Math., **32**, pp. 313–357.

V. A. Erma (1968). "An exact solution for the scattering of electromagnetic waves from conductors of arbitrary shape. II. General case", Phys. Rev., **176**, pp. 1544–1553.

L.C. Evans (1991). *Partial Differential Equations*, American Mathematical Society, Providence.

L. D. Fadeev (1956). "Uniqueness of the inverse scattering problem," Math. Revs., **18**, pp. 259.

L.D. Fadeev (1965). "Increasing solutions of the Schrödinger equation", Sov. Phys. Dokl., **10**, pp. 1033–1035.

V.V. Filatov (1984). "Construction of focusing transformations of transient electromagnetic fields", Sov. Geol. Geophys., **25**, pp. 89–95.

P. Filippi, D. Habault, J Lefebvre and A. Bergassoli (1999). *Acoustics: Basic Physics, Theory and Methods*, Academic Press, San Diego.

J.N. Flavin and S. Rionero (1995). *Qualitative Estimates for Partial Differential Equations*, CRC Press, Boca Raton.

A.S. Fokas, I.M. Gelfand and Y. Kurylev (1996). "Inversion method for magnetoencephelography", Inverse Problems, **12**, pp. L9–L11.

G.B. Folland (1992). *Fourier Analysis and its Applications*, Wadsworth and Brooks/Cole, Pacific Grove.

G.B. Folland (1995). *Introduction To Partial Differential Equations*, Princeton University Press.

G.E. Forsythe and W.R. Wasow (1960). *Finite-Difference Methods for Partial Differential Equations*, John Wiley, New York.

L. Fortuin (1970). "Survey of literature on reflection and scattering of sound waves at the sea surface", J. Acoust. Soc. Am., **47**, pp. 1209–1228.

J.N. Franklin (1974). "On Tikhonov's method for ill-posed problems", Math. Comput., **28**, pp. 889–907.

F.G. Friedlander (1958). *Sound Pulses*, Cambridge University Press.

B. Friedman (1956). *Principles and Techniques of Applied Mathematics.* John Wiley, New York.

A. Gamliel, K. Kim, A. I. Nachman, and E. Wolf (1989). "A new method for specifying nonradiating, monochromatic, scalar sources and their fields," J. Opt. Soc. Am., **6**, pp. 1388–1393.

E. Garber, S.G. Brush and C.W.F. Everitt (1995). *Maxwell On Heat And Statistical Mechanics*, Lehigh University Press.

W. Gautschi (1997). *Numerical Analysis: an Introduction*, Birkhauser, Boston.

D.B. Ge, A.K. Jordan and Lakhshman S. Tamil (1994). "Numerical inverse scattering theory for the design of planar optical waveguides", J. Opt. Soc. Am., **11**, pp. 2809–2815.

I.M. Gel'fand, M.I. Graev and N. Ya. Vilenkin (1966). *Generalized Functions*, vol. 5, Academic Press, New York.

D.B. Geselowitz (1970). "On the magnetic field generated outside an inhomogeneous volume conductor by internal current sources", IEEE Trans. Magnetics, **MAG–6**, pp. 346–347.

D.N. Ghosh Roy, L. Couchman and J. Warner (1997). "Scattering and inverse scattering of sound-hard obstacles via shape deformation", Inverse Problems, **13**, pp. 585–606.

D.N. Ghosh Roy, J. Warner, L. S. Couchman and J. Shirron (1998). "Inverse obstacle transmission problem in acoustics", Inverse Problems, **14**, pp. 903–929.

D. Gilbarg and N.S. Trudinger (1977). *Elliptic Partial Differential Equations*, Springer–Verlag, Berlin.

R.P. Gilbert, Y. Xu and P. Theijl (1992). "An approximation scheme for the three-dimensional scattered wave and its propagating far-field pattern in a finite depth ocean", ZAMM, **72**, pp. 459–480.

R.P. Gilbert and Y. Xu (1993). "An inverse problem for harmonic acoustics in stratified oceans", J. Math. Anal. Appl., **17**, pp. 121–137.

R.P. Gilbert, T. Scotti, A. Wirgin and Y. Xu (1997). "Identification of a 3D object in a shallow sea from scattered sound", Compte Rendues, Acad. Sci. Paris, serie IIb, **325**, pp. 383–389.

D. Givoli (1991). "Non-reflecting boundary conditions", J. Comput. Phys., **94**, pp. 1–29.

V.B. Glasco (1984). *Inverse Problems of Mathematical Physics*, American Institute of Physics.

H. Goldstein (1980). *Classical Mechanics*, Addison–Wesley, Reading.

G. H. Golub and C.F. Van Loan (1989). *Matrix Computations*, The Johns Hopkins University Press.

G.H. Golub and M. Heath (1977). "Generalized cross-validation as a method for choosing a good ridge parameter", Technometrics, **21**, pp. 215–223.

J.W. Goodman (1968). *Introduction to Fourier Optics*, McGraw–Hill Book Company, San Francisco.

R. Gorenflow and S. Vassella (1991). *Abel Integral Equations, Analysis and Applications*, Lecture Notes in Mathematics, vol. 1461, Springer–Verlag, Berlin.

I.S. Gradshtein and I.M. Rhyzhik: *Table of Integrals, Series, and Products*, Academic Press, San Diego, 1980.

J. Graves and P.M. Prenter (1978). "Numerical iterative filters applied to first kind Fredholm integral equations", Numer. Math., **30**, pp. 281–299.

H. Griem (1997). *Plasma Spectroscopy*, Cambridge University Press, Cambridge.

H. Groemer (1996). *Geometric Application of Fourier-series and Spherical harmonics*, Cambridge University Press, Cambridge.

C.W. Groetsch (1977). *Generalized Inverses of Linear Operators*, Marcel Dekker, New York.

C.W. Groetsch (1993). *Inverse Problems in the Mathematical Sciences*, Viewweg–Verlag, Braunschweig, Wiesbaden 1993.

C.W. Groetsch (1984). *The Theory of Tikhonov Regularization for Fredholm Integrals Equations of the First Kind*, Pitman, Boston.

G.T. Gullberg, R.H. Huesman, S.G. Ross, E.V.R. Di Bella, G.L. Zeng, B.W. Reutter. P.E. Christian and S.A. Foresti (1999). "Dynamic cardiac single-photon emission computed tomography", in *Nuclear Cardiology*, Eds. B.L. Zaret and G.A. Beller, Mosby, St. Loius.

N.M. Gunter (1968). *Potential Theory*, Ungar, New York.

V.V. Gusev and S.G. Potapov (1992). "Application of the finite element method to molecular quantum mechanics", Theoret. Expt. Chem., **27**, pp. 381–385.

B. Gustafsson, H.O. Kreiss and J. Oliger (1995). *Time Dependent Problems and Difference Methods*, John Wiley and Sons, New York.

S. Gutman and M. Klibanov (1994). "Iterative method for multidimensional inverse scattering problems at fixed frequency", Inverse Problems, **10**, pp. 573–599.

W. Hackbusch (1995). *Integral Equations: Theory and Numerical Treatment*, Birkhauser–Verlag, Boston.

J. Hadamard (1923). *Lectures on the Cauchy Problem in Linear Partial Differential Equations*, Yale University Press, New Haven.

T. Hagstrom (1997). "On high-order radiation boundary conditions" In: *Computational Wave Propagation* eds. B Engquist and G.A. Kreigsmann, pp. 1–21, Springer–Verlag, New York.

M. Hämäläinen, R. Hari, R.J. Llmoniemi, J. Knuttila and O.V. Lounasmaa (1993). "Magnetoencephelography-theory, instrumentation, and applications to noninvasive studies of the working human brain", Rev. Mod. Phys., **65**, pp. 413–497.

G. Hämmerlin and K.H. Hoffmann (1991). *Numerical Mathematics*, Springer–Verlag, New York.

P. C. Hansen (1998). *Rank-Deficient and Discrete Ill-posed Problems: Numerical Aspects of Linear Inversion*, SIAM Publications, Philadelphia.

E.J. Haug, K.K. Choi and V. Komkov (1986). *Design Sensitivity Analysis of Structural Systems*, Academic Press, Orlando.

S. Helgason: *The Radon Transform*, Birkhauser–Verlag, Boston, 1980.

G. Hellwig: *Partial Differential Equations*, Blaidsdell Publishing Company, New York, 1960.

H. von Helmholtz (1853). Ann. Physik and Chemie, **89**, pp. 354–377.

G. M. Henkin and R. G. Novikov (1988). "A multidimensional inverse problem in quantum and acoustic scattering," Inverse Problems, **4**, pp. 103–122.

G.T. Herman (1980). *Image Reconstructions From Projections: The Fundamentals of Computerized Tomography*, Academic Press, New York.

F. Hettlich (1995). "Frechet derivative in inverse obstacle scattering", Inverse problems, **11**, pp. 371–382.

F. Hettlich (1994). "On the uniqueness of the inverse conductive scattering problem for the Helmholtz equation", Inverse problems, **10**, pp. 129–144.

E. Hille (1964). *Analysis*, vol. 1, Blaisdell Publishing Company, New York.

B.J. Hoenders (1978). "The uniqueness of inverse problems". In: *Inverse Source Problems in Optics*, ed. H.P. Baltes, Topics in Current Physics, **20**, pp. 41–82, Springer–Verlag, Berlin.

Isaacson and Keller (1966). *Analysis of Numerical Methods*, John Wiley, New York.

V. Isakov (1990). "On uniqueness in inverse transmission scattering problem", Comm. Pure and Appl. Math., **15**, pp. 1565–1587.

V. Isakov (1990). *Inverse Source Problems*, AMS, Providence.

V.K. Ivanov (1962). "On linear problems which are not well-posed", Sov. Math. Dokl., **3**, p. 981.

V.K. Ivanov (1963). "On ill-posed problems", Mat. Sb., **61**, p. 211.

V.K. Ivanov (1966). "The approximate solution of operator equations of the first kind", USSR Comp. Math. Math. Phys., **6**, pp. 197.

J.D. Jackson (1975). *Classical Electrodynamics*, John Wiley and Sons, New York.

D.R. Jackson and D.R. Dowling (1991). "Phase conjugation in underwater acoustics", J. Acoust. Soc. Am., **89**, pp. 171–181.

M.A. Jawson and G.T. Symm (1977). *Integral Equation Methods in Potential Theory and Elastostatics*, Academic Press, London.

F. John (1960). "Continuous dependence on data for solutions of partial differential equations", Comm. Pure and Appl. Math., **13**, pp. 551.

C. Johnson (1987). *Numerical Solutions of Partial Differential Equations by the Finite Element Method*, Cambridge University Press, Cambridge.

D.S. Jones (1966). *Generalized Functions*, McGraw–Hill Book Company, New York.

D.S. Jones (1974). "Integral equations for the exterior acoustic problem", Q.J. Appl. Math., **27**, pp. 129–142.

D.S. Jones and X.Q. Mao (1989). "The inverse problem in hard acoustic scattering", Inverse problems, **5**, pp. 731–748.

D.S. Jones and X.Q. Mao (1990). "A method for solving the inverse problem in soft acoustic scattering", IMA J. Appl. Math., **44**, pp. 127–143.

A.K. Jordan and S. Lakshmanasamy (1989). "Inverse scattering theory applied to the design of single-mode planar optical waveguides", J. Opt. Soc. Am., **A6**, pp. 1206–1212.

F. Kang (1983). "Finite Element Method and natural boundary reduction", Proc. Intl. Cong. Mathematicians, Warsaw, pp. 1439–1453.

R.P. Kanwal (1997). *Linear Integral Equations*, Birkhauser–Verlag, Boston.

O.D. Kellog (1953). *Foundations of Potential Theory*, Dover Publications, New York.

A. Khinchin (1960). *A Course of Mathematical Analysis*, Hindustan Publishing Corp. (India), Delhi.

D. Kingcaid and W. Cheney (1991). *Numerical Analysis: Mathematics of Scientific Computing*, Brooks/Cole Publishing Company, Pacific Grove.

A. Kirsch (1993). "The domain derivative and two applications in inverse scattering theory", Inverse Problems, **9**, pp. 81–96.

A. Kirsch (1996). *An Introduction to the Mathematical Theory of Inverse Problems*, Springer–Verlag, New York.

A. Kirsch (1998). "Characterization of the shape of a scattering obstacle using the spectral data of the far field operator", Inverse Problems, **14**, pp. 1489–1512.

A. Kirsch (1998). "Factorization of the far-field operator for the inhomogeneous medium case and an application in inverse scattering theory", Inverse Problems, **15**, pp. 413–429.

A. Kirsch and R. Kress (1987). "An optimization method in inverse acoustic scattering". In: *Boundary Elements IX, vol. 3 Fluid Flow and Potential Applications*, Eds. C.A. Brebbia, W.L. Wendland and G. Kuhn, Computational Mechanics Publications, pp. 3–18, Springer–Verlag, Heidelberg.

A. Kirsch and R. Kress (1993). "Uniqueness in obstacle scattering", Inverse Problems, **9**, pp. 285–299.

R. Kittappa and R.E. Kleinman (1975). "Acoustic scattering by penetrable homogeneous objects", J. Math. Phys., **16**, pp. 421–432.

R.E. Kleinman and G.F. Roach (1982). "On modified Green's functions in exterior problems for the Helmholtz equation". Proc Royal Soc. London, **A383**, pp. 313–332.

N. S. Koshlyakov, M.M. Smirnov and E.B. Gliner (1964). *Differential Equations of Mathematical Physics*, North–Holland, Amsterdam.

R. Kress (1989). *Linear Integral Equations*, Springer–Verlag.

R. Kress (1998). *Numerical Analysis*, Springer–Verlag, New York.

E. Kreyszig (1978). *Introduction to Functional Analysis*, John Wiley and Sons, New York.

K.K. Kuo and M. Summerfeld (1984). eds., *Progress in Astronautics and Aeronautics*, **90**, AIAA, New York"

A.G. Kyurkchan, B. Yu Sternin and V.E. Shatalov (1996). "Singularities of continuation of wave fields", Physics–Uspekhi, **39**, pp. 1221–1242.

H.D. Ladouceur (1990). "An inverse problem in propellant combustion", in *Inverse Methods in Action*, Proc. Multicentennials Meetings on Inverse Problems, pp. 347–355.

E. Lalor (1968). "Inverse wave propagator", J. Math. Phys., **9**, pp. 2001–2006.

G.L. Lamb (1980). *Elements of Soliton Theory*, John Wiley, New York.

C. Lanczos (1949). *Linear Differential Equations*, Van Nostrand, London.

L.D. Landau and E.M. Lifshitz (1987). *Fluid Mechanics*, Pergamon Press, Oxford.

N. S. Landkof (1972). *Foundations of Modern Potential theory*, Springer–Verlag, Berlin.

L. Landweber (1951). "An iteration formula for Fredholm integral equations of the first kind", *Am. J. Math.*, **73**, p. 615.

K.J. Langenberg (1987). "Applied inverse problems for acoustic, electromagnetic and elastic wave scattering", in: *Basic Methods of Tomography and Inverse Problems*, pp. 127–470, ed. P.C. Sabatier, Hilger, Bristol.

H.P. Langtangen (1999). *Computational Partial Differential Equations*, Springer–Verlag.

H. L. Lass (1957). *Elements of Pure and Applied Mathematics*, McGraw–Hill, New York.

M.M. Lavrentiev (1967). *Some Improperly Posed Problems of Mathematical Physics*, Springer–Verlag, Berlin.

M.M. Lavrentiev, V.G. Romanov and S.P. Shishatskii. (1986). *Ill-posed Problems of Mathematical Physics and Analysis*, Am. Math. Soc., Providence.

L.P. Lebedev. I.I. Borovich and G.M.L. Gladwell (1996). *Functional Analysis: Applications in Mechanics and Inverse Problems*, Kluwer, Dordrecht.

J. Nédélec (2001). *Acoustic and Electromagnetic Equations*, Springer–Verlag, New York.

K.H. Lee, G. Liu and H.F. Morrison (1989). "A new approach to modeling the electromagnetic response of conductive media", Geophysics, **54**, pp. 1180–1192.

P. Linz (1978). *Theoretical Numerical Analysis*, John Wiley, New York.

B.A. Lippmann (1953). "Note on the theory of gratings", J. Opt. Soc. Am., **43**, pp. 408.

L.A. Liusternik and V.J. Sobolev (1961). *Elements of Functional Analysis*, Hindustan Publishing Corp. (India), Delhi.

R.J. Llmoniemi, M. S. Hämäläinen and J. Knuttila (1985). "The forward and inverse problems in the spherical head", In: *Biomagnetism: Applications and Theory*, pp. 278–282, Pergamon Press, Oxford.

J. Locker and P.M. Prenter (1980): "Regularization with differential operators. I: General Theory", J. Math. Anal. Appl., **74**, pp. 504–529.

A.K. Louis (1989). *Inverse und schlechte gestelte Probleme*, Teubner–Verlag, Stuttgart.

A.K. Louis and P. Maass (1991). "Smoothed projection methods for the moment problem", Numer. Math., **59**, pp. 277–294.

A.F.D. Loula, T.J.R. Hughes and L.P. Franca (1987). "Petrov–Galerkin forms of the Timoshenko beam problem", Comut. Methods Appl Mech. Engin., **63**, pp. 115–132.

M.A. Lukas (1980). "Regularization", In: *The Application of Numerical Solutions of Integral Equations*, eds. R.S. Anderssen, F. R. de Hoog and M.A. Lukas: Sijthoff and Noordhoff, Alphen aan den Rijn, The Netherlands, pp. 151–182.

R.K. Luneberg (1966). *Mathematical Theory of Optics*, University of california Press, Berkeley.

Iu.P. Lysanov (1958). "Theory of scattering of waves at periodically uneven surfaces", Sov. Phys. Acoust., **4**, pp. 1–10.

T.M. MacRobert (1967). *Spherical Harmonics*, Pergamon Press, Oxford.

J. Malmivuo and R. Plonsey (1995). *Bioelectromagnetism: Principles and Applications of Bioelectricity and Biomagnetism*, Oxford University Press.

J.T. Marti (1986). *Introduction To Sobolev Spaces And Finite Element Solution Of Elliptic Boundary Value Problems*, Academic Press, London.

V.P. Maslov (1968). " The existence of a solution to an ill-posed problem is equivalent to the convergence of a regularizing process", Uspekhi. Mat. Nauk., **23**, pp. 183–184.

T.D. Mast, A. Nachman and R. Waag (1997). "Focusing and imaging using eigenfunctions of the scattering operator", J. Acoust. Soc. Am., **102**, pp. 715–725.

W. McLean (2000). *Strongly Elliptic Systems and Boundary Integral Equations*, Cambridge University Press.

J.G. McWhirter and E.R. Pike (1978). "On the numerical inversion of the Laplace transform and similar Fredholm integral equations of the first kind", J. Phys. A: Math. Gen., **11**, pp. 1729–1745.

G. Meinardus (1967). *Approximation of Functions: Theory and Numerical Methods*, Springer, New York.

A. Messiah (1958). *Quantum Mechanics*, North–Holland, Amsterdam.

V.P. Mikhailov (1978). *Partial Differential Equations*, MIR, Moscow.

S.G. Mikhlin (1970). *Mathematical Physics: an Advanced Course*, North–Holland, Amsterdam.

R.F. Millar (1970). "The location of singularities of two-dimensional harmonic functions. I:Theory", SIAM J. Math. Anal., **1**, pp. 333–343.

R.F. Millar (1971). "Singularities of two-dimensional exterior solutions of the Helmholtz equation", Proc. Camb. Phil. Soc., **69**, pp. 175–188.

R.F. Millar (1973). "The Rayleigh hypothesis and a related least-squares solution to scattering problems for periodic surfaces and other scattering", Radio Science, **8**, pp. 785–796.

K. Miller (1970). "Least squares methods for ill-posed problems with a prescribed bound", **1**, SIAM J. Math. Analys., pp. 52–74.

L. Misici and F. Zirilli (1994). "Three-dimensional inverse obstacle scattering for time harmonic acoustic waves: a numerical method", SIAM J. Sci. Comput., **15**, pp. 1174–1189.

V.A. Morozov: *Methods for Solving Incorrectly Posed Problems*, Springer–Verlag, Berlin, 1984.

P.M. Morse and H. Feshbach (1953). *Methods of Theoretical Physics*, vol. 1, McGraw–Hill Book Company, New York.

J.C. Mosher, R.M. Leahy and P.S. Lewis (1997). "Matrix kernels for the forward problem in EEG and MEG", Proc. IEEE **ICASSP**, pp. 2934–2956.

C. Muller (1997). *Analysis of Spherical Symmetries in Euclidean Spaces*, Springer–Verlag, New York.

H. Murakami, V. Sonnad and E. Clementi (1992). "A three-dimensional finite element approach towards molecular SCF computations", Int. J. Quant. Chem.,

42, pp. 785–817.

D.A. Murio (1993).   *The Mollification Methods for Ill-posed Problems*, CRC Press, Boca Raton.

A. Nachman (1989). "Reconstructions from boundary measurements", Annals of Mathematics, **128**, pp. 531–576.

J.A. Nashed (1971). "Differentiability and related topics of nonlinear operators: Some aspects of the role of differentials in nonlinear functional analysis", In: *Nonlinear Functional Analysis*, eds. L.B. Rall, pp.103–309, Academic Press, New York.

F. Natterer (1986). *The Mathematics of Computerized Tomography*, Teubner–Verlag, Stuttgar.

F. Natterer (1984). "Error bounds for Tikhonov regularization in Hilbert scales", Appl. Anal., **18**, pp. 29–37.

A. Nayfeh (1973). *Perturbation Methods*, John Wiley and Sons, New York.

A.W. Naylor and G.R. Sell (1982). *Linear Operator Theory in Engineering and Science*, Springer–Verlag, New York.

J.C. Nédélec (1982). "Integral equations with nonintegrable kernels", Int. Eqs. Operator Theory, **4**, pp. 563–572.

T. Needham (1997). *Visual Complex Analysis*, Clarendon Press, Oxford.

R.G. Newton (1982).   *Scattering Theory of Waves and Particles*, Springer–Verlag, New York.

R.G. Newton (1985). "Relation between the Schrödinger equation and the plasma wave equation", Phys. Rev. A, **31**, pp. 3305–3308.

R.G. Newton (1989). *Inverse Schrödinger Scattering in Three-dimensions*, Springer–Verlag, Berlin.

Y. Nievergelt (1991). "Numerical linear algebra on the HP-28 or how to lie with supercalculators", Am. Math. Monthly, , pp. 539 - 544.

S.M. Nikolskii (1975). *Approximation of Functions of Several Variables and Imbedding Theory*, Springer–Verlag, New York.

A.N. Norris (1998). "A direct inverse scattering method for imaging obstacles with unknown surface conditions", IMA J. Appl. Math., **61**, pp. 267–290..

J. Ockendon, S. Howison, A Lacey and A. Movchan (1999). *Applied Partial Differential Equations*, Oxford University Press.

T. Oden and G.F. Carey (1982). *Finite Element: Mathematical Aspects*, vol. 4, Prentice–Hall, Englewood Cliff.

V.I. Oliker (1989). "On reconstructing a reflecting surface from the scattering data in the geometrics optics approximation", Inverse problems, **5**, pp. 51–65.

A. Papoulis (1965). *Signal Analysis*, McGraw–Hill Book Company, New York.

L.E. Payne: *Improperly Posed problems in Partial Differential Equations*, SIAM Publications, Philadelphia, 1975.

L.E. Payne (1960). "Bounds in the Cauchy Problem for Laplace's equation", Arch. Rational Mech. Anal., **5**, pp. 35–45.

R. Petit and M. Cadilhac (1966). "Sur la diffraction d'une onde plane par un réseau infiniment conducteur", C.R. Acad. SCi., Seri. B, **262**, pp. 468–471.

P.P. Petrushev and V.A. Popov (1987). *Rational Approximation of Real Functions*, Cambridge University Press, New York.

D.L. Phillips (1962). "A technique for the numerical solution of certain integral equations of the first kind", J. Assoc. Comput. Mach., **9**, pp. 84.

O. Pironneau (1984). *Optimal Shape Design for Elliptic Systems*, Springer–Verlag, Berlin.

R. Potthast (1998). "On a concept of uniqueness in inverse scattering for a finite nuumber of incident waves", SIAM J. Appl. Math., **58**, pp. 666–682.

M.J.D. Powell (1997). *Approximation Theory and Optimization*, Cambridge University Press, New York.

W.H. Press, B.P. Flannery, S.A. Teukolsky and W.T. Vetterling (1988). *Numerical Recipes In C: The Art of Scientific Computing*, Cambridge University Press, Cambridge.

A. Quarteroni and A. Valli (1994). *Numerical Approximations Of Partial Differential Equations*, Springer–Verlag, Berlin.

A.G. Ramm (1992). *Multidimensional Inverse Scattering Problems*, Longman Scientific and Technical, New York.

A.G. Ramm (1964). *Scattering By Obstacles*, D. Reidel Publishing Company, Dordrecht.

A.G. Ramm and A.I. Katsevich (1996). *The Radon Transform and Local Tomography*, CRC Press, Boca Raton.

J.W.S. Rayleigh (1897). "On the passage of waves through apertures in plane screens, and allied problems", Philosophical Magazine, **43**, pp. 259–272.

B.D. Reddy (1998). *Introductory Functional Analysis*, Springer–Verlag.

R.D. Richtmyer and K.W. Morton (1967). *Difference Methods for Initial value Problems*, Wiley–Interscience, New York.

T.J. Rivlin (1969). *An Introduction to the Approximation of Functions*, Dover Publications, New York.

R.T. Rocafeller (1970). *Convex Analysis*, Princeton University Press, Princeton.

P. Roman (1975). *Some Modern Mathematics for Physicists and Other Outsiders*, Pergamon Press, New York.

P. Roman (1965). *An Advanced Quantum Theory*, Addison–Wesley, Reading.

V. G. Romanov (1974). *Integral Geometry and Inverse Problems in Hyperbolic Equations*, Springer–Verlag, New York.

J.H. Rose, M. Cheney and B. De Facio (1985). "Three-dimensional inverse scattering: plasma and variable velocity wave equations", J. Math. Phys., **26**, pp. 2803–2813.

L.I. Rudin and S. Osher (1994). "Total variation based image restoration with free local constraints", Proc. IEEE, **ICIP94**, pp. 31–35.

L.I. Rudin, S. Osher and E. Fatemi (1992). "Nonlinear total variation based noise removal algorithms", Physica D, **60**, pp. 259–269.

P.C. Sabatier (1978). *Applied Inverse Problems*, Lecture Notes in Physics, vol. 85, Springer–Verlag, New York.

P.C. Sabatier (1990). "Modeling or solving inverse problems?" in *Inverse Methods in Action*, Proc. Multicentennials Meetings on Inverse Problems.

P.C. Sabatier (1991). "From one to three dimensions in inverse problems", pp. 1–23, in: *Inverse Problems in Scattering and Imaging*, eds. M. Bertero and E.R. Pike, Adam Hilger, Bristol.

J. Sarvas (1987). "Basic mathematical and electromagnetic concepts of the biomagnetic inverse problem", Phys. Med. Biol., **32**, pp. 11–22.

H.A. Schenk (1967). "Improved integral formulation for acoustic radiation problems", J. Acoust. Soc. Am., **44**, pp.41–58.

H. Schlichting (1979). *Boundary Layer Theory*, Springer Verlag, Berlin.

W. Schweizer, P. Fassbinder, R. Gonzalez-Ferez, M. Braun, S. Kulla and M. Stehle (1999). "Discrete variables and finite-element techniques applied to simple atomic systems", J. Comp. Appl. Math., **109**, pp. 95–122.

T.I. Seidman (1980). "Nonconvergence results for the application of least square estimation to ill-posed problems", J. Opt. Theory Appl., **30**, pp. 535–547.

J.R. Shewell and E. Wolf (1968). "Inverse diffraction and a new reciprocity theorem", J. Opt Soc. Am., **58**, pp. 1596–1603.

J. Shirron (1995). *Solution of Exterior Helmholtz Problems Using Finite and Ininite Elements*, Ph.D. Dissertation, University of Maryland.

J. Simon (1980). "Differentiation with respect to the domain in boundary value problems", Num. Funct. Anal., **2**, pp. 649–687.

V.I. Smirnov (1964). *A Course in Higher Mathematics*, vol. 4, Pergamon Press, Oxford.

F. Smithies (1937). "The eigenvalues and singular values of integral equations", Proc London Math. Soc., **43**, pp. 255–279.

F. Smithies (1958). *Integral equations*, Cambridge University Press, Cambridge..

J. Sokolowski and J.P. Zolesion (1992). *Introduction to Shape Optimization*, Springer–Verlag, Berlin.

A. Sommerfeld (1949). *Partial Differential Equations in Physics*, Academic Press.

I. Stakgold (1979). *Green's Function and Boundary Value Problems*, John Wiley and Sons, New York.

G.W. Stewart (1973). *Introduction to Matrix Computations*, Academic Press, Orlando.

O.N. Strand (1974). "Theory and methods related to the singular–function expansion and Landweber's iteration for integral equations of the first kind", SIAM J. NUm. Anal., **11**, pp. 441–459.

G. Strang (1988). *Linear Algebra and Its Applications*, Harcourt Brace Jovanovich, San Diego.

G. Strang and G.J. Fix (1973). *An Analysis of the Finite Element Method*, Prentice–Hall, Englewood Cliff.

S. Ström (1991). "Introduction to Integral Equations for Time-harmonic Acoustic, Electromagnetic and Elastodynamic Wave Fields", In: *Field Representations and Introduction to Scattering*, Ed. V.V. Varadan, A. Lakhtakia and V.K. Varadan, pp. 38–141, North–Holland, Amsterdam.

J. Sylvester and G. Uhlmann (1987). "A global uniqueness theorem for an inverse boundary value problem", Annals of Mathematics, **125**, pp. 153–169.

W. Szymczak, W. Guo and J. Rogers (1998). "Mine detection using variational methods for image enhancement and feature extraction", Proc. SPIE, **3392**, pp. 13–17.

R. Tabbar (2000). "Modeling the nano-scale phenomena in condensed matter physics via computer-based numerical simulations", Physics Reports 325, pp. 239–310.

N. Tabbara, B. Duchene, C. Pichot, D. Lesselier, L. Chommeloux and N. Joachimowicz (1988). "Diffraction tomography: contributions to the analysis of some applications in microwaves and ultrasonics", Inverse Problems, , pp. 305–331.

R. A. Tapia (1971). "The differentiation and integration of nonlinear operators", In: *Nonlinear Functional Analysis*, eds. L.B. Rall, pp. 45–102, Academic Press, New York.

A. Taylor (1964). *Introduction to Functional Analysis*, John Wiley, New York.

A.E. Taylor (1985). *General Theory of Functions and Integration*, Dover Publications, New York.

A.E. Taylor and W.R. Mann (1983). *Advanced Caulus*, John Wiley, New York.

A.N. Tikhonov and V.Y. Arsenin (1977). *Solutions To Ill-posed Problems*, John Wiley and Sons, New York.

A.N. Tikhonov and A.A. Samarskii (1963). *Partial Differential Equations of Mathematical Physics*, Holden–Day, San Francisco.

J. van Tiel (1984). *Convex Analysis*, John Wiley, Chichester.

E.C. Titchmarsh (1939). *Theory of Functions*, Oxford University Press, Oxford.

J.H. Tripp (1983). *Biomagnetism: an Interdisciplinary Approach*, eds. S.J. Williamson, G.L. Romani, L. Kauffman and I Modena, Plenum Press, New York.

Turkel (1983). "Progress in computational physics", Computers and Fluids, **11**, pp. 121–144.

F. Ursell (1978). "On the exterior problems of acoustics II", Proc. Cambridge Phil. Soc., **84**, pp. 545–548.

B.R. Vainberg (1966). "Principles of radiation, limit absorption and limit amplitude in the general theory of partial differential equations", in *Russian Mathematical Surveys*, **21**, pp. 115 - 193.

G.M. Vainikko (1987). "On the optimality of regularization methods", in: *Inverse and Ill-posed Problems*, eds. H.W. Engl and C.W. Groetsch: Academic Press, Boston.

J.M. Varah (1983). "Pitfalls in the numerical solution of linear ill-posed problems", SIAM J. Sci. Stat. Comput., **4**, p. 164.

I.N. Vekua (1953). "On completeness of a system of metaharmonic functions", Dokl. Akad. Nauk, SSSR, **90**, pp. 715–718.

G. Wahba (1977). "The approximate solution of linear operator equations when the data are noisy", SIAM J. Num. Anal., **14**, 651–667

G. Wahba and S. Wold (1975). "Periodic splines for spectral density estimation: the use of cross-validation for determining the correct degrees of smoothing", Comm. in Statist., **4**, pp. 125–141.

R. Wait and A.R. Mitchell (1985). *Finite Element Analysis and Applications*, John Wiley, New York.

Y. Wang and W. Chew (1989). "An iterative solution of two dimensional electromagnetic inverse scattering problem", J. Imaging Systems and Technology, **1**, pp. 100–108.

J. Warner, D.N. Ghosh Roy, J. Bucaro and L. Couchman (2000). "Inversion of penetrable obstacles from far-field data on narrow angular apertures", J. Acoust. Soc. Am., **107**, pp. 1111–1120.

H. Weyl (1919). Ann. Physik, **60**, p. 481.

G.B. Whitham (1974). *Linear And Nonlinear Waves*, John Wiley and Sons, New York, 1974.

D.V. Widder (1971). *An Introduction to Transform Theory*, Academic Press, New York.

C.H. Wilcox (1956). "An expansion theorem for electromagnetic fields", Comm. Pure and Appl. Math., **9**, pp. 115–134.

E.G. Williams (1999). *Sound Radiation and Nearfield Acoustic Holography*, Academic Press, San Diego.

S.J. Williamson and L. Kaufman (1989). "Biomagnetism", J. Magnetism Magnet. Mater., **22**, pp. 129–201.

G.M. Wing (1992). *A Primer on Integral equations of the First Kind. The problem of Deconvolution and Unfolding*, SIAM Publications, Philadelphia.

Y. Xu (1991). "An injective far-field pattern operator and inverse scattering problem in a finite depth ocean", Proc. Edinburgh Math. Soc., **34**, pp. 295–311.

Y. Xu, C. Mattawa and W. Lin (1999). "An inverse scattering method for underwater imaging", Report presented at SIAM Southeastern Atlantic Section of Meeting, University of Tennessee, Knoxville. See also Proc. ISAAC Congress, Fukuoka, Japan, 1999.

Y. Xu (1999). "Inverse acoustic scattering problems in ocean environments", J. Comput. Acoustics.

C. Yeh (1964). "Perturbation approach to the diffraction of electromagnetic waves by arbitrarily shaped dielectric obstacles", Phys. Rev., **135**, pp. 1193–1201.

D. Zidarov (1980). *Inverse Gravimetric Problems in Geoprospecting and Geodesy*, Elsevier, Amsterdam.

A. Zinn (1991). "The numerical solution of an inverse scattering problem for time-harmonic acoustic waves", in: *Inverse Problems and Imaging*, ed. G.F. Roach et al, Longman, London.

A. Zinn (1989). "On an optimization method for the full and the limited-aperture problem in inverse acoustic scattering for a sound-soft obstacle", Inverse Problem, **5**, pp. 239–253.

A. Zochowski (1992). *Mathematical Problems in Shape Optimization and Shape Memory Materials*, P. Lang, New York.

Printed and bound by CPI Group (UK) Ltd, Croydon, CR0 4YY

03/10/2024

01040418-0004